THE MODULAR HOME

The MOD

ULAR
HOME

Andrew Gianino

Storey Publishing

The mission of Storey Publishing is to serve our customers
by publishing practical information that encourages
personal independence in harmony with the environment.

Edited by Jeff Beneke
Art direction and design by Kent Lew and Cynthia McFarland
Text production by Erin Dawson
Cover photographs courtesy of Rob Coolidge, Architect: front cover right and bottom left; courtesy
    of The Home Store/Melanie Poisson: front cover top left and back cover.
Illustrations by Terry Dovaston and Associates and Glenco, Inc.
Indexed by Susan Olason, Indexes & Knowledge Maps

The information in this book is true and complete to the best of our knowledge. All recommendations
are made without guarantee on the part of the author or Storey Publishing. The author and publisher
disclaim any liability in connection with the use of this information. For additional information please
contact Storey Publishing, 210 MASS MoCA Way, North Adams, MA 01247.

Storey books are available for special premium and promotional uses and for customized editions. For
further information, please call 1-800-793-9396.

Printed in the United States by Von Hoffmann Graphics
10 9 8 7 6 5 4 3 2 1

Library of Congress Cataloging-in-Publication Data

Gianino, Andrew.
    The modular home / Andrew Gianino.
        p. cm.
    Includes bibliographical references and index.
    ISBN-13: 978-1-58017-526-5; ISBN-10: 1-58017-526-0 (pbk. : alk. paper)
    ISBN-13: 978-1-58017-576-0; ISBN-10: 1-58017-576-7 (hardcover : alk. paper)
    1. Modular houses. I. Title.

TH4819.P7G53 2005
643'.2—dc22
                                                        2004020308

Dedication

*To my wife and parents for their
unflagging love and support.*

TENTS

Acknowledgments

My editor, Jeff Beneke, did a remarkable job making the book more readable. Those of you who think the book is too long owe a special thanks to Jeff, since it would have been 25% longer without his sharp editing. I would also like to thank Sarah Guare, from Storey Publishing. Her patience and easy manner made the final revisions a surprisingly pleasant experience. Special thanks are also owed to Jan Gianino, my dear wife, and Bonnie Bandolin, my bookkeeper turned book reviewer, for their editorial assistance.

Several people gave me helpful criticism about the content of the book. I would especially like to thank Joe Muldowney of Manorwood Homes, Dave Wrocklage of Epoch Homes, Mark Kelly of the Hickory Consortium, and Michael Mullens of the University of Central Florida. Somehow these people found the time to read the entire book before it was edited. I would also like to thank several people from my company, The Home Store, for their feedback: Karen Markert, Chet Mitchell, James Kitchen, Dan Newbury, Layne Floyd, Brian Holmes, and David Walbridge.

The following gave me a better understanding of the options for financing the construction of a modular home: Mark Butler of Salem Five, Tracy Egloff of Northampton Cooperative Bank, Chris Eugin of Greenfield Cooperative Bank, Lori Grover of Greenfield Savings Bank, Scott Hanley of Citizens Bank, Denise Laizer of Easthampton Savings Bank, Jeff Smith of Florence Savings Bank, and Dave Williams of The Bank of Western Massachusetts.

I'd like to thank John Glenny of Glenco, Inc., Gordon Swift, and Layne Floyd for their invaluable assistance with the illustrations. I would also like to express appreciation to all of the modular manufacturers who provided photographs for the book. They are cited in the photo gallery section.

Finally, I would like to thank several people for their help in making The Home Store possible, since without their kind assistance I'd probably have returned to child psychology and this book would never have been written: my wife and her parents — Jan Gianino and Norma and Bill Robison; my sister and brother-in-law — Diane and Rusty Shattuck; two modular manufacturer presidents — Ed Langley and Barry Shein; two modular manufacturer general managers — Doug Stimpson and Joe Muldowney; two modular manufacturer sales representatives — Wil Sormrude and Janice Willits; two Home Store GCs — Doug Blowers and Layne Floyd; two bankers — Chris Eugin and Charlene Golonka; several subcontractors and vendors — Amherst Woodworking, Atlas Overhead Door, Czar Distributing, Deerfield Valley Crane, D.M.O. Construction, Mr. Gutter, Rugg Lumber, and Valley Concrete; one set crew — Doug Pease and his team; and all those Home Store employees who were with me through thick and thin for several years — Gordon Swift, Dick Wood, Brian Holmes, Karen Markert, Chet Mitchell, Jen Perry, Mark DeMarco, Mike Barna, Butch Dodge, Amy Szalony, Bonnie Bandolin, and Patrick Falla.

Introduction

WHEN I STARTED SELLING modular homes in 1986, I was a complete novice about home building. I had never constructed a home in my life, not even a tree house as a boy. As an undergraduate student, I studied political science — a far cry from construction. I then went on to earn my PhD in psychology, with a focus on child psychotherapy and infant social development. In short, I was trained to be a child therapist and infant researcher, not a home builder.

In 1986, my wife and I decided to invest some money into building a home for resale. When we discovered that all of the good builders were booked for the next 18 months and the available contractors were inexperienced yet charging the same as the good ones, we turned to modular homes. In our conversations with friends, we learned that many others were having the same problem. It was impossible to find a good builder to construct a quality home at an affordable price. We concluded that a better alternative than building one home was to become a representative of a modular manufacturer so we could sell homes to everyone who needed them.

At first, we did this as a part-time venture. Over the course of 18 months I came to realize that building homes was more enjoyable to me than infant research and less stressful than working with severely disturbed children and families. Given the challenges of selling 100 homes a year, I sometimes wonder how I came to that conclusion. My PhD in psychology didn't prepare me to be a builder, any more than if I had been a librarian, machinist, or attorney. I didn't know a 2×4 from a roof truss. In fact, I had never heard of either, nor had I known about modular homes until six months before I became a dealer.

Why should my early ignorance give you comfort about this book, rather than concern? Because after 18 years in business, I now know most of the things you should know about buying a modular home and probably don't. I learned these things while teaching myself, which I needed to do so I could help my customers.

When I first started talking to customers, I was stunned by how much we both needed to know. It seemed there were more technical terms in construction, more industry jargon, than in psychology, and that's saying a lot. In addition to understanding modular construction, I had to acquire some working knowledge of general contracting and the individual construction trades, from site work and foundations to plumbing, electrical, heating, and carpentry. I also had to learn about real estate, building permits, and construction loans. To realize how much there is to master, consider that it takes professionals years to become experts in each one of these fields.

My first few months of self-education were particularly stressful because I was supposed to be helping my customers carry out their responsibilities. Most of them, like most of you, knew very little about residential construction or modular homes. I was the one who was supposed to be the expert, and they couldn't wait several years for me to become one. I was reminded of what the medical interns used to say during their training at one of the hospitals where I worked: "See one! Do one! Teach one!" In other words, if you see a medical procedure done once, you should be able to do it yourself, which in turn qualifies you to teach it to others. Fortunately, this frightening concept is rarely practiced in the medical field. But I've certainly seen it adopted in the home-building field, where there isn't enough professional training. And I was one of the guilty practitioners, as I struggled during my first year to educate myself at the same time that I was helping my customers. The only comfort I could take was that few of my local modular competitors seemed to know much more.

In trying to educate myself, I discovered there were no good sources of information for a retail customer. Don O. Carlson, editor of *Automated Builder* magazine, had published an excellent book for professionals that proved very helpful to me. His *Automated Builder: Dictionary/Encyclopedia of Industrialized Housing,* which is widely read by modular manufacturers and builders, covers a lot of territory that my book does not explore. But I could not find anyone, even among modular manufactur-

ers, who had created substantive handouts that helped retail customers understand the process of designing and building a modular home. I assumed that some enterprising modular-home dealers would have produced educational materials to give them a competitive advantage. I quickly discovered, however, that most dealers, regardless of how good they are as contractors, are small companies without the time, inclination, or skills necessary to develop comprehensive educational tools for their customers.

Given the circumstances, the only reason I sold my first modular home was because my factory sales representative, Wil Sormrude, went beyond the call of duty and helped our salesperson explain to our customers what they were buying. Over the next couple of years I questioned Wil and others relentlessly. (When customers apologize to me for how many questions they ask, I tell them they have a right to be demanding, since they are investing a substantial part of their life savings in their new home. I also tell them that they don't compare to me as a relentless interrogator. Wil occasionally still reminds me that I was the most demanding customer he has ever met.)

It didn't take me long to realize that my customers' hunger for information presented me with an opportunity. As my wife and my mother kept pointing out, the best way for me to use my education to distinguish my company was to write a series of handouts on the more important aspects of building a modular home. My customers would then come to see my company as more informed and more helpful. If I could take this step further and develop sales systems that helped my customers understand what they were getting and what they were paying, I could shelter them from the risks that always attend building a new home. This, in turn, would save them money and provide them with a better overall experience, which would generate future referrals and subsequent sales for my company.

From the beginning, my handouts were written from the customer's point of view. I drew upon my memory of what it was like to build my first modular home without any relevant experience, training, or practice. I recalled what it was like to be told I had to make hundreds of decisions when I didn't understand my choices and couldn't find the time to learn them. I remembered what it was like to receive my first home and realize I had no idea what needed to be done to complete it. I felt strongly that if I were a customer building my first home, I wouldn't want to be rushed into a decision, since such haste would surely lead to regrettable mistakes. I also felt I'd be willing to invest the time researching what was involved in building a modular home because I would understand it was the only way to ensure that I got the most value for my money. I kept all these thoughts in mind when I began writing educational handouts. I also applied this understanding to the forms I created to document thoroughly what my customers were getting with their home. The handouts and forms certainly grew in length over the years, largely because they grew in insight, informed by my many painful mistakes and the lessons I learned from them. This book brings these handouts together, substantially expands upon them, and adds entirely new material.

When you build a modular home, three things need to happen. First, you need a modular dealer to design your modular plan, determine its building specifications, and price it. Second, you need a modular manufacturer to build the home as designed. Finally, you need a general contractor (GC) to put the home together. Depending on whom you select for these three functions, you could have one company provide all three of the functions or you could work with two or three different companies.

A few modular manufacturers sell their homes directly to customers and also complete the required GC work. Most manufacturers, however, sell their homes to modular dealers, who resell the homes to the public. When you purchase a modular home from a dealer, you have no direct contractual relationship with the manufacturer, just as you don't have one with the manufacturer of your car. Many dealers also complete the GC work, and in doing so provide "turnkey" modular services. This enables the customer to have one contract with one company to both design and assemble his home.

Some modular dealers, however, do not complete any of the GC work. Their customers must either hire someone to serve as their GC or try to perform the GC's job themselves. If you hire an independent GC, you will have to sign one contract with a modular dealer to design and price your home and a second contract with a GC to assemble it. Acting as your own GC is a more challenging alternative, requiring that you directly hire all of the necessary subcontractors for the job.

When I started in business, I didn't provide any of the GC services, nor did I recommend anyone who could provide them. My customers were completely on their own to put their home together. After a couple of years, I found an independent GC with modular experience to recommend to my customers. Six years later, I became a modular GC so that I could offer the full range of services my customers needed. These different experiences taught me a lot about the relative advantages and pitfalls of each type of arrangement for the customer.

I agree with those industry professionals who recommend that you purchase a modular home from a dealer who either is an experienced GC himself or works hand in hand with one. An exception would be if you are an experienced contractor or will be using an experienced contractor to complete the work. As I hope to make clear throughout the book, building a modular home requires a wealth of knowledge. This includes conventional construction knowledge, such as is required for site work, foundations, framing, and heating. It also includes the specific construction knowledge required to "button up" the modular units into a finished home, particularly the carpentry completion and mechanical hookups. Because this latter knowledge is so fundamentally important, I join those industry leaders who suggest that modular manufacturers sell their homes only to dealers who are able to provide complete turnkey services themselves or through another closely affiliated GC. Ultimately, the customers have the right to decide for themselves whether to use a modular dealer as their GC, but the modular dealer should at least be able to offer those services.

The reality, however, is that many manufacturers sell their homes to companies that don't provide construction services. It is not hard to understand these manufacturers' motivation. Each modular factory is set up to build hundreds of homes a year. A manufacturer is most successful when it keeps its production as close to capacity as possible, and it can more easily do this by selling to a wide variety of sources. Problems arise, however, when a modular dealer is unable to help his customers solve their GC problems. Although dealers with little or no construction experience can eventually acquire modular-construction experience, as I did, you may want to ask yourself if the dealers you are considering have already completed their education or will be doing so at your expense.

As this book makes clear, a customer who selects an independent GC, rather than a modular dealer who also serves as a GC, assumes greater responsibilities. For that reason, I describe the GC's responsibilities separately from the dealer's responsibilities,

even though the same company could do both. By discussing the dealer and GC functions separately, you can better understand what each of them must do to make the construction of a modular home a success. Should you work with a good GC who has limited modular experience or a good dealer with limited construction experience, you will be better equipped to enjoy their strengths and protect yourself from their weaknesses.

The use of the term *dealer* in this book will displease some industry professionals because it acknowledges the current, flawed situation in which some dealers do not offer construction services. But one of the goals of this book is to inspire some of the many good-quality stick builders to adopt modular technology and in so doing enlarge the ranks of those companies that are true modular builders.

I would also like to note that when I'm talking about the typical stick builder, I'm referring to a small custom builder, not a production builder who builds scores of practically identical homes in neighboring subdivisions.

When customers build a home for the first time, their experience is quite similar to parents having their first child. Friends and family tell prospective parents how hard it will be adjusting to the needs and schedule of their precious infant. They listen with excitement, sure that they understand and certain they'll be prepared. Few parents are truly prepared, as they later readily admit. You have to live through parenting to understand. You also have to experience the challenges of building a home to understand its demands and stresses. If you are wondering why I am being so candid, it's because I want your experience to be as free of surprises as possible. If this book does its job and prepares you for what lies ahead, you will be very glad when you've built your modular home, just as most parents are very happy when they've given birth to their first child.

I wrote this book primarily for those who hope to have some say in the design of their home, its building specifications and amenities. When you are done reading this book, you will know what to look for in a modular dealer, manufacturer, and general contractor. You will understand how to take advantage of the many design possibilities that modular construction allows. You will also understand which building specifications and optional features offer the best value. Should you need land or financing, you will know how to shop for it. If you are looking to build an addition to your current house, you will find out the many reasons why modular additions are such a good choice. And you will learn how to make all of these things happen on budget and on schedule.

Finally, it is my sincere hope that you will realize that no matter how much you learn from this book, your best course of action will be to find and work with an experienced, reputable modular dealer and general contractor.

1

Why Build Modular?

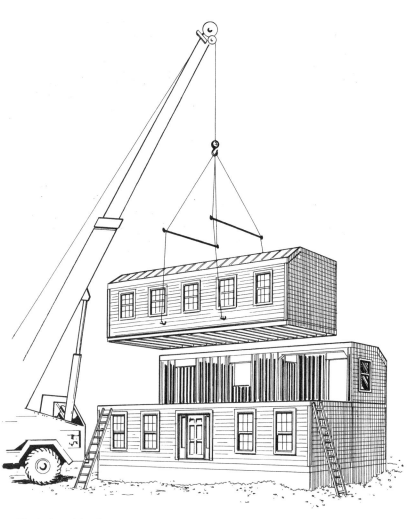

The last module of a two-story home being set on the foundation.

A MODULAR HOME IS BUILT out of boxes, called modules, which are constructed off-site, transported to a building lot, and assembled into a finished home. All of the materials, from framing, roofing, and plumbing to cabinetry, interior finish, and electrical, are identical to what you would find in a conventional "stick-built" home. The most striking thing about a modular home is not anything you can see, nor how it is made, but rather where it is made: in a modern factory designed to build good-quality homes.

Modular homes are one of four building systems that compete with stick-built homes. The other three building systems are precut homes, panelized homes, and manufactured homes. All four systems involve some form of factory work.

Stick-built homes are constructed from individual "sticks," such as 2×4s and 2×10s, that are delivered from a lumberyard and cut to size and assembled at the site. Since stick-built homes are constructed on-site, they are sometimes called "site-built" homes.

Precut homes are assembled out of the same basic sticks as conventional homes except that the correct pieces are prese-

lected and cut to size at a factory. Of the four building systems, precut "kits" provide the least factory assistance to the builder, since the factory does not assemble any of the building materials.

Panelized homes are constructed out of factory-built wall panels that are typically 8 feet tall and 4 to 40 feet long. Some panelized systems include plumbing, wiring, and insulation already installed ("closed-panel systems"), while others include only the framing and exterior sheathing ("open-panel systems"). The panels are assembled into a home by using a crane to set them onto the foundation.

Manufactured homes, known as mobile homes in the past, are similar to modular homes in that they are made of boxes. This similarity accounts for much of the mistaken impression that the two products are identical. The fact that both types of homes are manufactured in a factory also confuses people. The differences between modular and manufactured housing, however, are quite substantial. Manufactured homes are built only to preemptive federal codes governed by the U.S. Department of Housing and Urban Development (HUD), which among other things requires that they be constructed on a nonremovable steel chassis. Modular homes, along with stick, precut, and panelized homes, are always built in compliance with state and local building codes, which are much stricter than the federal codes. In addition, state building codes and local zoning regulations significantly limit where manufactured homes can be erected, but they do not constrain where modular homes can be built. Modular homes are manufactured, but that does not earn them the formal designation of manufactured homes. When I refer to *manufacturers* throughout this book, it should be understood that I am referring to manufacturers of modular homes.

Shown above is a crane lifting a panel into place on a panelized home.

Unlike a modular home, a panelized home is constructed from a series of factory-built panels.

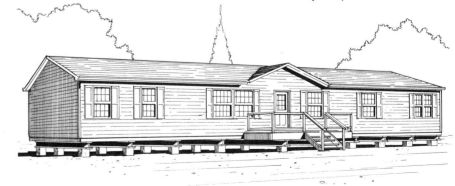

The Factory Advantage

All four building systems combine engineering know-how and factory-production methods to design and build more efficiently and with greater quality control. When done well, the efficiency results in lower costs and the quality control results in a better product. The idea of building homes, especially the components that make up a home, in a factory is not a recent phenomenon. Precut homes have been built in the United States since the 1890s. Americans began buying houses out of mail-order catalogs as soon as it was possible to ship the materials cross-country by railroad. Sears sold about 100,000 mail-order homes from 1908 to 1940. The use of production-line techniques again picked up after World War II and made a sizable contribution to reducing the housing shortage that developed after the war.

Even stick-built houses today use a growing number of mass-produced, factory-built components, including prehung windows and doors, roof trusses, interior moldings, drywall, and kitchen and bath cabinets. More and more aspects of home construction are being completed in factories because the factory environment helps to organize the construction process. By using automatic assembly equipment and repetitive assembly-line techniques, factories assemble component parts more efficiently and with greater consistency in product quality. This is true whether the components are assembled to make a window or an entire house.

Virtually all of the best products in the world, from computers and appliances to automobiles and planes, are manufactured in factories. That is why both consumers and industry professionals in Japan and Scandinavia consider the modular method of home building superior to site-built construction. This makes it ironic that the country that has led the world in the design and mass production of manufactured

On the modular factory's assembly line, materials are organized for greatest efficiency and quality control.

goods, the United States, took until the 1980s to embrace modular homes. Today, there is still a bit of a romantic notion that building a home on-site, piece by piece, is somehow superior. This belief lingers even though consumers would reject new appliances and automobiles that were built in someone's backyard, with the materials exposed to the weather and no one watching over the assembly.

Yet the romance with custom stick construction is starting to lose some of its bloom. According to the January 2004 issue of *Automated Builder* magazine, modular home sales grew by 34 percent in the previous four years, while production stick-frame sales grew by 16 percent. Many stick builders have converted to modular homes, driven in part by the severe shortage of skilled construction workers. This shortage is being caused by older, experienced workers retiring or choosing less physically demanding work and by younger people choosing other careers. In an ABC News study of 10,000 high school students that rated their interest in potential careers, the construction trades ranked 251st, right behind cowboy. In addition, those who are entering the trades are doing so with little formal training or mentoring. The problem is particularly severe in some trades, such as carpentry, where few companies have apprenticeship programs. A study by the National Association of Home Builders, which is made up primarily of stick builders, found that two of three builders are now forced to hire workers with skill levels below those expected for their jobs.

This labor shortage has eroded craftsmanship, driven up prices, and caused delays, shoddy construction, and unhappy homeowners. Frustrated by these problems, custom stick builders have turned to modular homes as a way to introduce some control into the building process. Modular manufacturers have, in turn, enticed them

by designing homes that meet the needs of builders' style-conscious customers.

Consumers in search of a custom-built home are also giving modular homes a more favorable look. Sometimes they turn to modular homes because they cannot get a stick builder to respond in a timely fashion. More frequently, superior quality, faster completion time, and better prices are the primary inducements, along with greater energy efficiency, extended warranties, and flexible design options. Customers who want high-quality finishes as well as high-quality construction increasingly understand that they can get both with a modular home.

The Quality Advantage

Many consumers assume that the primary advantage of modular technology is lower prices. That is only the third most important reason for building a modular home. Faster construction time is the second, and superior quality is the first. Here are some of the factors that produce the better quality that is the hallmark of a modular home.

The Workforce

A modular manufacturer does not hire carpenters, plumbers, electricians, roofers, and painters to build its homes, just as an automobile manufacturer does not hire mechanics to make its cars. In both industries, the manufacturer hires and trains people for the specific tasks required by factory production. The people skilled at hanging windows may not have learned to frame a set of stairs, and the people trained to frame the stairs may not be good at roofing, but both of them may be better at their jobs than the typical carpenter who has developed good overall skills but has not had the training or experience to master any one part of his craft. Factory-production workers are also likely to perform better because they are more closely supervised than the typical on-site construction worker.

Factory labor tends to be more stable and reliable than traditional construction crews. The workforce is mostly employees who receive good benefits, unlike most construction workers building a stick-framed home, who are usually small subcontractors without any benefits. Factory production is not delayed or halted because a couple of workers get sick or find another job. When the factory workforce becomes strained, the manufacturer can move workers who have been trained to perform multiple jobs into the vacated positions. When the manufacturer needs to replace or add workers, it has training and supervision systems in place that can bring its new workers up to speed quickly.

The Equipment

By design, modular factories are able to use larger, more powerful, and more sophisticated equipment than on-site construction. Because all main components are assembled with jigs, precision control is ensured. This means that every home is more likely to be built exactly as it was designed, with all corners square and all walls plumb.

The Framing Materials

The framing materials used by modular manufacturers need to be of very good quality to function with the precision equipment in the factory. Only the highest-quality, kiln-dried lumber can be used, because green, warped, badly knotted, or splintered pieces would slow down the finely tuned assembly-line jigs. This means that better materials, applied with tighter specifications, go into modular homes.

In the past, wood used in construction was of higher quality. It was more likely to have come from old-growth trees that produced clear lumber with few knots. Today, the lumber industry uses more unseasoned, juvenile lumber. A higher percentage of this lumber is green, which is more likely to dry unevenly, causing warping, twisting, and splitting. A modular manufacturer can reject lower grades of lumber without putting itself at risk for material shortages because it buys materials for multiple jobs, which requires it to closely monitor and maintain sufficient inventory. The manufacturer has personnel to inspect the materials and test them for moisture content and the time to submit a replacement order to the supplier if the materials are unacceptable. Knowing that a manufacturer will inspect materials more closely than a site builder and return materials when they do not meet the manufacturer's standards, most suppliers will make a greater effort to meet the manufacturer's needs for good-quality materials. The typical custom stick builder, in contrast, orders only enough materials for one job at a time from a supplier that is usually a few miles away from the site. He does not always have the time to inspect the materials before the subcontractors who will install them come to do their job. And the subcontractors cannot afford to wait for replacement materials when poor-quality materials are discovered. Consequently, the inferior materials are more likely to be used.

Climate Control

When materials are delivered to a modular factory, they are stored under cover, protected from the elements, and assembled indoors. Building in climate-controlled conditions enables manufacturers to avoid weather-related defects caused by rain and snow, strong winds, freezing temperatures, and searing heat. After a modular home is built at the factory, it is wrapped tightly with protective coverings for the journey to the site. This prevents unwanted moisture from saturating the lumber, drywall, and insulation. Keeping the moisture content down also reduces the chances for mold to grow. Within hours of removing the coverings, the set crew makes the home as weather tight as possible, even when the siding installation is delayed. This means that the interior materials are better protected from the weather.

A site builder's materials are exposed to the weather as soon as they are dropped off at the site. They continue to be subjected to the elements while the home is being framed. Only when the home is finally closed in, which can be weeks later, is it protected from inclement weather. Even good-quality materials swell when soaked in a rainstorm. Once wet framing materials dry out, they shrink, twist, and bend, causing problems such as bowed walls, drywall cracks, squeaky floors, and protruding nails or screws.

A climate-controlled factory environment offers two more advantages. First, it enables modular companies to avoid weather-related delays that prevent a customer from moving in on schedule. In fact, it allows homes to be built all year long. Second, it eliminates the poor workmanship that can result when construction crews have to work in either very cold or very hot conditions.

Building Codes and Standard Specifications

Modular homes are constructed to meet or exceed local and state building codes. There are several reasons why a manufacturer would exceed the building code. First, it has to build homes strong enough to be transported and set up, which is a factor that does not affect site builders. Second, if a manufacturer sells homes in several states, it may be more economical to build them all to the same code specifications, which means they would be built to the most demanding code. For example, some manufacturers use photoelectric smoke detectors, carbon-monoxide detectors, and housewrap for all of their homes even though only some of the states they serve

require them. Third, manufacturers can often purchase materials for a very good price in large volumes. Fourth, manufacturers realize that some customers are skeptical about the quality of modular homes, so they try to appeal to these customers by building their homes better than those of local site builders. This also allows them to distinguish themselves from their modular competition.

Many site builders follow a different logic. They build their homes to the minimum code standards required by the state or local building departments. For many site builders to succeed against their local competition, they have to cut costs by downgrading their building specifications.

Construction Standards

In the world of wood-frame residential construction, modular homes offer unparalleled strength. They are designed to be transported safely over long distances at highway speeds and lifted by a crane onto a foundation. Major components such as walls and floors are fastened together with both nails and special adhesive, with the adhesive providing a stronger bond than nails. Site-built homes are typically only nailed together.

Manufacturers employ several special techniques to strengthen the framing system at its stress points. For example, most manufacturers build their floor systems with a double perimeter band rather than the single band used by stick builders. This makes the floors exceptionally strong and rigid. Since they are built on perfectly square jigs, the floors are also perfectly square. The structure of two-story homes is also strengthened by installing both floor and ceiling joists on all modules. Consequently, the first floor of two-story homes has a ceiling that is independent of the floor of the second story above. A stick-built two-story home makes the first-story ceiling do double duty as the second-story floor. In addition to making for a stronger home, the modular method also helps reduce noise transfer between floors.

Many manufacturers also use metal plates to join the tops of intersecting interior and exterior walls. In addition, they use steel straps along the bottom and top of each side of the "marriage wall," which is where the two modules join together. This ties together the wall studs, the bottom plate of the wall, and the perimeter of the floor as well as the wall studs, the top plate of the wall, and the perimeter of the ceiling. The set crews then join each side-by-side module at its marriage wall with carriage bolts in the basement. The resulting basement carrying beam, which is made up of four or six members, is very strong. But strength is not its only advantage. Unlike the large dropped beams that divide most basements and reduce headroom, the integral carrying beam of a modular home sits flush to the ceiling. This is a significant advantage when installing the plumbing and heating systems, since the pipes and ducts can run directly across the basement

ceiling without obstruction from a beam.

Modular manufacturers bond drywall to the ceilings and walls with a sprayed-on adhesive that is considerably stronger than the screws or nails used by site builders. The only exception is one side of each interior wall, which is attached conventionally. (Since the adhesive dries instantly, it must be applied to the framing and drywall while they are touching. But it is not possible to reach the backside of the second wall after the first wall has been glued.) In addition to strengthening the framing, this method produces ceilings and walls that are more easily finished, and it eliminates the annoying "nail pops" that often plague stick-framed homes. The ceiling finish is also improved over stick construction by placing the ceiling joists crown-side up on the jig before attaching the drywall, which creates a flatter ceiling. Using larger sheets of drywall, which is impractical for most site builders, reduces the number of joints. The use by some manufacturers of ⅝-inch drywall on the ceiling eliminates the sagging you sometimes see in ceilings made with ½-inch

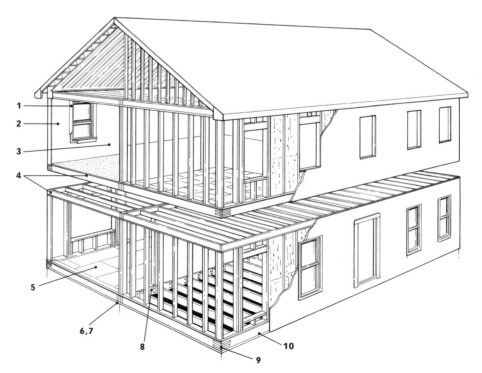

The construction techniques that strengthen a modular home include:

1. Window and door casings reinforced with metal splines in corners
2. Drywall bonded to wall studs and ceiling joists with super adhesive
3. Larger sheets of drywall used on walls and ceilings
4. Separate first-story ceiling and second-story floor systems on two-story homes
5. Floor sheathing glued and nailed
6. Flush basement ceiling with main carrying beam built into floor system
7. Double or triple perimeter joists used at marriage walls
8. Modules joined together at marriage wall with bolts in basement and attic
9. Corners of perimeter floor joists strapped
10. Double rim joists front and back of house

drywall, the standard in stick-framed homes.

There are many other details that make typical modular construction superior to typical stick construction. For example, modular manufacturers take extra steps to support plumbing pipes so they do not rattle or vibrate when used. To reduce the chance of the floor twisting, many manufacturers strap the corners of the perimeter floor joists. To reduce the possibility of

Speaking from Experience

THE TECHNIQUES UNIQUE to modular construction not only make homes strong enough to be transported and set on a foundation, but they also afford them more protection against any violent force. Some of the best stories about the structural durability of modular homes, involving homes that fell to the ground during the delivery or set, are ones the industry is most ambivalent about telling. They fear that customers will think these accidents happen frequently or that they might be expected to accept a damaged module, neither of which is true. Not to tell these stories does a disservice to the industry. What offers a more vivid testament than the image of a module sliding off a carrier as it exits an interstate at 30 miles per hour and rolls down a hill, turning over six times, surviving with only a few drywall cracks and a broken window?

Fortunately, my company has had only one such experience in 18 years. The house was a two-story Colonial with an attached in-law apartment. The two small modules for the apartment were shipped on the same carrier as two of the modules for the home. When my set crew lifted one of the two-story modules from the carrier, the carrier tipped over and the small module fell off the carrier and rolled over on its side. As shocked as I was by the accident, I was even more shocked by the condition of the module and my customer's response. The module survived with only a

few cosmetic cracks in the drywall, a broken pane of glass in a window, and an exterior door out of alignment. I apologized to my customer and said that I would build him a new module as soon as possible. He said that was not necessary, since the module was still in very good shape. I explained that I did not feel right having him take a damaged module, but he insisted. After the general-contracting work was completed on the home, none of us could tell that anything unusual had happened. Several years later, we have yet to encounter any problems with the module.

This sturdiness accounts for the many stories of modular homes surviving without serious structural damage after being subjected to hurricanes and tornadoes, something no stick-built house could withstand. It explains why a modular motel was found to be the only thing standing after Hurricane Hugo hit a North Carolina town, and why the Federal Emergency Management Agency (FEMA) gave the following report on Hurricane Andrew in Florida: "Overall, relatively minimal structural damage was noted in modular housing developments. The module-to-module combination of units appears to have provided an inherently rigid system that performed much better than conventional residential framing. This was evident in both the transverse and longitudinal directions of the modular buildings" (publication number FIA-22, February 1993, page 29).

squeaky floors, some manufacturers use nails and lag screws to fasten the interior walls to both the floors and exterior walls. Some manufacturers reinforce the corners of door and window moldings with metal splines, which keep the miter joints of interior trim tight.

Portability

Modular homes can be readily relocated from one building site to another. In fact, manufacturers and dealers often relocate a model home from a home show, where it was completely finished, to a model-home center, where it is reassembled, and then later to a customer's property, all without ill effect. Stick-built homes are occasionally moved, but generally only over short distances, at very slow speeds, and with a great deal of anxiety.

Energy Efficiency

One hallmark of good quality is energy efficiency. The most cost-effective way to increase energy efficiency is to reduce air infiltration, which is usually the single

biggest cause of heat loss. In fact, air leakage accounts for 25 to 40 percent of the energy used for heating and cooling a typical home. Failing to reduce the movement of air in and out of a home through all of the cracks and holes in and around the walls, ceilings, floors, windows, and doors can increase your energy bill even more than poorly insulated windows, walls, floors, ceilings, and doors.

Standard modular construction techniques tend to reduce air infiltration significantly better than site-built practices because they do a better job of sealing the breaches. For example, caulking along exterior sheathing seams and window flanges and hand-packing insulation around electrical fixtures help to eliminate infiltration of cold air in the winter and hot air in the summer. Sealing all penetrations to the building envelope, such as behind electrical outlets and switches on interior and exterior walls and around plumbing pipes at wall openings, also lessens air infiltration.

Modular manufacturers can be more successful at sealing a house because they

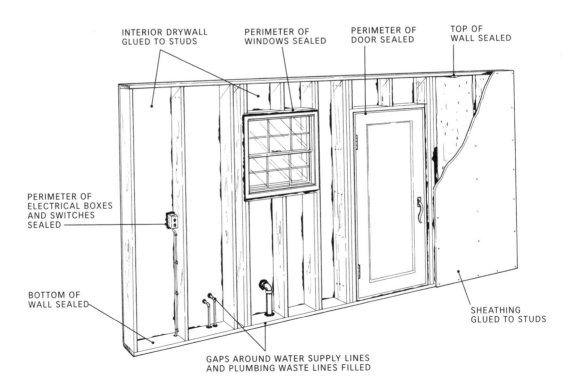

Standard modular construction practices, such as sealing air penetrations around electrical outlets and plumbing pipes, make a typical modular home more energy efficient than a typical stick-framed home.

INTERIOR DRYWALL GLUED TO STUDS

PERIMETER OF WINDOWS SEALED

PERIMETER OF DOOR SEALED

TOP OF WALL SEALED

PERIMETER OF ELECTRICAL BOXES AND SWITCHES SEALED

BOTTOM OF WALL SEALED

GAPS AROUND WATER SUPPLY LINES AND PLUMBING WASTE LINES FILLED

SHEATHING GLUED TO STUDS

have the luxury of building from the inside out. Site builders are compelled to build from the outside in. Building in a climate-controlled facility enables a manufacturer to install the drywall before the exterior sheathing. If a site builder did this and it rained, the water would damage the drywall. So a site builder needs to install the exterior sheathing before the drywall, which makes it impossible later to get behind the drywall to seal around electrical fixtures, plumbing pipes, windows, and other areas that allow air to penetrate the building envelope. There are special techniques and products that site builders can deploy to compensate for this shortcoming, but they add time and expense. Modular manufacturers need only to seal the breaches before they install the exterior sheathing, an easy task, to dramatically reduce air infiltration.

Evidence of the superior tightness of modular homes is readily available from the U.S. Environmental Protection Agency's Energy Star program. One of the goals of this program is to promote greater energy efficiency in new home construction. The tightness of the home is measured with what is known as the blower-door test. This test attempts to suck the air out of the house when all of the windows and doors are closed. The rate of air leakage, as measured by air pressure gauges, indicates the tightness of the house's shell. Tests performed by the Hickory Consortium for the U.S. Department of Energy showed that when a typical modular home is finished correctly by the set crew and general contractor, it does considerably better on the blower-door test than a typical stick-built home. While stick homes can be as energy efficient as modular homes, the stick builder needs to take special steps to achieve the energy efficiency obtained by a typical modular home.

The tighter construction of modular homes provides additional benefits. Closing off the air gaps in the walls retards the spread of fire, which allows occupants more time to exit their home safely and call the fire department. Reducing air infiltration also reduces unwanted moisture in the house, and it blocks bugs and rodents from gaining entry.

One potential downside to tighter construction is that stale air tends to build up over time. Fortunately, this can be easily remedied by adding supplemental information on ventilation. Some manufacturers now offer this as a standard feature. See page 114 for more detail.

Quality Control

The superior quality of modular construction is built on a foundation of strict compliance with state and local building codes. The first step in ensuring compliance requires a review of the proposed building plans drawn by the manufacturer. The state in which the home will be erected or a state-approved independent inspection agency completes this step. If the plans do not comply with the required building codes, they are sent back to the manufacturer for reengineering. Once they are approved, the manufacturer can build the home. Site-built plans rarely receive this much expert review since few local building departments have either the time or the engineering expertise.

Once construction begins, modular homes are subjected to far more inspections than site-built homes. At each stage, from framing to finishing, a manufacturer's quality-assurance inspector, armed with extensive checklists of performance standards, monitors the work for code compliance and craftsmanship. The electrical systems are checked to ensure that all components and wiring work as designed. The plumbing systems are examined to ensure that all pipes and connections are tight and in good working order. When problems are discovered, they are immediately corrected at the factory. The additional costs incurred by this quality-assurance system are more than compensated for by the reduction in service work in the field.

Specially trained inspectors from a state-approved independent agency complete an additional inspection of every home before it is shipped. They verify that each home is constructed in compliance with the state building code. These routine inspections are both planned and unannounced. When an inspector finds a code violation, the manufacturer is informed so it can remedy the problem and fix the breakdown in its quality-control program. If the modules are in compliance with the applicable building code, the inspector provides documentation in the form of a label affixed to the modules.

Modular factories use quality-control checklists to ensure that each home is built to the manufacturer's standards. This is a checklist used by Manorwood Homes of Pine Grove, Pennsylvania.

MODULAR HOME QUALITY ASSURANCE INSPECTION FORM

MODEL NO._____   Size _____

SERIAL NO._____   Style _____

DATE:_____   N/A = NOT APPLICABLE

Foreman, Q.C., or Designate to check "YES" if completed satisfactorily, "PART" if partially inspected with further inspection needed, checking "YES" once item is finished and inspected completely or "NO" if not satisfactory. Once item has been corrected, foreman, Q.C. or designate to check "CORR".

	PART	YES	NO	CORR
Drawings checked for PA Exempt Note				
(1) FLOORS				
Framing construction/Correct Fastening				
Water lines installed & supported				
D.W.V. installed & supported				
Heat duct assembly & support				
Joist Hangers Installed Properly				
Bridging Installed Properly				
Mandatory Checkpoint				
Decking Installed/Fastened/Gapped				
Vinyl installation: Correct Style/Color				
Serial number				
Checked by:				
Comments:				
(2-3) WALL CONSTRUCTION				
Exterior wall construction				
Marriage wall construction				
Interior wall construction				
Insulation: R-value/Stapled				
Drywall Screws Flush				
Glue Properly applied				
Checked by:				
Comments:				
(2-3) INTERIOR WALL SET				
Wall to wall fastening				
Fireplace installation (If app.)				
Stair Rise/Run (If app.)				
Checked by:				
Comments:				
(2-3-4-5) ROUGH PLUMBING				
Tub & shower installation; Correct Style/Color				
Furnace installation				
Water heater installation				
Vanity p-trap, water lines & vents				
Kitchen sink supply, drain, vent				
Gas line risers & shut-offs				
Gas line supports				
Gas line grounding to chassis				
Penetrations sealed/firestopping				
Checked by:				
Comments:				
(2-3) Exterior & Marriage Wall Set				
Exterior & marriage wall to floor fastening				
Column straps (If app.)				
Wall to wall fastening				
Checked by:				
Comments:				

	COMPLIANCE			
	PART	YES	NO	CORR.
(4) ROOF ASSEMBLY				
Truss construction & spacing				
Truss to edge rail fastening				
Ridge beam proper size & fastening				
Ceiling board fastening				
Vapor barrier installation				
Overhang construction/Flip eaves				
Gabel Overhang				
Checked by:				
Comments:				
(4) ROOF SET				
Roof to sidewall fastening				
Roof to marriage wall fastening				
Roof to interior wall fastening				
Roof to end wall fastening				
S/L Attic Floor Decking: Size____ # of pcs.__				
Checked by:				
Comments:				
(2-3-4-5) ROUGH WIRING				
Box installation correct size				
Smoke detector location				
Panel box installation (in home)				
Wire protection				
Securement and support of wires				
Ext. wall wiring/protection				
Penetrations sealed/firestopping				
Checked by:				
Comments:				
(5) Back Paneling				
Drywall Nailed/Glued/Screws Flush				
Drywall Perimeter/Inside Nailed				
Mandatory Checkpoint				
All cuts &corners taped				
Checked by:				
Comments:				
(5-6) ROOFING & ROOF WRAP				
Roof insulation & eave venting				
Penetrations sealed/firestopping				
Roof sheathing installation				
Dormer construction (If app.)				
Shingle Underlayment				
Drip edge				
Shingle: Correct Style/Color				
Shingle installation				
Ridge Blocks (OSB or Plywood)				
Roof Wrap plastic correct				
Proper sticks and staples used				
Checked by:				
Comments:				
(5-6-7) Sand & Paint				
House sanded				
House painted				
Checked by:				
Comments:				

APPROVED

APR 1 3 2004

PFS CORP.

Customers are often impressed by the fact that factory construction involves a series of inspections to ensure that good building practices are being followed. Local building inspectors of stick-built homes are not required to inspect for quality workmanship, only code compliance. If something is done poorly but meets the building code, the inspector must pass it. The addi-tional inspections are a big advantage of building a modular home.

Warranty

Most custom stick builders provide the minimum warranty required by state law. In most states this is a one-year warranty on all features. Most modular manufacturers provide an extended warranty for little or

---

MODULAR HOME QUALITY ASSURANCE INSPECTION FORM

(5-6) SIDING	COMPLIANCE			
	PART	YES	NO	CORR.
Sheathing installation (Proper Gap)				
Mandatory Checkpoint				
Window installation/Sealed				
Exterior door installation/Sealed				
Siding installation: High staples/Loose fit				
Soffit installation: Not screwed too tight				
Siding: Correct Style/Color				
Bath Exhaust				
Checked by:				
Comments				

(5-6-7) Frame Set				
Correct Size Carrier				
Home set on carrier properly				
Home secured to carrier				
Checked by:				
Comments:				

(8) FINISHED ELECTRICAL				
Light fixture installation				
Exterior receptacle				
Range hood installation				
Crossover connections (If app.)				
Visual Polarity (If app.) Exhaust Fans				
Fireplace				
Fluorescent lights				
Dropped Panel Box Completion (if needed)				
Checked by:				
Comments:				

(7-8-9) FINISHED PLUMBING				
Water closet installation Correct Style/Color				
Water inlet through floor				
Island Installation Soaker Tub/Whirl Pool				
Showerhead Installation/Secured				
Vanity Bowl Installation Correct Style/Color				
Furnace, w/h flue, combustion air intake				
Checked by:				
Comments:				

(8-9-10) FINAL FINISH				
Furnace fire stopped & flue connection				
Water heater fire stopped & flue connection				
Optional appliances installed/secured				
DWV schematic & components (If app.)				
Carpet (Correct Style and Color)				
Heat Registers/Boots				
Cabinets: Installation/Style/Color				
Countertops: Installation /Style/Color				
Bath Fixtures Correct Model/Color				
Appliances: Correct Model/Color				
Interior Doors and Trim				
Checked by:				
Comments:				

**Mandatory Checkpoint-Mandatory Inspection: Unit must be signed off by Q.C. or designee before next work process is performed. **

(9-10) TEST/LABEL/INSPECT	COMPLIANCE			
	PART	YES	NO	CORR.
Operational/Polarity Test				
Continuity Test				
Dielectric Test				
Supply Line Test				
Drain Line Test				
Gas Test high pressure				
Gas Test Low Pressure				
Install GFI				
Panel Box Labeled				
Access panels installed/labeled				
Panel box labeled				
Mandatory Checkpoint				
Drywall/Paint Inspected				
Checked by:				
Comments:				

(10) Final Wrap/Inspection				
Final close up and wrap				
Amp Connectors covered & secured				
House Lagged properly				
Checked by:				
Comments:				

(9-10) QC Labeling				
DATA PLATE INSTALLED				
STATE LABEL INSTALLED				
THIRD PARTY LABEL INSTALLED				

Note: All comments and problem clarifications to be written in space below or on the back of this page.

APPROVED

APR 1 3 2004

PFS CORP.

no additional charge. In addition to a one-year warranty for workmanship and materials, the typical program includes a two-year warranty for plumbing, electrical, and heating problems and a 10- or 15-year warranty against structural defects. A nice feature of these warranties is that you can transfer them when you sell your home, which adds to its resale value.

Modular manufacturers tend to use more brand-name products, which provide good quality and come with their own comprehensive warranties. Manufacturers of these established brands are likely to be around to provide warranty coverage.

The wholesale distributors who supply the modular manufacturers add another layer of warranty assistance, since they provide strong support to companies that do volume purchases. Few custom stick builders can command as much attention from their suppliers as a modular manufacturer who purchases materials for hundreds of homes a year.

Modular and Do-It-Yourselfers

Many people want to build their own home but are afraid of getting in over their head. This book does not encourage you to take the do-it-yourself route. In fact, it strongly recommends hiring a professional GC with modular experience. If you are going to take the leap, however, modular construction is a much better way to act as your own GC while building your first home than stick building, primarily because there is less work to do and it is more defined and manageable. Also, if you want to complete some of the construction tasks yourself, you are less likely to be overwhelmed by the work because the tasks are more limited in scope. This book will not turn amateur home builders into modular professionals, but it will help you make the right decisions and stay on budget and on schedule.

Planning Discipline

Building a modular home compels the customer and the dealer to complete most, if not all, of the planning steps before beginning construction. That is because the manufacturer cannot begin construction without knowing what it is going to build and how much it is going to charge. This forces the customer and dealer to make final decisions about the home's design and its building specifications.

Some customers building a custom stick home complete their planning tasks with the help of an architect, others with the help of a builder. But it is much easier to skip some of these steps when stick building. To some extent, customers and builders can make things up as they go along. These planning steps sometimes get short shrift when stick building because it takes a large investment of time on the part of both the customer and the builder, just as it does when building a modular home. When pressed for time, both parties often prefer to start construction and postpone the meetings and decisions to a later date. Modular dealers would be just as prone to do this if they could get away with it. The planning discipline imposed by modular construction is an advantage for customers, because as most banks will tell you, poor planning almost always leads to cost overruns.

For example, customers who do not spend enough time thinking through the design of their home often discover some design flaws once construction has started. One week they notice that the foyer closet is too small. The next week they decide that it is better to make the dining room smaller so the kitchen can be larger. The following week they realize that the door to the study will be swinging into a bookcase. When they make discoveries like this, they have the choice of accepting the layout as is or asking the builder to make the necessary changes. This might include tearing down the walls he has already built. If they decide to make the changes, all of the additional costs are passed on to the customer through expensive change orders.

Delayed planning can also cost a customer money when "allowances" are used to cover the cost of materials that have not yet been selected, such as cabinets and flooring. This is sometimes done so that construction can be started before every selection has been made. The builder assigns a dollar amount for those items yet to be specified. The allowances are adjusted, often higher, depending on the cost of the customer's final decisions. If you build a modular home, you and your dealer will not have many opportunities to procrastinate on your decisions, and you will not be subject to as many unbudgeted expenses.

The Time Advantage

Building a modular home will normally save you time compared with building a conventional stick-framed house. Quicker construction time saves you money and reduces the stress of home building. The faster you build your home, the sooner you will be able to enjoy it and dispense with the worries of construction.

Construction Time

When a manufacturer is building without a production backlog, it delivers a house five weeks after the customer has finalized the plans and obtained a building permit and financing. The general contractor begins the site work and completes the foundation while the manufacturer is building the house. If the house is small and uncomplicated and requires little on-site construction, and the GC is not too busy, it will be completed in about three weeks. Thus, a prospective homeowner can move in eight weeks after finalizing the preliminary steps.

In most regions of the country, the fastest stick builders require at least 13 weeks to build a small, uncomplicated custom house. Many need several weeks longer. Even the most efficient stick builder is significantly limited by small subcontractor crews. His systems are no match for the modular manufacturer's assembly line and large workforce. Consequently, the simplest stick-built house completed by the most competent custom builder is likely to take at least five weeks longer than a comparable modular home. A typical two-story house that includes a garage, porch, and deck will take a few weeks longer for both the modular builder and the stick builder.

A larger, more complicated modular home, such as a 7,500-square-foot two-story, will take considerably longer to build. If the design requires substantial custom finish work and multiple site-built structures and the GC has a substantial backlog, it might take 16 weeks or even longer to complete the home after it is set on the foundation. Adding the five weeks required by the manufacturer to build and deliver the house produces a total of 21 weeks. A good custom stick builder is likely to take at least a couple of months longer than a modular builder to complete such a project, since he will have considerably more work to do from scratch than the modular builder, who will have most of the house built at the factory.

You might have heard of stick builders completing homes in two months. These are production stick builders constructing cookie-cutter homes in their own very large subdivisions. In a sense, their speed is also a product of an assembly-line system. They bring the factory to the field by having each construction trade go from house to house, repeating what it had done at the last house. They offer fewer designs and less customization than even the most limited modular manufacturer.

During peak building periods, whether seasonal or during construction booms, the total time required to build both modular and stick-built homes increases, often substantially. The modular manufacturer's lead time grows and the modular GC's completion time increases. But a good custom stick builder's backlog also grows and his completion time similarly increases. The typical modular manufacturer, however, is better able to respond to greater demand by boosting production than the typical small custom builder. In addition, the modular GC's typical scope of work is substantially less than the scope of a custom builder, so he is much less dependent on overbooked subcontractors and can therefore get things done more efficiently.

The Cost Advantage

Modular manufacturers are usually able to build homes more affordably than custom stick builders for the same kinds of reasons that automobile manufacturers are able to build a new car for far less than a mechanic building the same car in his garage. These reasons include:

• **Volume purchasing.** Buying materials in bulk allows modular manufacturers to secure significant discounts.

Modulars Are Not for Everyone

Not every style and design of home can be built economically by a modular manufacturer. The most significant limiting factor is the size of the modules that can be driven from the factory to the job site. Federal, state, and local laws limit how wide and long each module can be. Fourteen-foot-wide modules are the most popular, and most conventional house designs can be comfortably built with these modules. Now that many manufacturers are building 16-foot-wide modules, even more conventional designs can be built without sacrifice in layout or style. However, there are many contemporary designs that are too expensive and impractical to build as a modular home. If you prefer such a design, you will need to have it stick-built.

One rule of thumb that puts this constraint in perspective is that if a design cannot be built as a modular home, which means it cannot be built out of rectangular boxes, it will likely be more expensive to build regardless of the type of construction used. In short, designs with multiple bump-outs and roof angles are always more expensive to build. If you want to get the most home for your money, you will likely want a design that can be built as a modular home even if you ultimately decide to build it as a site-built home.

Although most building sites can take delivery of a modular home, there are some locations that require a redesign of the house plan into smaller modules or enough extra site work that building a modular home is not practical. Narrow approaching roads with hairpin turns, lots on the side of steep hills, and very narrow properties can pose challenges. The only way to know if a building lot can comfortably receive a modular home is to have a modular dealer visit it.

- **Lower labor costs.** Experienced factory workers are well paid, with good benefits, but they do not command the wages paid to professional carpenters, plumbers, electricians, and painters.

- **Assembly-line efficiencies.** Organizing the construction process with inventory controls, power tools and equipment, and quality-control systems enables the factory crews to produce a home with greater efficiency than a handful of construction workers building in the field.

- **Less material loss.** Inclement weather does not damage the materials.

- **Less pilferage and vandalism.** Factories are easier to secure than building sites.

- **Less waste.** By working with standardized dimensions, modular manufacturers are better able to make optimal use of materials and avoid waste.

When material scraps are generated, manufacturers reuse them when they can. Many companies use pieces of drywall to brace the backs of their drywall seams as an effective way of reducing cracks. They also recycle their waste, which gives them a direct financial return, reduces landfill costs, and helps protect the environment.

The likelihood of saving money on a modular home grows as the size of the home grows. Likewise, the financial benefits increase for a modular home that is easy to build, has few upgrades, and requires no unusual work by the GC. That is why larger two-story homes that avoid the most expensive optional features are often the best deals for a customer. Most people who build larger homes do not receive all of this benefit, however, because they select more costly designs and upgrades, such as complex roofs, hardwood and tile floors, solid-surface countertops, and upscale cabinetry.

A custom modular home is not always less expensive than a comparable custom site-built home. It depends on the construction costs that are typical in the region. In areas with high labor costs, modular homes tend to be less expensive than site-built homes. Consequently, modular homes compete very favorably in most suburban areas. In fact, modular homes can be quite inexpensive when built in communities with exceptionally high construction costs, such as Greenwich, Connecticut. In sparsely populated rural areas with low site-built construction costs, modular homes may not offer any cost advantages (although they can still offer the other advantages discussed in this chapter). This may change if the shortage of skilled labor continues to result in rising labor rates, as has already happened in many areas.

To compare the costs of modular versus site building, you need to make sure you are comparing apples with apples. For a modular house, add the dealer's price to the general contractor's price for finishing the house at the site. In an area with average construction labor costs, you can expect to save about 5 percent over a comparable site-built house for the part of the house built by the manufacturer and GC. You will not save any money on the other construction work, since the GC will charge the same amount to clear the land, drill the well, install a septic system, install a driveway, build a foundation, and landscape the property regardless of the type of house being built. He will also charge the same for building a garage, deck, porch, and other site-built structures.

The cost to deliver a modular home to the customer's site can often be significant. Buying from a nearby factory can reduce the cost. But if the closer factory is located in an area that has substantially higher labor costs than the other factories serving your area, you probably won't save money.

The construction costs for a modular home are always fixed in advance of construction. Most modular dealers and manufacturers offer guaranteed prices for long periods of time, even in a market of moderately rising material and labor costs. To secure this price lock, customers need to authorize their home to be built, sign off on all plans and specifications, and have their order accepted by the manufacturer. This process greatly reduces the possibility of cost miscalculations and overruns in the future.

Many site builders have in their contracts a price-escalation clause that remains in force until the house is finished. In a typical construction season, prices are unlikely to rise by more than the overall rate of inflation. But every few years there are some dramatic price increases, usually caused by sudden spikes in demand due to natural disasters or building booms. If material prices were to rise sharply during construction of a stick-built home, the customer might be responsible for a big jump in costs.

To be sure, even when building a modular home, the GC may include a price-escalation clause in his contract. But this would not apply to materials used at the factory in building the house. It would affect only those materials used for site-built

Speaking from Experience

OVER THE YEARS, we've built a few modular homes that were designed by architects. In each case, the finished product was among the best-designed homes we've built. The suggestion to build the contemporary Cape shown on page 154 as a modular home came from the architect, Robert Coolidge, AIA, of Branford, Connecticut. Rob had first thought about renovating my customer's existing home, but quickly realized that building new would make more sense. When they calculated what the local custom stick-builders would charge and how long they would take, they decided to take a closer look at modular construction.

Rob, my customer, and I spent a lot of time going back and forth with design ideas, discussions about how modular homes are constructed, and general cost information. We had several meetings in which Rob showed me some preliminary drawings, and I told him what we could and could not do and what it would likely cost. He then used the information to refine his design.

Since Rob had little prior experience with modular homes, I helped educate him about the size of the modules and the structural requirements. Rob and my customer benefited from a trip to AvisAmerica, which manufactured the modules. On the visit, they saw how the homes were built and spoke directly with the engineering department. The brainstorming with the engineering staff was very helpful. The education was not one-sided, however, since Rob taught me about the custom possibilities for modular homes. I still think I learned the most.

When we began working together, we had hoped to build the two stories out of four modules. But in the end we decided to build the second story, including the entire roof, on-site. This required a lot of on-site work, but the finished product was exactly what we all had envisioned. Anyone who is unfamiliar with the possibilities of modular construction would swear the home was built by a custom stick builder. But attractive design and quality construction were the only things it had in common with good stick construction, since it was built faster and for less money.

features such as garages, porches, and decks. Consequently, the overall impact on the cost would be smaller than if the clause applied to the entire house.

Savings with Modular Financing

Many customers save even more money on the construction financing of their modular homes than they save purchasing it. As will be discussed at length in chapter 9, a construction loan disburses funds as work is completed. You are required to make interest payments on the outstanding balance each month. The significantly shorter construction phase for a modular home means fewer monthly interest payments to service the loan. This will considerably reduce your financing costs.

Lenders like modular homes because customers are unlikely to incur large cost overruns, the home will be built on schedule, and the quality is predictable. The construction inspections are also easier to complete for a lender, since a modular home is mostly built by the first inspection.

Saving on Insurance

Insuring a modular home after it is built is identical in coverage and cost to insuring a site-built home. However, insuring a modular home during construction will save money compared to insuring a site-built home due to the shorter construction time.

The shorter construction period also lessens your exposure to the typical risks that attend construction sites, such as vandalism and the pilferage of construction materials. Vandalism is further curtailed because the modules can be secured more rapidly than a site-built home. The ability to quickly secure the modules also makes it more difficult for someone to steal construction materials. Pilferage is further reduced because of the size of the modules; you cannot walk off with a module in the way you can carry away a few boards of lumber. Completing the home more quickly also reduces your biggest financial risk, that of a personal injury to a contractor working on the job or a neighborhood child playing around the home after hours.

Reselling a Modular Home

Customers often ask whether a modular home will be worth as much as a comparable site-built home when it is first built and later when it is resold. As far as professional bank appraisers are concerned, the answer to both questions is yes. When a bank appraiser assesses the value of a site-built home and a modular home that are built to the same specifications and located in the same neighborhood, she applies the same appraisal rules to both homes and comes up with the same value. Likewise, most people shopping for a new home evaluate a house with their eyes in terms of its perceived quality. Few people have any idea who built a house or how it was built, and most really do not care. Once a house is constructed, its resale value is determined by how it appears to potential buyers, not its pedigree. As long as you select a good house design, equip it with desirable amenities, have it built by a quality-conscious manufacturer and general contractor, and locate it in a desirable community, you will do very well on resale with both appraisers and customers. The same holds true for stick-built houses. In other words, neither form of construction has an advantage when it comes time to resell.

Designs and Amenities

In the late 1970s, modular homes had the reputation of being affordable, as long as you were satisfied with mediocre finish quality and limited design choices. Today, modular homes are still affordable, but with first-rate quality and a wide selection of creative designs. Modern assembly lines are producing unique custom designs that were not possible a few years ago.

All modular manufacturers have an array of standard plans that are traditional in function and style. Many manufacturers supplement their standard offerings with more modern designs that could pass for a contemporary architect's creation. Many of them, in fact, have been created by architects, as that profession has begun to explore the possibilities of modular design. Manufacturers have learned to use wood and steel beams to open up floor plans, and they have borrowed from both classical and contemporary elevations to give their homes attractive exteriors. Steeply pitched roofs, vaulted and cathedral ceilings, reversed gables, and angle bay towers are just a few of the design elements that grace modular homes today. When complex house plans cannot be built completely at the factory, a few of the cutting-edge manufacturers build as much of the house as possible at the factory, with the rest to be completed on-site. Robert Coolidge's design, discussed on page 22, is an example. As the cost of site-built homes increases, you can expect to see more of this hybrid construction.

Many manufacturers are willing to help people design a plan that expresses their personal preferences. You may start with one of their standard plans or a design of your own. Computer-aided design (CAD) has made it possible to prepare custom plans quickly and accurately. As long as it can be built economically in their factory, most manufacturers will consider building it.

The design work begins with the dealer, but the modular manufacturer completes the engineering when it prepares production plans, which detail the actual construction of a house. The combined design and engineering services provided by the dealer and the manufacturer are substantially greater than what is provided by custom site builders, who are usually small and without the necessary time and resources. Most dealers and manufacturers charge very reasonable fees for their services. When a customer needs detailed design assistance to construct a site-built home, he often must obtain it from an architect, who rightfully will charge thousands of dollars for the work.

Manufacturers routinely customize a home's interiors as well as exteriors according to a customer's tastes. Skylights, central vacuum cleaners, whirlpool tubs, and bay and bow windows are just as common in modular homes as in site-built homes. Hardwood floors, solid-surface countertops, and cedar siding can be installed either at the factory or on-site. After a home is set on the foundation, the GC can customize it with a porch, balcony, deck, sunroom, or garage. In short, modular homes can be built to be every bit as luxurious as stick-built homes.

Modular construction has become the construction method of choice for the quality- and cost-conscious house buyer. The best-built modular homes are better built than the vast majority of stick-built homes. Modular homes offer better materials, state-of-the-art construction technology, superior finished quality, and time and money savings. And you gain all of these benefits without sacrificing either design or amenities.

2 Selecting a Dealer

SHOPPING FOR A MODULAR home means, first and foremost, finding a modular dealer who offers the house you want at a fair price and can back this up with good support. Many customers find the process to be difficult and confusing. It is not like shopping for a new car or refrigerator. There are no consumer magazines offering comparisons and recommendations. Unless you know several people who have already built modular homes and can offer some advice based on their experiences, it may be hard to take the first step.

Part of the difficulty rests in the fact that dealers (and all house builders, for that matter) offer different levels of craftsmanship and service, different materials and techniques that typical customers know little about, price variations that seem to make little sense, and a package of choices that can be more overwhelming than enticing.

This chapter is addressed to customers who want to build a modular home but do not know where to start. You may or may not have a building lot, or the financing finalized, or a budget or specific house style in mind. But you will need to find a dealer. In this chapter you will learn what a modular dealer's job is, or what it should be, and how to evaluate and compare dealers based upon their craftsmanship, service, price, and products.

You should begin the process of selecting a dealer as soon as possible. It can take many weeks to identify candidates, verify credentials, and receive and review estimates. There are many things you can do first when shopping for a dealer. You can check out the candidates' homes at model-home centers or open houses. Examining some actual products may give you an insight into the level of quality offered by each dealer. For similar reasons, you can visit each modular factory if it offers tours.

Instead of focusing on the finished product, you can design your house plan or determine your building specifications by shopping on the Internet and looking

Modular Home Resources

The Modular Building Systems Association
Phone: 717-238-9130, Web site: www.modularhousing.com

National Association of Home Builders
Phone: 800-368-5242, Web site: www.buildingsystems.org/3.html

through books on house building. You need to determine both of these before you can ask modular-dealer candidates to provide an estimate. Yet you may not be able to complete your house design or list of preferred specifications unless you start meeting with dealers, since you will need their assistance to determine what they can do. Whichever way you begin, you will not want to spend a lot of time with any one dealer until you have some confidence he can provide the services you need.

The best way to identify prospective dealers is to speak with customers who have built a modular home. The most reliable recommendations come from people you know. If you do not know anyone who has built a modular home, you can still find useful guidance from a list of previous customers that the dealer should be willing to share with you.

Other potential sources of recommendations include the local chapter of the National Association of Home Builders (NAHB) and local building inspectors, general contractors, lenders, realtors, architects, and attorneys. If local resources do not provide you with a sufficient list of potential dealers, use the Internet. Internet searches have a major advantage over the Yellow Pages, another good source of prospective dealers, because the Web sites of manufacturers and dealers often contain a wealth of useful information. The Modular Building Systems Association and the National Association of Home Builders both have Web sites that list many of the modular manufacturers. Since you will not know much about the candidates that the Internet and Yellow Pages produce, you will need to spend more time checking their credentials. Whether you contact a modular manufacturer via the Internet or a phone call, the manufacturer is likely to put you in touch with its dealer in your area. Ideally, you will select at least three dealer candidates, each representing a different manufacturer.

Dealers and Manufacturers

Shopping for a modular dealer also means shopping for a modular manufacturer. To make the right decision, you need to evaluate both the dealer's services and the manufacturer's homes. Even when the dealer and manufacturer are the same company (that is, when the manufacturer sells its homes through a company-owned retail center), you will want to evaluate the services separately from the homes.

Most manufacturers, however, sell their homes through independent dealers. In this situation, it can sometimes be difficult for a customer to judge where the contributions and responsibilities of one end and those of the other begin. Visiting model homes and talking with past customers will help, but your actual experience of a manufacturer's attributes will be largely filtered through the dealer.

In fact, customers seldom discover the true differences among manufacturers. For example, quality differences are sometimes difficult to identify when comparing model homes that have been dressed up to look their best. Model homes mislead customers into thinking most manufacturers offer equal quality. Customers can have a difficult time sorting out whether price differences for a similar plan are due to the dealers' retail prices or manufacturers' wholesale prices. In addition, it often seems as if different manufacturers are willing and able to build similar plans. Most dealers work with their manufacturer to help their customers build any of the standard modular sizes and designs typical of the industry. Customers might conclude that what really matters when designing a home is how much design assistance the dealer provides. Indeed, customers often determine that the most important factors when shopping for a modular home have to do with the dealer's services. It is common for customers to feel they are buying a home from a dealer without regard to his manufacturer's abilities. The personal contact provided by dealers reinforces this perception, making modular dealers the face of the industry.

Realizing that most customers attach greater importance to the role of the dealer than to the manufacturer, many dealers feel comfortable selling more than one manufacturer's homes. Some even change manufacturers every year or two. They realize that most customers are buying a home from them because of the personal support and services they provide rather than a manufacturer's product quality and services. Conventional builders feel the same. They know customers buy their homes not because they use a particular brand of window or cabinet or because they buy from a certain supplier. Customers purchase their homes because they provide superior craftsmanship and services for a fair price or because they offer a superior location.

There are exceptions to this categorization. They almost always involve a modular manufacturer who has been able to build an identifiable brand name for itself. This tends to occur most often in communities close to the manufacturer's factory. It is also enhanced when top-quality modular dealers forge long-standing relationships with a manufacturer. In these situations, the manufacturer's reputation grows along with the dealer's. The power of a dealer, however, is particularly evident when the dealer breaks this bond and selects another manufacturer, causing the manufacturer to adopt a new dealer. In this case, new customers are more likely to follow the established dealer (and his new manufacturer) than the old manufacturer (and its new dealer). This reinforces the feeling of many dealers that they are more important than the manufacturer.

The fact that customers and dealers sometimes depreciate the contribution of the manufacturer does not mean this is the wise thing to do. A smart customer will want to consider the real differences among man-ufacturers in specifications, craftsmanship, and warranty service. It is not that the manufacturers' contributions are more important than their dealers'. It is just that you need to get the best that you can afford from both. You can do this only if you know what each is responsible for contributing.

Determining a Budget

If you have not yet completed your financing or obtained a building lot, you will need to do both soon. Begin by talking to a mortgage lender so you can determine how much you can borrow. This helps you avoid two common mistakes: setting your sights on a home that is more expensive than a lender will approve and purchasing one that is less than you actually can afford. Ask the lender to tell you in writing how much it will lend, what the fees and other expenses will be, and how much you will need to invest as a down payment. You should not commit to a lender, however, until you know your dealer's and GC's policies. See chapter 9 for more details on financing.

Next, look at real-estate advertisements and call a few realtors to find out what a typical building lot costs in the communities you are considering. To see what you will get for your money, visit a few of these lots. See chapter 6 for more details on building lots.

Once you know how much money you will have and how much the land will likely cost, you can determine a budget. Subtract the price for the land from the total amount of money you can spend (loan amount plus your deposit less lender costs) and use this price as the maximum you can afford for all costs associated with building a new home. You should create a contingency fund by allocating at least 3 percent of your total budget (you would need 5 to 10 percent if you were stick building). What is left will have to cover both the modular home and GC work.

Of course, you could start with a price from a modular dealer, since this will tell you how much money you have left to spend on land and how much money you need to borrow. Clearly, what you learn at each step will affect your evaluation of the others. Consequently, regardless of which tasks you take on first, you need to make some progress on all fronts soon after you begin.

Customers are most likely to receive the best that a manufacturer has to offer when the manufacturer has a committed relationship with its dealer. That is why most modular dealers form a close affiliation with one or two companies. They may change manufacturers every few years for one reason or another, but they prefer to do business with the same company or two because it affords them and their customers several advantages. Other dealers choose not to form an alliance with any particular manufacturers. Instead, they switch from manufacturer to manufacturer every time they sell a home, selecting whichever manufacturer is willing to give them the biggest discount for the home. Sometimes they will pass along the savings to you; sometimes they will keep it for themselves. Either way, this dealer strategy can cause problems.

If a modular dealer is loyal to a manufacturer, the manufacturer is going to go out of its way for the dealer and his customers when there is a warranty-service need. It is not that a manufacturer that sells one home a year to a dealer will ignore its warranty obligations. But it will not assume the financial burden of correcting a problem unless it is convinced of its responsibility, especially if it heavily discounted the house to get the dealer's sale. This may seem unfair to you, but businesses in every industry give preferential treatment to their most loyal customers.

Modular dealers often need the assistance of their manufacturer's engineering and sales departments. This is particularly true when a customer wants to build a custom design or select some nonstandard features. Manufacturers may not always be able to do what the dealer and customer ask, but they will make an extra effort to help a loyal dealer. Sometimes this extra consideration is a big reason customers are able to get the house of their dreams.

It is a challenge for anyone, not just a novice customer, to master all of the details (such as floor plans, standard features, optional selections, and prices) for one manufacturer. It is impossible to master them for several manufacturers at the same time. Each manufacturer will provide the dealer with its basic building specifications, but there will always be many details that are not contained in their lists. Dealers learn of them only by working closely with a manufacturer over a period of time.

When a dealer works with a new or little-used manufacturer, he does not always know what he is selling, since he does not always know what specifications are included in a given package. Likewise, the manufacturer may not know what the dealer thinks he is buying. Manufacturers and dealers learn a lot about each other's expectations and preferences through each new home they sell. Many dealers have their own standards that they expect the manufacturer to meet, but there is little opportunity for this information to be shared when a dealer jumps from manufacturer to manufacturer. For example, when a dealer orders a 9-in-12 roof, he might assume that he will receive storage trusses, which will give his customer usable attic space. Consequently, he may include it in his contract with the customer. When it comes time to order the home with a new manufacturer, however, the dealer may not think it necessary to inform the manufacturer of this specification, since it was standard with his previous manufacturer. If the dealer turns out to be mistaken, the customer won't feel any better knowing that the dealer will be obligated to remedy his mistake.

Dealers and General Contractors

A dealer's primary job is to design, price, and order a modular home from the manufacturer. The job of turning it into a livable home is that of a general contractor (GC). Most dealers also function as GCs or work closely with affiliated GCs, and this is a good route for you to follow. Acting as your own GC or working with one who does not have experience with modular homes can be risky, since you and your GC will not know the particular requirements of a modular home. If you are intending to select an independent GC, you should ask each modular dealer to identify the specific construction tasks required to complete your home, including the plumbing and electrical hookups, heating installation, and carpentry button-up. (See chapter 7 for a thorough discussion of the general contractor's responsibilities.) The dealer's written information should reduce the likelihood that you and your GC will underestimate the scope of work. Should the dealer not have the knowledge to provide this information, ask yourself what else he does not know, and who will provide the answers to your GC's technical questions.

As indicated earlier, an underlying industry problem is that few manufacturers or other industry professionals have developed systems to educate dealers, general contractors, or consumers about the modular-specific areas of construction, namely scope of work, procedures, and warranty obligations. Only a few dealers have developed their own comprehensive educational programs for consumers who want to act as their own GC or hire an independent contractor without modular experience. Several manufacturers and dealers have created checklists to enumerate these issues, but as anyone who has ever used construction checklists can attest, they do not do the job by themselves. For one thing, few dealers, GCs, or customers feel they have the time to study checklists. Second, reading alone is not the preferred method of self-education for most people when they are learning how-to assembly tasks, such as finishing a two-story vaulted foyer so that movement between modules, and thus future drywall cracks between floors, is

minimized. Illustrations, photographs, videos, and, best of all, professional hands-on training are all better teaching tools. For those GCs who have been building modular homes for several years, the poverty of educational tools is not much of a problem. But if you are trying to teach yourself or a GC new to modular construction without some assistance from a dealer, you will have an uphill climb. You will be much more vulnerable to misunderstandings and mistakes, which are likely to cost time, money, and stress. If you are in this situation, it would be worthwhile for you or your GC to take the very good two-day course in modular construction taught by The Penn College Modular Housing Training Institute (MHTI; phone 570-327-4768). Even if you don't use your modular dealer as your GC, you should give strong preference to those dealers who have the technical construction knowledge required to assemble a modular home.

Assessing Craftsmanship

The best way to evaluate the craftsmanship offered by modular dealers and manufacturers is to walk through one or more of their homes. No amount of words of reassurance from a dealer will be as helpful as seeing the quality with your own eyes. Most dealers will offer you a chance to visit a home in one or more of these situations:

- A home at the end of the manufacturer's assembly line
- A home immediately after it is set
- A finished home

Since each of these situations is different, it is important that you know what each situation tells you so you are not misled by what you see.

A HOME AT THE END OF THE MANUFACTURER'S ASSEMBLY LINE

Viewing a home at the end of an assembly line is the best way to learn about the manufacturer's level of craftsmanship and quality control apart from the contribution of a general contractor's finish crew. Any one home on any given week, however, may not be representative of what the manufacturer's typical home looks like. This is because all modular factories have some variability in the quality they produce, depending on the design of the home, which of their crews work on the modules, and whether they are behind or ahead of schedule. In addition, most manufacturers spend from one to three days fine-tuning the appearance of the modules after they come off the production line and before they are delivered to the dealer. This means you need to see a home after it has undergone this cosmetic makeover to see what it will look like when it is ready for delivery.

A HOME IMMEDIATELY AFTER IT IS SET

The delivery and set of a modular home will always impose minor stresses on the structure of each module. These will inevitably produce some visible symptoms such as small drywall cracks and slight misalignments of moldings and doors. Although these symptoms are easily fixed by a competent general contractor, it is worth knowing what affects their frequency. Longer modules with long open spans, where the walls between modules have been removed to make a larger room, are especially vulnerable to these symptoms. This will be less true for those manufacturers that sheathe the outside of the marriage wall, which joins two modules, or use some other comparable technique to strengthen the structure. Those manufacturers that temporarily brace the open spans during delivery also tend to have fewer symptoms. You will see more of these symptoms when a manufacturer uses old carriers with weak suspensions to deliver the modules.

Difficult delivery routes and rough site conditions will have the same effect. How the set is conducted, especially how the set crew picks up each module, will also affect the number of drywall cracks and misalignments. Using more straps for longer modules will provide greater support and thus reduce the stress on the modules. Consequently, when you view a home immediately after it is set, you will not necessarily know whether any drywall cracks and misalignments were caused by the manufacturer's construction, one of the other factors, or a combination of the two. On the other hand, should you see a home with few drywall cracks and misalignments, you can feel reassured about the work of all of the respective players.

A FINISHED HOME

The finished quality of all modular homes is a product of both the manufacturer's and the general contractor's efforts. Not surprisingly, some manufacturers deliver better quality than others. The same is true for GCs. A modular home will exhibit the best quality when both the manufacturer and the GC are top-notch. But you might be surprised what a very good GC can do with a home manufactured with below-average quality, or how poorly a well-built home can look when buttoned up by an ineffectual GC. When you view a finished home, whether a customer's home or a model home built for a manufacturer or modular dealer, you need to consider who served as the general contractor, and whether you will be using the same one to complete your contracting work. If you are considering using the same GC and you like the quality, you will not need to be as concerned with whether the home was delivered with this quality or it was created by the GC. On the other hand, if you do not find the quality acceptable, it probably will not matter to you who was responsible, since you are likely to look elsewhere for a dealer and GC.

It gets more complicated when you like

the quality but are not planning to use the same GC. Your eyes will not be able to tell you whether the manufacturer built the home with good quality or the GC did a superior job improving what the manufacturer delivered. You may be able to learn more if you can talk with the customer, but this will not be possible if you are looking at a model home. If the quality is less than you expect, you will have to look closely to determine who was responsible for the unacceptable workmanship. If the entire home appears to suffer from poor workmanship, chances are good that both parties share the responsibility. If you are not hiring the same GC who worked on the home or homes you inspected, it is a good idea to visit a couple of the manufacturer's homes at the factory or immediately after they're set.

The Dealer's Role

Finding a dealer who will sell and deliver a well-built modular home is not a difficult task. Finding one who is committed to providing a full range of services beyond that can be more of a challenge.

To order, deliver, erect, and service a modular home requires a significant effort by many people. It also requires someone to coordinate these efforts. Consider that each of the following has responsibilities that must be coordinated to build a home.

• The dealer's staff: to design a floor plan, select building specifications, and price the package

• The manufacturer's sales and engineering departments: to ensure that the dealer's order is understood and executed correctly

• The realtor: to find a building site

• The building inspector: to issue the required permits and approvals

• The lender: to provide financing

• The installation crew and crane: to set the home on the foundation

• The general contractor: to oversee the construction

• The manufacturer's service crew: to complete the warranty work

• The dealer's service crew: to assist with the warranty work

Because of the complexities involved in selling and building a modular home, most manufacturers do not, and will not, act as the dealer or general contractor. As they readily admit, they do not sell their homes directly to retail customers because it is too difficult and risky. To protect themselves, many manufacturers require the dealer's retail customers to sign a form acknowledging that the manufacturer did not sell them their home. This is to ensure that the retail customer will not be able to hold them accountable for mistakes or misdeeds by the dealer.

Automobile and appliance manufacturers do not ask you to sign such a form. What is so different about modular homes? The simple answer is that modular homes are not self-starting or self-installing. You cannot turn the key and drive away or plug them in and use them.

Like home building in general, modular construction is a complicated business. Today's modular dealer rarely sells a fully predesigned product. In the past, manufacturers limited their dealers to 20 or 30 engineered plans that allowed only simple changes. The dealer's primary job was to help the customer select one of the plans and then personalize it with the limited number of optional selections and colors that were offered.

Today, most modular homes are custom designed, with most of the design work done by the dealer and most of the engineering work by the manufacturer. Many dealers, and even some manufacturers, will tell you they do not build the exact same floor plan more than once in any given year. All manufacturers offer many optional selections, and most companies offer items not listed in their price books. With customers, dealers, and manufacturers continually pushing the envelope, customization puts a premium on an experienced sales, design, and engineering staff. Since you will be unlikely to obtain the necessary modular expertise or secure a manufacturer's engineering assistance on your own, you will need the assistance of a competent dealer.

In addition to not selling a fully designed product, the dealer is not selling a finished product. To make a modular home into a finished product, a general contractor must orchestrate the "button-up" efforts of carpenters, plumbers, electricians, heating contractors, and others. When manufacturers were primarily mass-producing "two-box" ranches with no clear-span openings, the GC did not have much to figure out or do. That is why phrases like *zip up* and *button up* arose to describe the finish process, even if they oversimplified the work involved. Now that modular homes have become more than simple ranches, the finish work has increased in complexity. Many of the more complicated designs require substantially greater construction knowledge and time to complete.

The increase in complexity has led to an increase in misunderstandings among manufacturers, dealers, and GCs about what is included by the manufacturer and what is required by the GC. The claim by some manufacturers that 90 percent of the home is completed at the factory is certainly a contributing factor. This is misleading, even if it makes for a nice headline. Not only are most novice retail customers confused, but so are experienced dealers and GCs. The problem would be manageable if the manufacturers always created fully documented assembly instructions for each custom plan. But this seldom happens, and the dealer is obligated to help the GC figure out what is required. This makes it necessary to select a dealer who understands both modular design and modular completion. If your dealer cannot provide this service, you and your GC can get in deep trouble very quickly.

In fairness to manufacturers, they make no pretensions to providing these services unless they have a dealership directly servicing the area. They know they cannot provide the direct assistance and local coordination required. They usually are too far away, and they have neither the staffing nor the retail and construction skills. Recognizing their limits, they insist on selling through regional representatives. They knowingly count on their dealers' salespeople and contractors to provide the necessary customer services. You would be wise to follow their lead and seek out a competent dealer.

Sales Assistance

While shopping for a modular home, you will be somewhat dependent on the knowledge, availability, and patience of the dealer. You want a dealer who will take the time to educate you about your many responsibilities and who will help you carry them out.

First and foremost, you need the dealer to provide a general understanding of the projected cost and schedule. He can begin at your first meeting by giving you a ballpark price. Although a ballpark price will not give you the level of detail you need to select a dealer, it will give you a reasonable idea whether you can afford to build with a particular dealer and how his prices compare to those of other dealers.

At a later meeting, the dealer should be able to give you complete itemized pricing that you can take to the bank, literally and metaphorically. To assess the likelihood that the dealer can provide this level of detail, ask to see a copy of his typical estimate. If it is one or two pages long, he is not providing the detail you need to protect yourself. This does not mean that he is unwilling to provide you with more detail, but you will need to ask.

If you find yourself over your budget when you receive a detailed estimate, ask the dealer to help evaluate your options. An experienced dealer will know which changes will maximize savings while requiring the fewest compromises. He will know what you can do with the house design to reduce its cost and which upgrades can be left out now because they can be added later when you will have a little extra money. A responsible dealer will try to help you build your home with the features you need most. Then he can help incorporate some of the additional features you want most. There is likely to be some give and take among your needs, wants, and budget. What will not help you is a dealer who is always encouraging you to build a house that exceeds your practical needs and your budget.

Once you begin discussing a ballpark price, you will also want the dealer to map out a ballpark timeline for the entire project. He should include the time it will take him to design and price your home, the time it will take his manufacturer to build it, and the time it will take a typical general contractor to complete the contracting work. He should also be forthright about the many kinds of obstacles that can derail the schedule, such as the time you need to make final decisions, the time required to obtain a construction loan and building permit, and the manufacturer's production schedule.

It might take you more than one meeting, but you will need to determine if the dealer has the knowledge and experience to help with your home design and selection of specifications, including optional features. Historically, residential architects have provided these services, since they are the best trained to do so. If an architect does not fit your plans or budget, however, you can take advantage of the fact that there is less to manage when building a modular

home and so complete these planning steps with the assistance of your dealer. If you work with an architect, have him help you select a dealer, since they will need to work closely together.

One way to assess a dealer's product knowledge is to observe whether he can tell you his manufacturer's standard features and optional selections without having to look them up every time you ask a question. Any dealer who exhibits this level of uncertainty does not know his manufacturer's product very well, which may be because he has not made the appropriate effort or sold many of the manufacturer's homes. A dealer should be able to provide detailed pricing for his standard offerings without having to contact his manufacturer for assistance.

Most customers have a lot of questions, no matter how much detail a dealer is able to offer in his presentation. So it is important to find a dealer who has the expertise, temperament, and time to answer questions. When a dealer is not able to answer a question, he should offer to find an answer and get back to you as promptly as possible. If a dealer lacks the knowledge or patience to provide you with the information you need to make an informed decision, he is probably not someone you should trust to build your home.

Design Assistance

Few customers are willing to take a modular dealer's standard plan and build it without modification. Whether they customize a standard plan or complete a custom design, customers need the help of a dealer who understands both the design constraints and the possibilities of today's modular construction. Such a dealer will also know how to make a custom plan that cannot be built as a modular home into one that will give you the essential features you want.

If you are building a custom design, assess each dealer's design skills by asking to see some of his custom designs, ones ideally done from scratch. Ask him to demonstrate his experience with the style and size of your preferred modular design. For example, if you are building a T-Cape with a cathedral ceiling and a finished second floor that crosses the T, ask the dealer if he has built anything like this before, since it has some requirements that are not typical of most modular homes.

As important and necessary as this level of design assistance is, it can be quite costly to a modular dealer in time and effort if you are requesting a custom plan drawn from scratch. You can receive this design assistance for free if you find a dealer who provides the help as a way of demonstrating his modular expertise and service commitment. Should you receive such assistance, you will want to give it appropriate weight when you are ready to make your decision. Some dealers, however, will not provide the necessary design assistance for free unless you make a commitment to purchase a home from them. This can pose a problem, since you will need a reasonable approximation of your plan to show each candidate so that he can provide you with a reliable cost estimate. If you find yourself in this situation, you will need to pay one of the dealers, a home designer, or an architect to complete a draft of your plan.

Attention to Details

Most people outside of the construction business do not realize how many details are involved in building a home. Nor do they realize how difficult it is to juggle so many details without making mistakes. That is why "attention to details" is such an important service, and one of the features of a good dealer. A dealer who is a stickler for details will help you stay on schedule and make sure that you receive the home you want. Some dealers are better with details than others. While you may not enjoy people who are obsessive in your personal life,

you will really appreciate a dealer with this characteristic.

Communication

Miscommunications, and the misunderstandings they cause, are common in construction projects. It may be asking too much to suggest that the problem can be avoided entirely, but there are steps that customers can take to keep it to a minimum.

First and foremost, a customer should have his dealer and GC put everything in writing. "Everything" here refers to much more than an order form and a contract. Many small details get discussed at sales meetings that do not necessarily find their way into these formal documents. Read a contract and you will likely find that it excludes "oral representations," which is another way of saying that if a detail was discussed and even agreed to verbally but did not find its way into some form of documentation, it has no legal validity.

One way you can learn a lot about a dealer's facility with details is to observe how he documents your conversations with him. Consistent and clear documentation is critical because every meeting between a dealer and a customer will generate a lot of discussion about a home's floor plan, exterior appearance, building specifications, features, and colors. Meetings also will include discussions about the building site, scheduling, banking, and budget. The only way customers can be sure of getting the house they want, on schedule and on budget, is if the dealer listens, understands, and documents all decisions and then follows through on them. Sales meetings can be a bit chaotic, with discussions being sidetracked and interrupted and decisions agreed to and then discarded after reconsideration. Putting all decisions reached during each meeting into writing is the best way to ensure that they do not get lost in the shuffle.

The dealer does not, and should not, bear all of the responsibility for maintaining this written record. The customer has an even greater interest in seeing that each and every detail relating to his new house comes to fruition. Customers can sometimes send mixed messages to a dealer, saying they want one option one day and a different one the next day. A conscientious customer, therefore, is one who never has to say to a dealer, "But don't you remember the conversation we had about . . . ?" If the conversation was important, put it in writing.

A dealer also has a professional obligation to communicate effectively. He should not resort to industry jargon without making sure that you understand it, and he should explain the important details fully. If you find yourself with a dealer you cannot understand, even after you ask him to clarify, you should find another candidate. If you ignore this advice and buy a home from this dealer, you should probably be prepared to believe him when he says, "Don't you remember I told you that?"

Complete Drawings

Before committing to a dealer, make sure he will provide you with a full set of house drawings. He should first provide you with a set of preliminary drawings based on your initial design and then follow these up with a second draft of preliminary drawings after you have revised the first. When you have made your final changes, he should turn the preliminary drawings into permit drawings, which will be given to the building department when you apply for a permit. The fee for preparing these drawings should be included in the base price of your home. Some dealers will provide you with additional drafts of your preliminary drawings for no additional fee, while others will ask you to pay for them. Find out how many sets of drawings are included by the dealer for no additional charge.

Each set of drawings should include elevation, electrical, and HVAC (heating, ventilation, and air-conditioning) plans in addition to the floor plans. The elevations depict all four sides of the home. Viewing these drawings is a good way of learning how the outside of your home will appear. The electrical drawings enable you to determine if your home will include sufficient receptacles, lights, thermostats, phone jacks, and cable jacks and if they are all in convenient locations. The HVAC layout allows you to see if baseboard-heating units or forced-air ducts are properly located.

Some dealers provide these additional drawings when they complete their preliminary drawings. Others routinely postpone creating them until they complete their permit drawings. A customer should insist on having the additional drawings completed before the permit drawings are made. That way, the customer will be able to make changes in the drawings before the permit is issued. Any changes made after the permit has been issued are likely to require a new permit.

Complete Specifications

A modular home is made up of hundreds of materials. Each material is selected and then installed by its manufacturer according to a set of specifications. For example, windows may be selected because they are maintenance-free, vinyl framed, double pane with low-e glass, and tilt for easy cleaning. Each manufacturer starts with a set of standard specifications for those components that are part of every home. Some of the components are described in terms that combine details about the materials and the installation, such as that the exterior wall is made up of 2×6 studs spaced 24 inches on center. Other components are specified in terms of a product brand, such as Merilatt Classic cabinets. Some manufacturers will add details about a product's features, such as that the kitchen faucet is made by Moen and comes with a single-lever handle.

Each dealer, with the help of his manufacturer, has an obligation to tell the customer the basic specifications for all of the important materials used in a home. You may think you are receiving this information from a dealer when he gives you his manufacturer's standard building specifications, but there are always many, many details that are not contained in these lists. In fact, the potential level of detail is far greater than most customers realize. Consider, for example, the many specifications you might want to know about your windows:

- Brand
- Warranty on glass and frame
- Size for each application (first floor, second floor, kitchen, bathroom)
- Even or uneven sash
- Operation (single or double hung, casement)
- Tilt-in function
- Screens (half, full)
- Interior sill (picture frame, extended)
- Interior jamb (paintable, stainable)
- Interior sash (vinyl, paintable, stainable)
- Exterior jamb (vinyl, aluminum, wood)
- Exterior sash (vinyl, aluminum, wood)
- Exterior color (white or other specified color)
- Color of screens and window hardware
- Installation of side-by-side windows (mulled, separated by framing)
- Window grills (interior, integral)

Compare this level of detail to the information that most modular dealers provide their customers: "double-hung windows, with screens." This is the written information that the dealers receive from their manufacturers, and for some customers this level of detail may be enough. Even if you want more information than this, it is unlikely that you care equally about each of the specifications listed above, and you may

not care at all about some of them. For example, you may care a lot about whether the window tilts, since you do not wish to climb a ladder on your two-story home to wash the windows. You may also care that the sash is wood, since you intend to stain it. But you may not care whether it comes with an extended windowsill or whether the exterior is wrapped in vinyl or aluminum. Someone else will almost certainly have different concerns and priorities.

You will want to know what is included, but also what is not included, with each of the products you purchase. For example, when you select an optional whirlpool, it might not occur to you to ask if an optional water heater is available to warm the recirculating water unless the modular dealer informs you that one is not included. You might learn that one is not available for the whirlpool you selected, or you might choose to forgo it because of the additional cost. Either way, at least you will know what you are getting, and perhaps have the option of making the choice yourself, rather than leaving it up to the dealer.

You will also want to know about features the dealer does not offer but that you might reasonably expect to be included. A typical example is hardware for the kitchen and bathroom cabinet doors and drawers. The cabinet manufacturers seldom provide the hardware as a standard feature, since they prefer to allow customers the chance to personalize their selection. Some modular manufacturers provide this hardware on their own, but others do not. If the manufacturer used by one of your dealer candidates does not, the dealer should volunteer this information. His silence would not be golden to your ears if you discovered this after your home was built.

It is tempting to think that you can avoid the demand on your time required to obtain and review the many component specifications for your home by relying on what you see in a model home. A model will show you what you will get if you build the same design, if it displays the manufacturer's most current products, and if you select the exact same products on display, including any costly upgrades. This rarely happens. Few modular dealers can afford to exhibit more than a couple of designs, and the probability is that you will not build the exact same plan as the model. Not many dealers have the time or money to update their models every time the manufacturer changes products, which happens at least once a year for some products and sometimes more frequently. It is not uncommon for a model home to show cabinets, windows, and plumbing fixtures that are no longer available. In addition, most dealers dress up their model with attractive upgrades that may not be in your budget. Finally, you may select an optional feature that is not on display in the model, since it is impractical for a dealer to display all of his manufacturer's product choices, even if he has multiple models that he updates regularly. A model home is unlikely to show you all of the products that you will get when you purchase a home.

In fairness to modular dealers and manufacturers, it is impossible for them to anticipate which details are important to you. They might want to put the burden back on you by suggesting it is your obligation to inform them of which details are important. But you are unlikely to know this until a dealer points out what he and his manufacturer are planning on giving you as part of their standard specifications and optional selections. Request that each dealer provide you with all of the important details in his written proposal. With this information in hand, you can complete a fair comparison of all the modular dealers.

Assessing the Building Site

If you do not yet have a building lot, you may need some help selecting one. At the very least, you will want someone to tell you which of the lots you are considering will be easy and affordable to build on and which will be difficult and expensive. Modular dealers who lack general-contracting experience may not be competent to recommend one lot over another, in which case you should seek the assistance of a GC.

However, you will need to rely on a dealer if you have concerns about whether a modular home can be delivered to a building lot you already own or are considering purchasing. This requires that the dealer visit the site, ideally before you sign a contract with him.

Before the site work begins, the dealer should meet with the customer, GC, and excavation contractor at the site to develop a plan for delivery day and set day. He should document his recommendations. These should include a drawing indicating where the modules will be stored for the delivery and where the crane and modules will be positioned for the set. After the initial site work is completed, the dealer should return to approve the excavator's work, and, if necessary, give additional instructions to ensure that the site is ready. If any of the candidates tells you he is unable or unwilling to assist your GC and his excavator in these ways, you should ask yourself if he has the required competencies and attitude to build your home.

House Set

Responsibility for hiring the set crew and crane should always be left to the modular dealer or manufacturer. A customer should refuse to hire the set crew and crane, even if a dealer promises it will save substantial money. There are several reasons for this. First, the set procedures require a great deal of specialized knowledge, skill, and teamwork that a crew acquires only through training, supervision, and experience. Second, because of the size and cost of the modular units, as well as the risks associated with the set procedure, whoever sets a home has substantial liability. Third, if a set is done poorly, the GC's job will be more difficult and the quality of the finished home may suffer as a result. Finally, the goal of a dealer who asks the customer to hire the crane and set crew is to hold the customer responsible for any problems with how the house goes together. Since he neither built nor set the home, the dealer can disclaim responsibility for the problems. It is best to avoid dealers who operate this way.

Product Samples

Making product selections is difficult without samples or brochures, and making confident color selections is difficult with small samples. A dealer should have samples or catalogs of his manufacturer's standard offerings and optional features, and he should have color samples large enough for you to make sound judgments. If a dealer does not have all of the samples you want to inspect, try to find the information on the product manufacturer's Web site. Think twice about selecting any feature you have not seen in person or in a brochure.

Warranty Service

Most modular dealers rely on their manufacturer to perform warranty-service work, even though their customers may live far from the manufacturer. Manufacturers want to correct their own problems and dealers often do not have the manpower to complete the work. Since most manufacturers have two or three service crews for each factory, they must rotate their visits through each of their dealer's territories. This may mean that the service crew will not make it to your area until a couple of weeks after you request service. When the request is made soon after your home has been set on the foundation, this delay is unlikely to pose a problem for the GC. But if the request is made just before the GC intends to finish his work, he may be delayed waiting for the service crew to visit. Some manufacturers handle this situation better than others. The best way to assess the warranty service of a manufacturer is to talk to past customers about their service. See chapter 10 for a more complete discussion of warranties.

A few of the larger modular dealers supplement their manufacturer's service crews with crews of their own. This usually gives them a significant advantage in providing timely service. Larger dealers that purchase many homes a year from a manufacturer often have another advantage. Their manufacturers are more likely to provide consistent and prompt service, since they want to keep their volume dealers happy. Smaller dealers are sometimes able to compensate for their size by providing more personal, hands-on assistance.

Model-Home Center

Modular dealers with a model-home center make a strong long-term financial commitment to their business. Their investment in one or more model homes can be thought of as a statement that they intend to be selling modular homes for the foreseeable future. This does not necessarily make them better dealers, but it can be comforting to know they will be there for you if you have a warranty-service problem.

Availability

The best modular dealers meet with new customers, complete detailed estimates, and coordinate their activities with those of the manufacturer, general contractor, lender, and building inspector. When they also serve as the GC, they complete detailed estimates, find subcontractors, supervise construction, and provide warranty service. To keep up with their responsibilities, they usually have to finish several different tasks each day. When they work without partners or employees, making them a sole proprietor, they accomplish this by limiting the number of homes they build, usually to one or two at a time. They also limit their customer meetings to scheduled appointments. Larger dealers provide these services by having several people do the different tasks. They are more likely to have a model-home center, often staffed by more than one person and open several days a week.

The size of the dealership, however, does not necessarily determine the level of service it provides. A large company that is always busy and understaffed for its volume will often fail to meet a customer's needs. A sole proprietor who makes himself available to his customers might actually give better service. On the other hand, many sole proprietors are unable to keep up with their responsibilities. Sometimes the first thing that goes is their return phone calls to you. This can be particularly troublesome when you are comparing dealer candidates or after your home is complete. Be particularly wary of a dealer who sells modular homes as a part-time job and has a full-time career in another business.

Discuss with each candidate how available he will be. Pay close attention to your personal experience. If you have a difficult time meeting with a dealer before you buy a home, it might only get worse once he has taken your order.

The order lead time of the dealer's manufacturer can also figure into your assessment of his availability. Some manufacturers have a backlog several months long at the same time that others can deliver a home in seven weeks. The longer backlog might be due to a large, multiunit project that is taking up several weeks of the manufacturer's production schedule or it may be due to the popularity of its homes. A shorter backlog may be due to the manufacturer's greater production capacity or the unpopularity of its homes. In other words, you will

not always be able to make a judgment about a manufacturer's quality by the length of its lead time.

Some customers use lead time as a decisive factor, even when the difference in time between their preferred modular dealer and the faster company is only a few weeks. An unscrupulous dealer can take advantage of this by promising a delivery time that cannot be met. By the time the customer realizes this, it is too late to back out. Even when the delivery differences among dealers are real and your personal situation makes this an important consideration, weigh carefully the relatively small additional cost and inconvenience of the delay against the advantages of purchasing a home from the dealer of your choice. These advantages include the many years you will enjoy the superior building specifications and craftsmanship. They can also include a large array of services before and after the sale that can make your experience of building a home a pleasure rather than a disappointment.

Experience

Having a modular dealer who is readily available will not be so helpful if he has limited experience with the profession's many responsibilities. You will particularly value a dealer's experience if you are designing a custom plan with multiple upgraded features. An experienced dealer is more likely to find creative design solutions that will please your taste while preserving your budget.

The only way a person working by himself can acquire the necessary modular experience is if he personally works in the industry for several years. In a larger company, no one individual other than the founder may have the overall knowledge of an experienced sole proprietor, but the staff's cumulative experience may exceed the individual's. This will only be true, however, if the larger company nurtures its staff with good training and supervision and is in turn rewarded by its employees with loyalty and low turnover. Again, size helps only if it is used well.

The best modular dealers provide all of the services discussed above. Be confident that a dealer can deliver these services before you decide to make him a finalist among your candidates.

How to Interview a Dealer

It will take a few meetings with each prospective dealer to learn enough about his building specifications, craftsmanship, price, and customer services to make a decision. During this time, look into each dealer's background, modular experience, and company size. Begin each meeting with a plan for acquiring this information. Use

Questions for the Dealer

Most customers will not need to ask every dealer each of these questions. Rather, this is a list of questions from which customers can pick and choose those that apply to their specific situation. If, for example, you have a friend who has built a house with the same dealer, you may already know quite a bit about the dealer's experience and reliability. If the dealer has also functioned as a general contractor for a number of years, you may not need to grill him about his GC experience, but you will want to ask him if he is experienced building the specific type of home you are considering.

- ☐ How long have you been selling modular homes?
- ☐ How many homes do you build a year?
- ☐ What type of homes do you sell mostly?
- ☐ Will you build a custom design if I bring one to you?
- ☐ Will you help me create a custom plan?
- ☐ Do you have other employees? If so, what do they do? How will they help me?
- ☐ Will you personally help me from start to finish? If not, can I meet the person or people who will?
- ☐ Which modular manufacturers' homes do you sell?
- ☐ How long have you been selling each of them?
- ☐ How do they compare in terms of:
  - ☐ Quality?
  - ☐ Standard building specifications?
  - ☐ Optional features?
  - ☐ Warranty service?
  - ☐ Price?
  - ☐ Production lead time?
- ☐ Are there other differences of note among the manufacturers?

- ☐ Which manufacturer would you use if you were building for yourself?
- ☐ How long do you anticipate it will take before you are ready to put my home in the manufacturer's schedule?
- ☐ Do you anticipate the manufacturer's lead time changing between now and the time we put my home in its schedule?
- ☐ What experience do you have as a general contractor?
- ☐ What experience do you have with helping customers prepare a site for the delivery and set?
- ☐ Who sets your modular homes?
- ☐ How can I contact you in the future?
  - ☐ Cell phone?
  - ☐ Work phone?
  - ☐ Home phone?
  - ☐ E-mail?
  - ☐ Fax?
- ☐ If I have a warranty problem after the home is delivered but before I move in, what do I need to do to get the problem fixed?
- ☐ How long can I expect it to take?
- ☐ Will you take responsibility for the manufacturer's warranty problems if the manufacturer does not?
- ☐ If I have a warranty problem after I move in, what do I need to do to get the problem fixed?
- ☐ How long can I expect it to take?
- ☐ When you sign a contract with a customer, do you have a "legalese" section that states the terms and conditions of the contract? If so, can I have a copy to review?
- ☐ How much of a deposit do you require?
- ☐ Under what circumstances is the deposit refundable?

the checklist on page 40 to help guide you through each interview.

Visiting a dealer's model home or a home under construction offers a good opportunity to ask questions. While small talk can help break the ice and allow customers to get a sense of how comfortable they feel with the dealer, it is important that they also choose the questions they ask carefully.

Throughout the interview, take note of the dealer's personality and how well it fits with your own. If you do not feel comfortable with the dealer or do not like his answers to your questions, find a gracious way to tell him candidly that you will not need his services. Do not string him along pretending he is still one of your candidates just to get him to complete a quote you can use for comparison. This is unfair to the dealer, and it is not in your best interest. The only estimates that are useful are those that come from dealers you are considering.

After completing this initial screening of dealers, and while awaiting written estimates, have each dealer's insurance company mail you an insurance binder; it is best to receive it directly from the insurance company, since it is fairly easy to fake the forms. This is the only way to ensure that the dealer has a current policy. Make sure that the coverage includes sufficient liability insurance and workers' compensation; ask your insurance agent for the recommended coverage. This will protect you if the dealer or one of his subcontractors is not fully insured and someone is injured on your property. You are also protected if there is significant damage to your property.

While waiting for the estimate, you may want to investigate the dealer's credentials. First, contact the Better Business Bureau and local consumer affairs office to see if any complaints have been filed against the dealer, and if so, whether they were handled satisfactorily for the customer. Second, ask the state attorney general's office if there have been any civil suits filed against the dealer. Third, ask the dealer for the names of his commercial bank, suppliers, and subcontractors. If he is in good standing with all of them, he should not object to you checking the references. Finally, ask the dealer to give you at least four customer references, and then contact each of them. Most people are happy to tell you about their experiences with a dealer, good and bad. If a dealer uses more than one manufacturer,

Questions for Previous Customers

☐ Are you happy with your home?

☐ Did the dealer do what he promised in a timely manner?

☐ Did the dealer hold to his price? Were there any overcharges?

☐ Were there extra charges because the dealer's allowances were too low to get you what you wanted?

☐ How did the dealer respond when he was asked to make changes? Were the changes documented and fairly charged?

☐ How good was the dealer's choice of building specifications, such as windows and framing?

☐ Did the dealer change the specifications from what was agreed to or expected?

☐ How good was the manufacturer's workmanship?

☐ Did the manufacturer take care of all of the little details?

☐ Did the dealer meet his deadlines?

☐ Were there any surprises? If so, did they cost you more money?

☐ Did the dealer help you prepare your site for the delivery and set?

☐ What did you think of the dealer's set crew and crane company?

☐ Did they do all that you thought they would do?

☐ What did you think of their workmanship?

☐ Did they work in a safe and efficient manner?

☐ Did the set crew or dealer clean up the site and get rid of all of the trash when the set was done?

☐ How well did the dealer honor his warranty?

☐ Did the manufacturer provide prompt service?

☐ Did you have any disagreements with the dealer? If so, how were they handled?

☐ Was the dealer easy to reach while your home was under construction?

☐ Would you buy another home from the dealer?

ask to speak to customers who had homes built by the manufacturer you are considering. Ideally, two of the references will have built in the last year and two will have built at least two years ago. Recent customers will be better able to tell you how effective the dealer was during the sales process, while past customers will tell you how well the house held up and the dealer honored his warranty obligations. Use the checklist on page 41 as a guide when questioning customer references.

Do not automatically reject a dealer based solely on one unfavorable reference unless the incident is egregious. Any dealer who has been in business for long is likely to have a few dissatisfied customers.

Obtaining Estimates and Comparing Prices

It usually takes more than one meeting with a modular dealer to get a complete price estimate that details all of the specifications and itemizes all of the prices. This is particularly true when the first meeting takes place at a dealer's open house, since you will probably be sharing the dealer's time with other customers. Even if the dealer is available for an extended meeting, you might be wise not to commit to a full-length appointment until you have reason to believe he deserves to be a finalist. Instead, ask him to provide you with a ballpark price.

You might be tempted to get a ballpark price by asking a modular dealer for a "price per square foot" for his homes. Many a dealer will be happy to answer this question, since it usually takes no effort to throw out a number. This is the wrong question, however, since the answer is almost always misleading and it will not help you compare dealers.

To calculate the cost of building a modular home, you must begin with the price of the home and then add in the price of the GC work. When you ask a dealer for a price, you need to be clear about whether the two of you are talking about one or both of these prices. Although any price based on square footage alone is problematic, a price per square foot for just the home is the most misleading price you can receive.

Price per square foot is misleading in part because the cost to build a home varies considerably depending on the design. For example, one-story homes are usually more expensive per square foot than two-story homes, if you keep everything else the same. The most important reason is that a one-story home of any size requires twice the foundation and twice the roof as a two-story home with the same square footage. In addition, modular designs that require more on-site construction will usually cost more per square foot than designs that are almost completely built at the modular factory, since the additional on-site construction will cost more per square foot than the part built at the factory.

Perhaps surprisingly, a one-story home can cost more to build per square foot than a two-story of the same size. This is because the one-story requires twice the foundation and twice the roof size, increasing the cost per square foot.

Square-Foot Price: One-Story vs. Two-Story		
1,650-SQUARE-FOOT HOME	**T-RANCH ONE-STORY**	**TWO-STORY**
Modular base price	$75,600	$70,500
Foundation (including excavation, drainage, basement insulation)	$13,000	$7,500
House plus foundation	$88,600	$78,000
Price per square foot	$54	$47

Comparing Square-Foot Costs

	MODULAR HOME ONLY	GC COMPLETION INCLUDED	ADD $36,000 IN OPTIONAL FEATURES	ADD $73,000 IN OPTIONAL FEATURES	ADD $122,000 IN OPTIONAL FEATURES
Price per square foot	$44	$75	$90	$105	$126
Modular base price (2,420 sq. ft. Whately 1 two-story)	$106,500	$106,500	$106,500	$106,500	$106,500
Excavation, foundation, button up, and misc. GC tasks	$0	$75,000	$75,000	$75,000	$75,000
OPTIONAL FEATURES					
Oak floor in living room and dining room	$0	$0	$3,300	$3,300	$3,300
Ceramic tile in foyer	$0	$0	$1,200	$1,200	$1,200
Misc. upgrades, such as cable jacks	$0	$0	$1,000	$1,000	$1,000
Whirlpool tub replacing soaking tub in master bath	$0	$0	$1,000	$1,000	$1,000
Cherry kitchen cabinetry	$0	$0	$2,300	$2,300	$2,300
Two-story walk-out bay in living room and master bedroom	$0	$0	$4,200	$4,200	$4,200
Cherry bath cabinetry	$0	$0	$0	$750	$750
Ceramic tile floor in kitchen and breakfast nook	$0	$0	$0	$3,100	$3,100
Ceramic tile floor in three baths	$0	$0	$0	$1,400	$1,400
Oak floor in three bedrooms	$0	$0	$0	$5,300	$5,300
42" oak stair treads and custom oak railings	$0	$0	$0	$2,400	$2,400
9-in-12 roof with storage trusses	$0	$0	$0	$0	$4,620
9' ceiling on first floor	$0	$0	$0	$0	$1,900
Wood windows	$0	$0	$0	$0	$2,000
Stained six-panel pine doors and stained moldings	$0	$0	$0	$0	$4,400
ADDITIONAL EXCAVATION WORK					
Septic (in lieu of town sewer)	$0	$0	$0	$8,000	$8,000
Well (in lieu of town water)	$0	$0	$0	$5,000	$5,000
Difficult site — trees, blasting, long driveway, additional fill	$0	$0	$0	$0	$20,000
OPTIONAL GC SERVICES					
24' x 24' two-car garage	$0	$0	$17,500	$17,500	$17,500
9-in-12 roof with storage trusses on garage	$0	$0	$2,200	$2,200	$2,200
10' x 18' deck	$0	$0	$3,500	$3,500	$3,500
Warm air heat and central A/C replace hot-water baseboard	$0	$0	$0	$8,000	$8,000
1,600-square-foot paved driveway	$0	$0	$0	$3,000	$3,000
44' x 6' front porch	$0	$0	$0	$0	$8,800
Masonry chimney	$0	$0	$0	$0	$7,500
Total cost	**$106,500**	**$181,500**	**$217,700**	**$254,650**	**$303,870**

Note: This chart contains approximate retail prices typical of modular dealers and GCs serving the Northeast as of this writing. Some areas within the Northeast have considerably lower prices, some considerably higher. There is even more variability across the country. It is possible to add another $75 per square foot to the price with additional features such as solid-surface countertops, cedar siding, decorative interior and exterior moldings, brass plumbing fixtures, upgraded lighting, and custom-painted walls.

You might try to make this question work by asking for a price per square foot for a given style of house. But there are many types of each style of home. Compare a typical one-story ranch to a large contemporary one-story with multiple rooflines and attic storage. The second type of design is significantly more expensive to build. Also, larger homes for any given style cost less per square foot than smaller homes of the same style, since there are many costs that do not go up appreciably when you make the house bigger. For example, almost all homes have only one kitchen and heating system, both of which get spread out over more square feet in a larger home. You might try to avoid this problem when asking for a price per square foot by being even more specific about the floor plan you prefer. But this will work only if the dealer has an idea of what type of specifications you expect.

You can add $10 to $100 per square foot to a home by upgrading from the basic standard specifications to a large number of expensive optional features, as the table on page 43 demonstrates. For example, stained cedar siding costs much more than vinyl siding, imported tile floors are substantially more expensive than vinyl, granite countertops come at a much higher price than laminate countertops. Yet none of these options affects the square footage of the house. Also, a home with a masonry fireplace, whirlpool tub, and central air-conditioning is more costly per square foot than one without these amenities, as is a home with a garage, porch, and deck. If you go through the entire house upgrading every product and adding amenities in this way, the price of the home, and therefore the price of the home per square foot, will rise significantly.

Site conditions can also significantly impact the cost of a home, regardless of its square footage. Installing a well and septic system, for example, usually costs substantially more than connecting to municipal water and sewer. So does building on a heavily wooded, hilly lot with a winding 200-foot driveway compared to a flat lot with no trees and a 50-foot driveway.

For these and other reasons, you can take most any house type and come up with a price per square foot that will be meaningless unless you know exactly what you are getting. You can avoid this problem by providing the dealer with a specific floor plan, preferred building specifications, level of amenities, and the site conditions on your lot. Once you do this, however, you are no longer pricing your home per square foot. There is just no good reason to ask a dealer for this price.

The Ballpark Estimate

A competent modular dealer can prepare a meaningful ballpark estimate in less than an hour if he knows something about the floor plan, preferred building specifications, level of amenities, and actual site conditions. The more complete the information, the more accurate the price is likely to be. Since it will be a ballpark price, it will be missing most of the specific details that go into a new house. The dealer will account for these details by making some assumptions. His assumptions, for example, may include the use of one of his standard floor plans as a starting point. The dealer may provide an allowance for optional features of a certain amount, or he may itemize an estimate for several of the more popular options the customer shows an interest in, such as upgraded kitchen cabinets and hardwood floors in the living room and dining room. You should ask the dealer to explain these pricing assumptions and how his estimate will change if they prove incorrect.

If he is including a ballpark estimate for the services of a general contractor, which he should do even if he does not provide those services, he should explain these assumptions as well. For example, he may

Ballpark Budget

But Don't Take To The Bank[1]

Customer ___TOM WATERS___ Date ___5-15-04___

Land Cost.....___ALREADY OWN___.. _____

Modular Home Base Price ....Specify Plan ___WHATELY 1___ | 106,500

   Series.......Heritage ( ) ...... Custom ( ) ...... Leader ( ) ..... New Horizon ( ) ....Mainstreet ( )

Modular Home Optional Features[2] .... | 16,750

Siding	_STD_	Exterior	_DECORATIVE MOLDINGS 600_
Roof	_ARCHITECTURAL 350_	Interior	_OAK STAIR TREADS 750_
Floors	_WOOD DR, LR; TILE FOYER 4500_	Fireplace	_GAS 2500_
Heat	_WATER BASEBOARD 850_	Whirlpool	_NO_
Cabinets	_CHERRY KITCH 2200_	Misc.	_ALLOWANCE 5,000_
Other			

General Contracting Services .... | 75,000

   Site-Work....... Excavation (✓) Town Sewer (✓) Town Water (✓) Foundation (✓)

   Button-up ....Interior Carpentry (✓) Exterior Carpentry (✓) Plumbing (✓) Heating (✓)

             Electrical (✓) Misc. _____ | _____

             Other _____ | _____

   Septic ___—___ Well ___—___ Trees + Other Site-Work ___—___ | _____

General Contracting Optional Services[3]

Two-Car Garage	_W/ ARCHITECTURAL SHINGLES 17,200_		17,200
Mudroom/Breezeway	_—_		
Second Story of Cape	_—_		
Porch	_—_	Deck _10' x 18' 3,500_	3,500
Masonry Fireplace	_—_	Air Conditioning _—_	
Misc.	_ALLOWANCE_		4,000
Other	_—_		

Total Cost.. | 222,950

[1] Our Ballpark Budget is designed to provide you with an estimate of our typical construction costs. It will help you decide whether to meet with us again to determine your costs more accurately and in detail. **We will provide you with a comprehensive quotation, which you _will_ be able to take to the bank with confidence, when we complete our Modular Purchase Order and General Contracting Services quotes (see below).** As you might guess, these quotes may be higher or lower than our Ballpark Budget, depending on your actual selections, the project's scope, the additional information you provide, and the conditions of your land.

[2] We offer hundreds of optional features in addition to those listed above. Examples include cathedral and 9' ceilings, lighting fixtures, circle and elliptical windows, exterior decorative moldings. We will give you a itemized list of these costs when we complete our Modular Purchase Order quote for you. You can make an appointment with us to complete your Modular P.O. quote at your convenience. (Note: we do not typically include appliances, such as refrigerator, dishwasher, stove, microwave, or washer, dryer.)

[3] Our General Contracting Services quote covers the **"site work"** done to prepare your land to receive the modular home plus the **"button-up"** work needed to complete your home after it is set on the foundation. In addition to the tasks listed above, it includes many smaller tasks, such as a bulkhead to the basement, basement insulation, trash removal, house cleaning, painting of exterior doors, etc. Our quote will provide you with an itemized list of each of these costs. However, our quote cannot include all possible costs. In addition, it will specifically exclude certain items you might need to budget for, such as a survey, perc test, septic design, utility fees, permit fees, landscaping, or paved driveway. Before we can complete our General Contracting Services quote, we will first need to look at your land.

Copyright Jan. 1994 by The Home Store, Inc.   Revised July 20, 2004

A dealer should be able to give you a relatively quick estimate of what a modular home might cost if you tell him a little about what you are considering building. The form above gives examples of the types of costs you need to consider.

assume that the property has municipal water and sewer along with overhead electric, all of which could change when the building lot is purchased. His estimate for the general contractor should also include any site-built structure that is being considered, such as a garage, porch, or deck. The sample shown on page 45 provides the minimal details that should go into the ballpark estimate.

Detailed Estimates

After selecting your modular dealer finalists, the next step is to solicit a detailed written estimate from each so you can complete an apples-to-apples comparison. If you are like most customers, you will not have selected a definite house plan and set of specifications when you requested the ballpark quotes. In this case you will first need to determine these and present them to each of the candidates.

If you decide on a plan that is copyrighted by one of the dealers or his manufacturer, ask the other dealers for their own comparable design, one that is similar in essentials such as basic style, square footage, number of bathrooms, and kitchen size. Since the plans will not be identical, you will need to take the differences into account. If you use your own custom design, one or more of the dealers may prefer to build it in a different way from the others. It will be important to learn this fact so that you can weigh the differences in design along with the other information.

When deciding on building specifications, you will likely find that each dealer offers a few different standard specifications and optional features. Your job will be to make them comparable. Once all of the dealers receive equivalent plans and specifications, they will be in a position to complete their written estimates.

Evaluate each dealer's estimate alone and then in comparison with the others.

Make sure that the dealers used the correct house plan. The estimate should specifically reference the plan by style and dimensions. Ideally, a copy of the plan will be attached. Having this verification will prevent you from considering an estimate that was completed on the wrong house plan. This mistake can happen when you are considering two different plans with the dealer and he confuses them.

The total price for each estimate should include a base price and a subtotal of the optional prices. The base price should cover the standard specifications for your house plan, and the optional prices should include the upgrades you select. You may find the costs for applicable taxes, delivery, crane, and set included in the base price or itemized separately. Since these costs will total thousands of dollars, be sure they are specified in writing in one place or the other.

If you use one of the dealer's standard plans with few modifications, the base price will be very easy to understand. The price will refer to the entire house plan you are building. If you make significant modifications to a standard plan, the base price may not include all of the components to the plan. For example, if you add 4 feet to the length of a standard plan, the added price may be shown as optional rather than included in the base price.

For a true custom design, the dealer can still generate a base price by starting with one of his standard plans. He would first select the standard plan closest in style, size, and features to your plan. Then he should compare the materials and features in the two plans that most affect the price, such as the quantity of windows, doors, cabinets, and bathrooms. Finally, he would adjust the base price of his standard plan to account for the differences. Alternatively, he or his manufacturer could create a special base price for your custom plan.

Customers are sometimes seduced by a

low base price. A base price that includes the most inexpensive specifications offers you the best value only if you can accept the specifications as they are. If another modular dealer offers a higher base price for a set of specifications that better meet your needs, you can compare the two companies only by upgrading the price of the more affordable company to match the specifications of the other and evaluating the adjusted total price.

Although a low base price does not necessarily equate to the best value, it can be advantageous when working within a tight budget. As long as the dealer complements his less expensive specifications with a wide range of potential upgrades, he affords you the opportunity to choose which specifications you want to include. A dealer who packs his standard specifications with more expensive materials offers a superior basic package but not necessarily a superior value. If you cannot afford all of the more expensive features, his top-of-the-line standard specifications are of little use to you. You are better served by a dealer who is willing to start with affordable specifications that you can upgrade according to your priorities.

When reviewing an estimate, customers tend to pay more attention to the optional upgraded specifications that incur an additional price than to the standard specifications included in the base price. This is understandable, since the optional prices are itemized and readily identified for closer scrutiny. They are also discretionary and easily eliminated for budgetary purposes. But if you are going to make a true comparison among estimates, you need to compare all specifications, whether included as standard or optional. This is particularly true when two or more estimates offer a similar price and you have upgraded only a couple of items.

For instance, let's say you have limited your upgrades to architectural shingles and a better level of carpet in the living room with three dealers and two of them have a similar total price. You may think there are no substantial differences in price between those two.

But if one dealer includes a steeper roof and nicer cabinets in his standard specifications, he is offering the better value. The only way you will notice this, however, is to read which specifications are included for no charge as well as which carry additional costs. In other words, equal prices do not necessarily mean equal value. It is also why your bottom line is not just the total price. Like all other consumer products, you need to learn what you are getting to understand who gives the better value.

After reviewing the estimates, you may decide to forgo a manufacturer's standard and optional selections so that your modular dealer or general contractor can supply a locally purchased alternative that you prefer. Many customers do this with the finished flooring, especially hardwood and tile, and siding, particularly wood and fiber-cement board. Some customers even ask a dealer to supply custom cabinets. If you take this path, a dealer may have you select a specific product with a fixed price or he may allocate an allowance that you can apply to a future selection. If you opt for an allowance, the final cost to you will be determined by the actual cost of your selection. Using an allowance is a reasonable practice as long as it includes enough money to get what you want. The only way to know this is to do some preliminary shopping for the item at the vendors used by the dealer.

Reviewing the Estimates

After reviewing the estimates, ask the best candidates to revise their proposals as necessary so you have comparable estimates. With the corrections in hand, weigh the price differences among companies against any differences in their specifications and their floor plans. Factor in the information you collected about the craftsmanship and customer service of the respective dealers.

When one dealer's estimate is significantly lower than another's, ask the higher-priced dealer to explain and justify his higher price. If there are two or more higher-priced dealers, take your questions to each of them. This way you can obtain the assistance of two or more modular home professionals in evaluating the lower-priced competitor's proposal. If that dealer's proposal is a worthy competitor, you will motivate the more expensive dealers to consider some concessions.

If you plan to try and get a higher-priced dealer to make price concessions, be prepared to convince him that a competitor's offer is equivalent in features and service. Few competitors will take your word for it, just as you would not take at face value everything the various dealers and GCs tell you. Show the dealer the competitor's written estimate. The more expensive dealer might be able to demonstrate that the low-priced competitor's proposal is inferior to his. You may be disappointed to learn this, since it will remove the incentive for the dealer to lower his price, but it will help you accomplish your ultimate goal: to get the best value, which is a combination of price, features, and service, and not just the cheapest price.

What do you do if the more expensive dealers tell you that they cannot match the lower price and stay in business? Most professional consultants suggest that you ignore the lowest estimate, if it is substantially lower, on the assumption that something is wrong with it, even if you cannot identify what it is. In most cases, you will receive inferior craftsmanship and customer service, and you will pay for these for years to come. If you receive only two estimates, however, you will not know with confidence if the lower one is fair. That is why you should try to get at least three estimates.

One explanation for a below-market price is that the dealer has not earned a reputation that merits normal profits. Since such a dealer still needs to make a living, he may limit his time selling modular homes so that he can earn his primary income from another job. Or he may increase his sales volume. Either strategy may enable him to compensate for his low unit profit margins, but both can create problems for his customers.

A dealer who sells homes on a part-time basis will only be available to service customers part-time. This may not be as big a problem during the shopping phase as it could become after delivery. When the general contractor has a problem or you have a warranty issue, the dealer may be unable to respond in a timely fashion.

A dealer who cranks up the sales volume in an attempt to compensate for minuscule profit margins may not be able to afford support staff. And a dealer without support staff will quickly become overextended in trying to help multiple customers. More important, this kind of dealer may not have the financial resources to help a customer with an expensive warranty problem. If he has a financial setback, he may not have the cash reserves to remain in business. Customers should be cautious about giving a deposit to a dealer who is in danger of going out of business.

Some modular dealers are able to sell their homes at below-market prices by purchasing them from the manufacturer that is offering the lowest price at the time. Although this can occasionally be a good business practice, it can cause problems for the customer. For example, a manufacturer selling a modular home at a desperately low price, because it has few sales, may not offer good craftsmanship. Since every dollar the manufacturer saves on labor is another dollar it can use to pay its bills, it may be compelled to lay off skilled workers and supervisors. A customer may need to visit the factory under these circumstances to see firsthand the current level of quality. Finally, a manufacturer in this financial situation poses a risk to both the customer and the dealer once the dealer sends the customer's deposit to this financially struggling manufacturer.

On the other hand, a modular dealer and manufacturer who insist on maintaining a comfortable profit margin are companies that are more likely to be in business for a long time. A dealer and manufacturer who stay in business will be able to stand behind their work and care about their reputations. No reputable dealer or manufacturer is going to sell a home for less than it has to. Dealers and manufacturers who offer bottom-of-the-barrel prices usually have a reason for doing so that will not benefit customers in the long run.

In the future, the claim that you cannot obtain significantly better prices from one dealer without sacrificing quality, customer service, and peace of mind may be disproved. Currently, most modular dealers are independent from their manufacturers. When more manufacturers and dealers join forces, the result will create business efficiencies that will lower their costs enough to reduce prices significantly. This could happen by manufacturers becoming dealers, dealers becoming manufacturers, or manufacturers and dealers forming partnerships. But such efficiencies require eliminating a lot of redundant overhead, which will be difficult to achieve without closer ties between dealers and manufacturers. There is no evidence today that any modular companies have achieved these cost advantages. When they do, it will usher in the golden age of modular home building.

Then stick builders will have an even more difficult time competing on quality, service, and price.

Preparing a Contract

In the next step, the customer asks the dealer to turn his estimate into a formal agreement. This requires that he add contract legalese to his list of detailed specifications and itemized prices. Before proceeding, have an attorney review the dealer's contract. If for any reason you are unable to have your attorney review the contract before you decide to move ahead, insert a contingency into the contract stipulating that your attorney must approve it or you can cancel the contract. The customer and his or her attorney should pay particularly close attention to the following points while reviewing the contract.

Change Orders
The contract should spell out the dealer's change-order policy. This should indicate when changes can and cannot be done, who can authorize them, what administrative fees apply, how they are paid, and how much delay they will cause.

Delivery and Set Delays
The contract should state the additional fees and other liabilities the customer will incur if he does not take delivery and pay for the home on the agreed-upon date. The language should relieve the customer of responsibility for delays due to the dealer or "acts of God."

Price Adjustments
The cost of building materials can increase between the time an order is placed and the time the home is delivered and paid for. The contract should indicate when the customer is responsible for covering that increase.

Payment Terms

The contract should spell out how the dealer expects to be paid. It should include the deposit requirements and the terms of the final payment. Some dealers insist on receiving payment when the house is delivered (COD). Others also accept payment after the house is set on the foundation, as long as payment is made by a lender that writes an "assignment-of-funds letter." In this letter, the lender commits to paying for the home immediately after an authorized person inspects it, usually within a day or two of the set. The dealer may not have a lot of discretion with these alternatives, since he will have to comply with his manufacturer's payment terms. The contract should state the type of payment that is required. Most dealers will insist on a bank or certified check. The contract should also indicate who the check needs to be made payable to. See chapter 9 for details.

For customers who are paying COD, some dealers will require an additional deposit to ensure that they have sufficient funds to pay the manufacturer to return the home to the factory should the customer not pay on delivery. This may seem absurd, but it does happen, and the dealer needs to protect himself.

Unloading and Demurrage

Unless the dealer is preparing the site for delivery and set day, he will want to hold the customer responsible for any additional costs if the site is not properly prepared by the customer's GC. This is understandable. The costs can mount rapidly when the manufacturer's drivers are forced to stay beyond their scheduled time and a bulldozer or tow truck is needed to assist the transporters, carrier, or crane. The additional costs can be even more substantial if poor site preparation or an improperly installed foundation prevents the crane and set crew from completing the set on schedule. When the

dealer prepares the site, however, it is his responsibility to do so properly.

Insurance

The manufacturer or its subcontractor will deliver the modules. The contract should spell out the manufacturer's responsibility for insuring the modules during delivery and when the risk of loss and need for insurance passes to the customer. The dealer or manufacturer should be responsible for insuring the activities of the crane company and set crew, if they are completing the set.

Reservation of Rights to Make Product Changes

Manufacturers often change the products that go into their homes. Sometimes that happens after an order is placed but before the home is delivered. For example, they may change the brand of vinyl window or the bathtub model. They may do this because they can no longer obtain the original product or because of building code requirements. Sometimes a change is made to upgrade the product. Manufacturers usually state that they will include only replacement products that are equal to or better than those listed in the original order. Customers and their attorneys should decide how to handle this. A reasonable position is that the dealer must inform the customer in writing of any change.

Warranty

The contract should declare what is covered and what is excluded in the warranty for both materials and workmanship. Ideally it should specify the list of standards the customer and the dealer can apply to resolve any disagreements. One popular set of standards is set forth in the *Guidelines for Professional Builders and Remodelers*, published by the National Association of Home Builders. If the modular home includes an extended warranty, its set of construction standards could be used. The warranty should state the length of the coverage, and the dealer should provide the customer with copies of all supporting documents. See chapter 10 for more information on warranties.

Placing the Order

Once an agreement has been reached, the customer should sign the contract and give a deposit. The dealer can then place the order with the manufacturer and initiate the first draft of preliminary drawings.

Contract Contingencies, Cancellation, and Authorization to Build

Customers need to complete several tasks before they are ready for a modular home to be built. The most important task, which usually takes at least a couple of months and sometimes much longer, is to select the building specifications and finalize the house plans. Finalized specifications and plans are necessary both for a manufacturer to build a home and for a customer to complete a financing application and apply for a building permit. Obtaining a building permit in turn involves multiple additional steps that can take several weeks. It is best if a customer starts all of these steps as early as possible. However, a customer cannot take these steps very far without first signing a contract with the dealer, since the dealer cannot afford to create any of the final drawings until a customer has first ordered a home.

A customer could conceivably order a home only to be later denied a permit or financing. This situation puts a customer in a "catch-22" position. The customer cannot build a home without financing and a permit, but the customer cannot get either of these without entering into a contract to build a home. One way around this is to separate the preliminary step of ordering a home from the irreversible step of building it. A modular dealer can accomplish this by having a cancellation clause in his contract that allows the customer to cancel the order if she is unable to meet any specified contingencies, such as obtaining financing. If a customer is unable to meet a contingency, she notifies the dealer in writing, and the dealer refunds the customer's deposit, less a reasonable fee for his services and house drawings, which the customer gets to keep. If and when the customer is ready to proceed, the customer authorizes the dealer in writing to direct the manufacturer to build the home.

This two-step process eliminates two risks: the risk of losing all of the deposit if the customer is forced to cancel the order and the risk that the dealer will build a home that the customer cannot buy. You should insist on similar language in your dealer's contract, unless you are sure you will be able to build your home.

3 Designing a Home

ONCE YOU HAVE DECIDED TO build a modular home, you will want to select a plan that meets your needs and expresses your personal style. Most modular dealers try to steer you to one of their standard offerings, and most customers can usually find at least one standard plan that comes close to meeting their needs. Dealers like to work from standard plans because they save them time and simplify their job, since the design, engineering, and pricing have already been done by the manufacturer. Standard plans are engineered to be built economically, the dealers can get their customers a price immediately, and customers benefit accordingly. Some dealers offer a few standard plans with big discounts, but the usual trade-off is that you can make only a few minor changes to these designs. Typically, you cannot move walls or modify the kitchens or the bathrooms.

When standard plans do not work, ask dealers if they will help you design a custom home. If you already have a plan in mind, find out if it can be built as a modular. Many plans drawn up for stick-built houses cannot be built with modular construction, although dealers may be able to come up with a very similar design. If your plan can-

not be built as a modular home, ask dealers to show you plans offering similar features.

Starting from Scratch

Some customers do not have any idea of the type of house they want to build. What then? You could work with an architect, although finding one with extensive modular experience may be difficult. When that's the case, look for an architect who is open to receiving assistance on modular construction and a dealer who is willing to provide it. Fortunately, with the growth of the modular housing industry, more and more architects are developing skills in modular design.

Whether you work with an architect or not, there is nothing more helpful when you are struggling with what to build than to visit model homes. Seeing a home and walking through its rooms will tell you more about what it would be like to live in that design than looking at its drawings.

Looking at Plan Books

All modular dealers have a book of standard plans. Each plan in the book typically includes a floor plan and an exterior elevation. The floor plan shows the location and size of each room, while the elevation provides an idea of what the finished house will look like. When looking through a plan book, do not be misled by the pairing of floor plans and exterior drawings into thinking that you cannot make adjustments. In fact, each floor plan can have a multitude of exterior looks, and each exterior look can be applied to many different floor plans. For example, all homes can have a garage and porch, even if the artist has not included them in the drawing. Likewise, you can adjust the slope of the roof, add dormers and decorative gables, and opt for oversized roof overhangs if you choose, regardless of what you may see in the drawing.

As the examples show, each plan can have a simple, unadorned look and a complex, ornate look, as well as many looks in between. Remembering this frees you up to consider some interesting plans that have been paired with what are unattractive elevations to your eye. It will also motivate you to take a second look at some desirable elevations that are matched with unworkable plans. A practical way to do this when you are looking at floor plans is to cover up the elevation drawings with a piece of paper. Otherwise you will find your eyes continually drawn to the elevation plans as you turn the pages.

When you find a plan that appeals to you, imagine living in the house. Visualize walking through it, entering first through the front door and then through the other exterior doors. Think about traffic flow and the location of various rooms. Imagine greeting guests and hanging up their coats. See yourself coming in from the car with a bag of groceries, or your children returning from their play in the backyard. Visualize placing your groceries on a countertop or table before putting them away. Make sure you have ample cabinets and closets in the convenient places; as best you can, count the cabinets and closets, noting their size.

Imagine serving a meal at the table, and what you will see when eating. Consider whether the children's or guest bedrooms are too close to or too far from the master bedroom. Would you have to walk through one main room to reach another room? Are the halls too long? Think about the views through all windows.

Finding the Right Plan

Before looking through a book of floor plans, determine which features are most important to you. Answering the following questions will help you clarify your priorities.

☐ What do you like about the floor plan of your current home? What would you change?

☐ What types of floor plans have you liked in other homes, including model homes and homes of family and friends?

☐ What type of home will fit best in your new neighborhood?

☐ What type of home will work best with the topography of your lot?

☐ What design will allow you to take advantage of the sun?

☐ What is your ideal budget? What is the most you can spend, leaving 2 or 3 percent aside as a contingency fund?

☐ Do you need all of the space finished right away, or will an expandable plan work best, such as an unfinished Cape or raised ranch?

☐ Do you prefer one-story or two-story living?

☐ How many bedrooms and bathrooms do you need?

☐ Do you prefer an informal family room separate from a more formal living room?

☐ Do you prefer an informal eating area ("nook") separate from a more formal dining room?

☐ Do you need a study or home office?

☐ Do you want the laundry on the first floor, on the second floor, or in the basement?

☐ Would you like an exercise room?

☐ What other rooms would you like to have?

☐ Are you counting on a walk-in closet or pantry?

☐ How big a kitchen would you like?

☐ How big would you like the rooms in your house to be? (Measuring all of the rooms in your current home as well as in model homes and recording this for future reference will help you immensely when designing a new home.)

Each of these elevations of the Whately series illustrates how a two-story home can be dressed up. The one at the top left is the typical standard elevation for a two-story. The other three are enhanced by adding some of the following: a taller roof pitch or a hip roof, a decorative gable on the roof, a porch, a garage, and dormers. The fronts on the first story of these three plans are enhanced with front-door side-lights, window mantels, and a pediment over the front door. The elevation at the bottom left is dressed up with a double window topped with a half round window on the second story. The elevation on the bottom right is improved by placing a transom window over the front door.

Whately № 1

The Whately 1 is a 27' 6" x 44' two-story with a living room, dining room, breakfast nook, kitchen, three bedrooms, and two and a half baths. Its signature feature is a vaulted foyer with a second-story balcony wrapped in railings. The master bath includes a 6-foot garden tub and a 5-foot shower. The plan shows two popular options, a walk-out bay in the family room and a second sink in the master bathroom. It also demonstrates how to attach a mudroom and two-car garage.

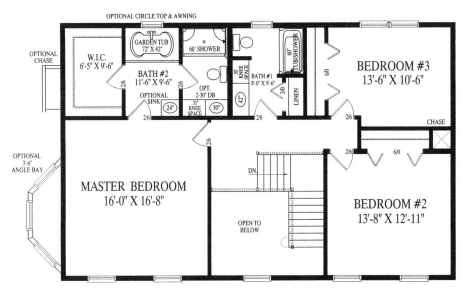

SECOND FLOOR

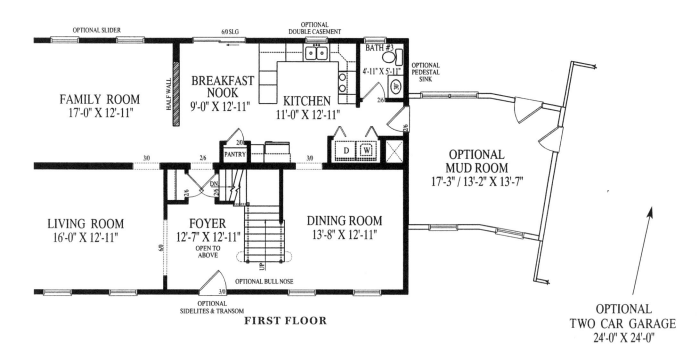

FIRST FLOOR

Whately № 2

The Whately 2 rearranges the Whately 1 to have both a great room and a fourth bedroom without enlarging the home. The great room is created by removing the wall between the family room and the living room, and space for the fourth bedroom is found by rearranging the master and hall bathrooms.

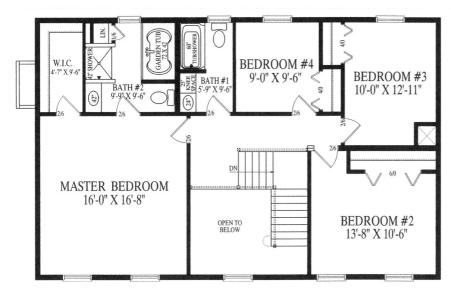

SECOND FLOOR

FIRST FLOOR

Whately № 3

The Whately 3 also rearranges the Whately 1 to have a fourth bedroom, but it does so while retaining the master and hall bathroom layouts. This is accomplished by replacing the vaulted foyer with a second-story bedroom. The downstairs is reconfigured to put the dining room next to the kitchen and the family room and living room on opposite sides of the house. In addition, the laundry room is combined with the half bath. The plan also creates a direct hallway to a garage bonus room.

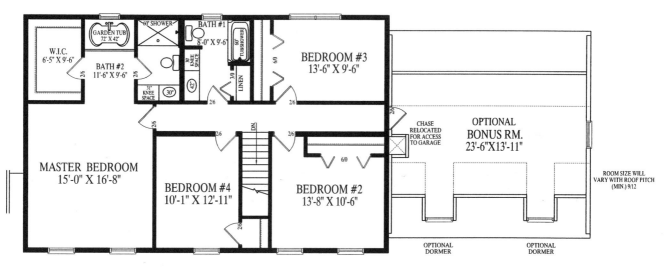

SECOND FLOOR

FIRST FLOOR

Whately № 4

The Whately 4 adds 2 feet to the length of the Whately 2, making the home 2,530 square feet. The plan is rearranged to offer an L-shaped kitchen with an island, a separate laundry room, and a half bath in the foyer. The second floor combines four bedrooms with a wrap-around vaulted foyer. The redesign of the second floor creates a sitting area, which in turn can provide access to a set of stairs to the attic. The plan also shows how to add a first-floor master bedroom as well as an alternative second-floor bathroom design.

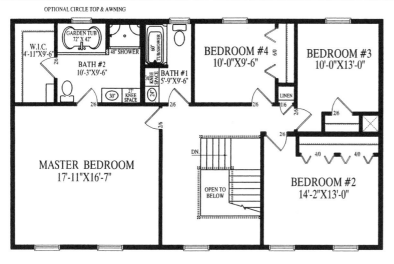

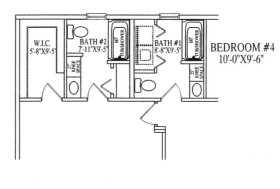

SECOND FLOOR

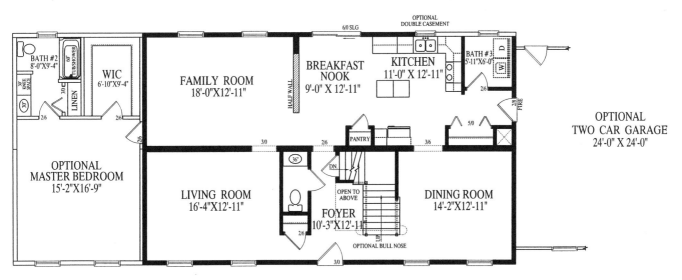

FIRST FLOOR

Whately № 5

The Whately 5 adds 4 feet to the Whately 2, making the home 2,640 square feet. This creates a larger master bath, master bedroom, and great room.

SECOND FLOOR

OPTIONAL
COMPUTER LOFT

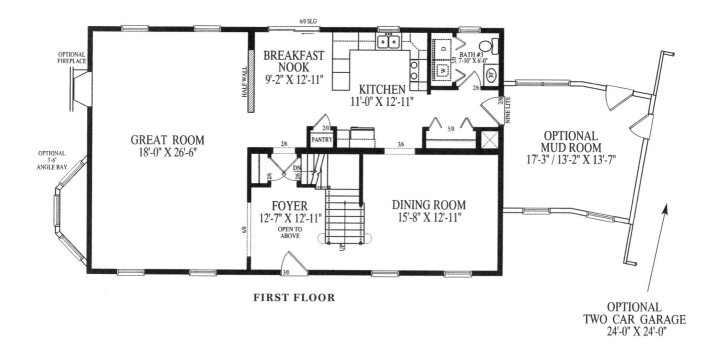

FIRST FLOOR

OPTIONAL
TWO CAR GARAGE
24'-0" X 24'-0"

Whately № 6

The Whately 6 adds 4 feet to the Whately 3 to make a fifth bedroom or study. It incorporates the great room of the Whately 2 and the laundry room of the Whately 4. An interesting feature of the Whately 6 is a bump-out in the front of the house, which is created by jogging the rooms one foot in on both sides of the foyer. Since this change eliminates the entrance from the foyer to the dining room, a more formal dining room is created by building a full wall on that side of the stairs.

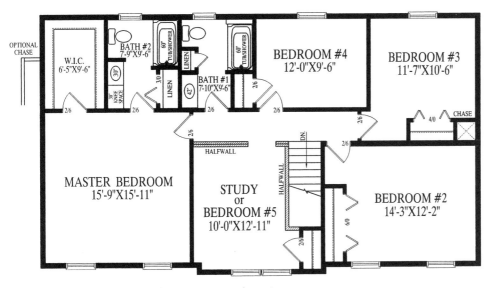

SECOND FLOOR

FIRST FLOOR

Each of these elevations of the Sugarloaf series illustrates one of the many ways a T-ranch can be dressed up. The one at the top left is the typical standard elevation for a T-ranch. The other three were enhanced by adding some of the following: a garage, porch, walk-out bay, picture window, chimney, and taller roof pitch. Notice how the garage in the bottom right elevation changes the character of the house by having its gable roof face front, creating an H-look.

Sugarloaf № 1

The Sugarloaf 1 is a 1,392-square-foot T-ranch with a 24' x 24' living area wing and a 24' x 34' bedroom wing. The plan has a living room, dining room, kitchen, three bedrooms, and two baths. The Sugarloaf 1 and Sugarloaf 3 are almost identical except that the width of the Sugarloaf 1 is 24' rather than 27'6". This makes the main rooms in the house about 2 feet narrower than those in the Sugarloaf 3, but it also makes the house more affordable for those customers who don't need the extra space.

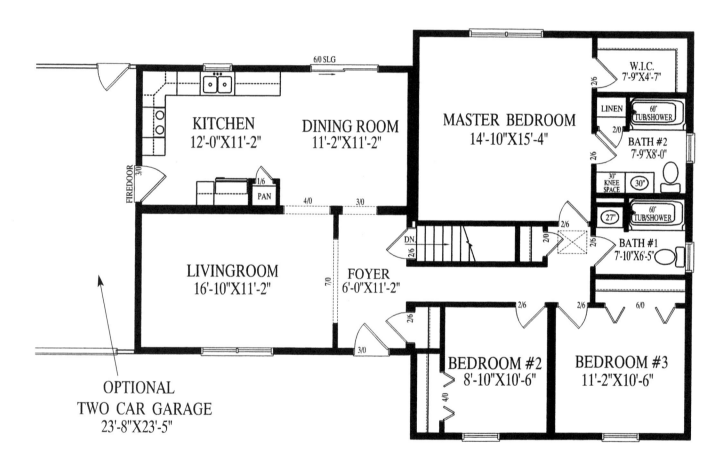

Sugarloaf № 2

The Sugarloaf 2 is the same square footage as the Sugarloaf 1 even though it includes a half bath and a larger living room and laundry room.

Space for the additional rooms is created by taking 2 feet from the bedroom area and adding it to the living area. This makes the master bed-

room and bathroom smaller and eliminates the walk-in closet in the master bedroom.

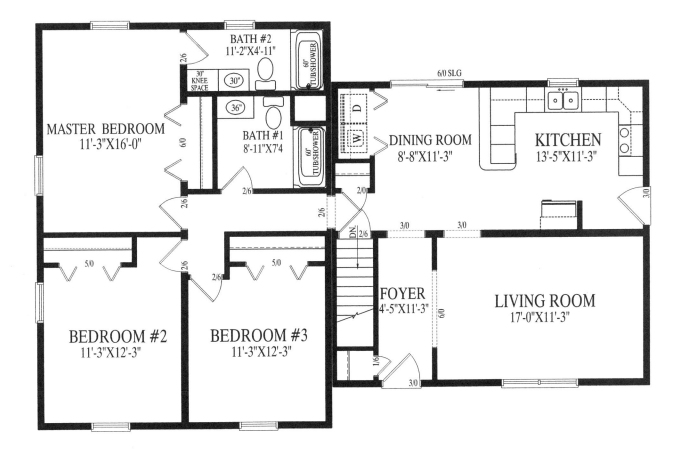

Sugarloaf № 3

The Sugarloaf 3 adds 4 feet to the width of the Sugarloaf 1, making the home 1,595 square feet. This makes all of the main rooms almost 2 feet wider. The additional width is used to make a U-shaped cabinet layout in the kitchen, a laundry closet in the hall bathroom, and a double closet in bedroom 3. The plan shows two popular options, a second sink in the master bathroom and a site-built walk-out bay on the rear of the dining room.

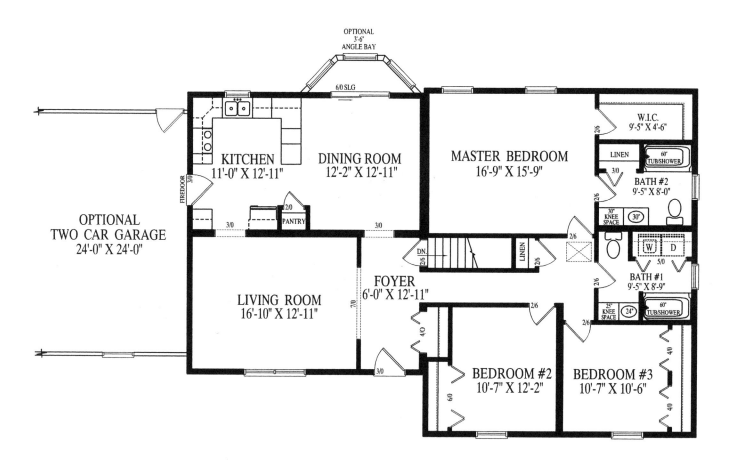

Sugarloaf № 4

The Sugarloaf 4 adds 2 feet to the length of the bedroom wing to make the home 1,650 square feet. The additional length is used to add a fourth bedroom. How-ever, this requires elimi-nating the tub/shower from the master bath-room, making it a half bath.

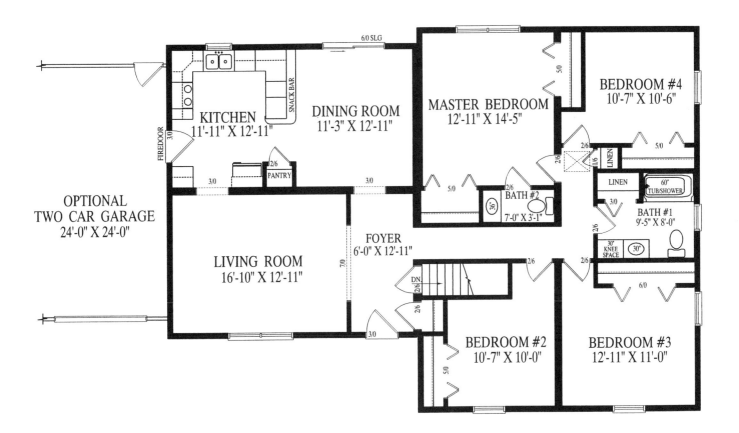

Sugarloaf № 5

The Sugarloaf 5 is designed to illustrate how to make a standard plan, the Sugarloaf 3, into an accessible home based on universal-design princi-ples. For example, it has wider halls and doorways as well as a 5-foot turning radius in the kitchen and bathrooms. By moving the dining room to the outside wall, this plan also makes it possible to add a factory-built walk-out bay on the gable end of the module.

Sugarloaf № 6

The Sugarloaf 6 adds 4 feet to the kitchen and living room wing of the Sugarloaf 3, making the home 1,705 square feet. The additional space is used to add a half bath-room and laundry closet next to the kitchen as well as to enlarge the living room. Moving the laundry room from the hall bath-room makes it possible to add a spacious linen closet. Some customers leave the laundry closet in the hall bathroom and use the additional kitchen closet as a second pantry or a coat closet.

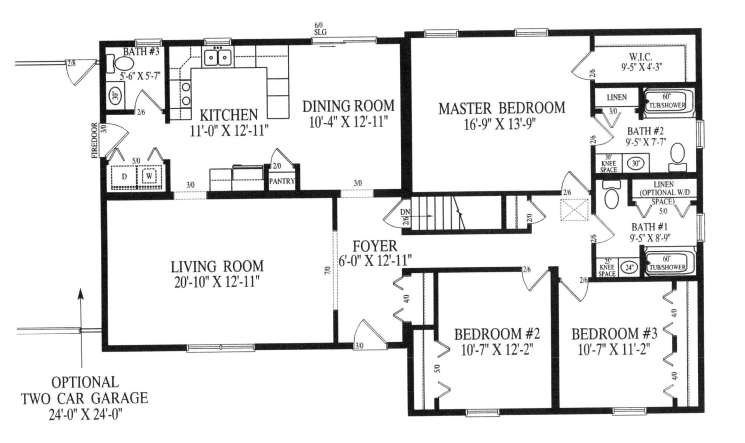

BATH #3
5'-6" X 5'-7"

KITCHEN
11'-0" X 12'-11"

DINING ROOM
10'-4" X 12'-11"

MASTER BEDROOM
16'-9" X 13'-9"

W.I.C.
9'-5" X 4'-3"

LINEN

60"
TUB/SHOWER

BATH #2
9'-5" X 7'-7"

30"
KNEE
SPACE

LINEN
(OPTIONAL W/D
SPACE)

BATH #1
9'-5" X 8'-9"

25"
KNEE
SPACE

60"
TUB/SHOWER

D   W

PANTRY

FIRE DOOR

LIVING ROOM
20'-10" X 12'-11"

FOYER
6'-0" X 12'-11"

DN

BEDROOM #2
10'-7" X 12'-2"

BEDROOM #3
10'-7" X 11'-2"

OPTIONAL
TWO CAR GARAGE
24'-0" X 24'-0"

6/0
SLG

Sugarloaf № 7

The Sugarloaf 7 adds 4 feet to the bedroom wing of the Sugarloaf 3. The space is used to create a fourth bedroom while preserving a fully functional master bathroom. The walk-in closet, however, is sacrificed and the hall bathroom is considerably smaller.

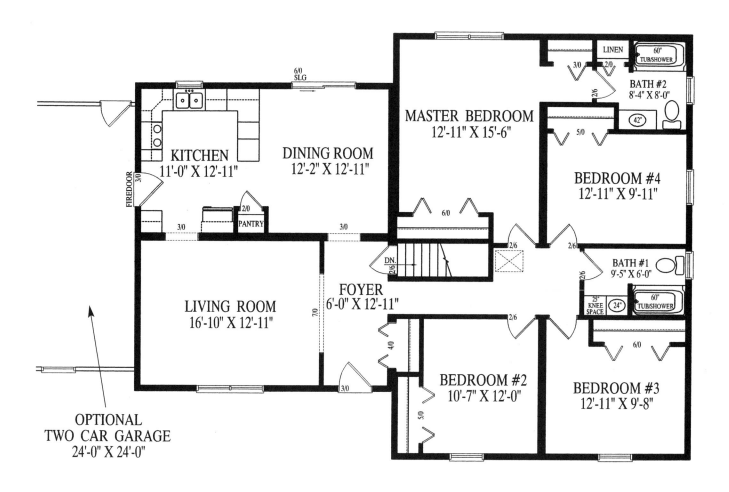

KITCHEN
11'-0" X 12'-11"

DINING ROOM
12'-2" X 12'-11"

MASTER BEDROOM
12'-11" X 15'-6"

LINEN

60"
TUB/SHOWER

BATH #2
8'-4" X 8'-0"

42"

BEDROOM #4
12'-11" X 9'-11"

PANTRY

FOYER
6'-0" X 12'-11"

DN.

BATH #1
9'-5" X 6'-0"

25"
KNEE
SPACE

24"

60"
TUB/SHOWER

LIVING ROOM
16'-10" X 12'-11"

FIREDOOR

BEDROOM #2
10'-7" X 12'-0"

BEDROOM #3
12'-11" X 9'-8"

OPTIONAL
TWO CAR GARAGE
24'-0" X 24'-0"

Changing Plans

As you examine books of house plans, keep in mind that each of them can be changed to fit specific needs. For example, all six of The Home Store's Whately plans (see pages 56-61) started with the Whately 1 and all seven of the Sugarloaf plans started with the Sugarloaf 3. The optional plans were created by making relatively minor and affordable modifications. You can do something similar to virtually any dealer's standard plans.

The Whately 2 was created when a few customers said they really liked the Whately 1 except that they needed a great room and four bedrooms. This was accomplished by removing the wall between the family room and the living room, and rearranging the master bathroom and the hall bathroom on the second floor. The Whately 3 was created to keep the second-floor bathrooms the same size as in the Whately 1 while adding a fourth bedroom. This was accomplished by closing off the vaulted foyer. The downstairs was also reconfigured to put the dining room next to the kitchen and the family room and living room on opposite sides of the house. In addition, the laundry room was combined with the half bath. The Whately 4 added 2 feet to the length of both floors. It also created an L-shaped kitchen with an island, added a separate laundry room, put the half bath in the foyer, and showed how an optional first-floor master bedroom could work. The second floor combined four bedrooms with a wraparound vaulted foyer, which created a sitting area. The Whately 5 added 4 feet to the Whately 2, which enabled a larger master bath, master bedroom, and great room. The Whately 6 added 4 feet to the Whately 3 to make a fifth bedroom or study. It also created a bump-out look in the front of the house. Study the plans carefully and you can see several other small changes in each of them.

Simply lengthening a plan can help create the home you are looking for. Adding square footage is often beneficial and sometimes necessary when moving walls around. You can also add bump-outs on the end of a home, as in the Whately 1, 3, and 5. These structures, which can be angled or rectangular and one or two stories tall, add a lot of natural light, enlarge the space, and enhance the interior and exterior appearance of a home. In addition, adding a wing to the end of a home, as was done with the Whately 4, can create a first-floor master bedroom suite. With any of the Whately plans, you can also build a mudroom wing between the main home and the garage.

The Home Store's Sugarloaf series illustrates most of the same design opportunities as the Whately series. It also shows how you can widen a home to enlarge all of the rooms. For example, the Sugarloaf 3 adds 3 feet 6 inches to the width of the Sugarloaf 1, which makes the kitchen, dining room, living room, and all three of the bedrooms about 2 feet bigger. Conversely, the Sugarloaf 1 shows what can happen when you reduce the width of a home; in fact, the Sugarloaf 1 was created from the Sugarloaf 3. The Sugarloaf 5, which was designed for The Home Store by the late Ron Mace, FAIA, when he was the director of Center for Universal Design at North Carolina State University, was specifically designed to illustrate how to make a standard plan, the Sugarloaf 3, into an accessible home based on universal-design principles. The Sugarloaf 5 design was awarded a Building Innovation in Homeownership Award by the U.S. Department of Housing and Urban Development in 1996. It has wider halls and doorways as well as a 5-foot turning radius in the kitchen and bathrooms; the actual model home exhibits more than 30 other universal-design features. For more discussion on universal design, see pages 91–94.

As the Whately and Sugarloaf plans illustrate, adding a fireplace and making a U-shaped kitchen into an L-shaped kitchen with an island are routine changes, as is adding or removing a whirlpool tub in the master bathroom. Many of these changes necessitate other changes. For example, you might need to borrow space from an adjoining room to make the master bathroom big enough to fit the whirlpool tub. And some changes will not work. For example, you cannot build a 60-foot-long home without any walls in the middle of the house to carry the weight of the roof. Also, some changes will be expensive, such as widening a home from 27 feet 6 inches to 31 feet 6 inches, although the benefit is almost always greater than the cost.

You can take one type of home and make it into an entirely different type. For example, you can turn a ranch into a Cape Cod by adding a steeper roof, a set of stairs, and rough mechanicals to the second floor. You can also make many Cape Cod plans that have an unfinished second story into a chalet by adding beams in the ceiling, railings for a loft, and trapezoid windows.

Although it requires a lot more effort and creativity, you can also make some two-story plans into a T-ranch by placing the second story perpendicular to the first story. Since most dealers have a limited selection of standard plans, these little tricks can sometimes help a dealer expand the possibilities. Many dealers charge a small fee for making the changes, since they require additional drawing time. There will also be charges or credits for items added to or omitted from the standard plan.

Size Limitations

A modular home is created when one or more modules are transported to a building site and assembled on a foundation. Each modular section is a semi-independent structural unit, essentially a box that is built to interconnect with other modules. Whereas "sticks" are the basic building unit in stick construction and walls are the basic building unit in panelized construction, modules are the basic building unit in modular construction. Modular design, engineering, and construction work because many home designs can be divided into modular sections.

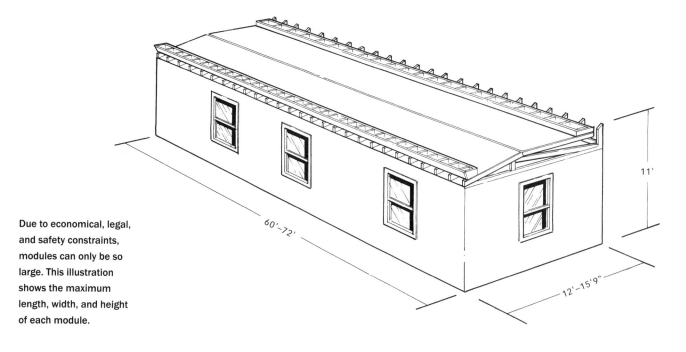

Due to economical, legal, and safety constraints, modules can only be so large. This illustration shows the maximum length, width, and height of each module.

Modular construction, like all construction, has design constraints. The limits to what can be built are a function of a module's maximum width, length, height, and structural capability. The limits are themselves determined more by what can be safely, legally, and economically built and transported than by what a manufacturer can fabricate. It is technologically possible to build almost anything as one or more modules, regardless of size. But delivering two 30-foot by 40-foot modules from the factory to the job site would be another matter, as would lifting them onto the foundation. In spite of these constraints, the design possibilities for modular homes are countless.

Width

Most modular manufacturers build modules in at least three widths, typically 12 feet, 13 feet, and 13 feet 9 inches. Some companies also build widths of 14 feet 9 inches and 15 feet 9 inches. The widths can vary by a couple of inches among companies, depending on the size of a manufacturer's production jigs. Maximum widths are determined by the federal and state transportation regulations as well as by each factory's production system. For special needs, such as an existing foundation or a zoning issue, many manufacturers can build to a slightly different width for a modest charge.

Length

Most manufacturers will build modules up to 60 feet long. Some companies will build up to 72-foot-long modules, although many states will not permit these extra-long modules to be delivered. The production line setup and the length of a manufacturer's carriers also play a role in what a company can build.

Height

Federal, state, and local regulations limit the maximum height of any vehicle and its

cargo, usually to 13 feet 6 inches. They also regulate the minimum height of any object, fabricated or natural, that overhangs a road, such as a bridge, wire, or tree limb. Any exceptions to the minimum height, such as with an older bridge, have to be clearly posted. These maximum and minimum height restrictions help prevent damage to low-hanging objects and vehicles, and their cargo, including modular homes.

Complying with these restrictions means that each module cannot exceed the maximum shipping height. Since the measurement is taken with the module sitting on top of its carrier, the height of the carrier,

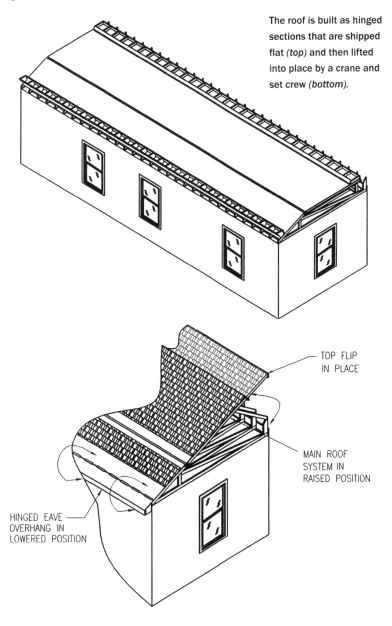

The roof is built as hinged sections that are shipped flat *(top)* and then lifted into place by a crane and set crew *(bottom)*.

TOP FLIP IN PLACE

MAIN ROOF SYSTEM IN RAISED POSITION

HINGED EAVE OVERHANG IN LOWERED POSITION

which is typically 2 feet 6 inches, also counts in the calculation. This limits the actual height of the module to about 11 feet. Those modules that do not contain any part of the roof, typically the first-floor modules of a two-story, seldom approach this maximum height. Those modules that support sections of the roof must be engineered to comply with the restrictions, since the height of the modules with the roof in an upright position exceeds the maximum restrictions. An essential requirement of most modular engineering is to design the roof systems so they lie flat on top of the module during delivery. Each section of the roof is fabricated into two or more components, which are hinged to the module and each other. Once the module is on the foundation, the set crew uses the crane to lift and unfold the roof to its correct height.

Sometimes the roof design makes the sectional hinging impossible, either because the design cannot be hinged or because the resulting height would exceed regulations. In those situations, the manufacturer will build one or more of the roof components as panels, or it will build a module without any of the roof components installed on top of the module. The roof will then be constructed on-site from either sticks, panels, or specially fabricated modular-roof sections.

Combinations

Most modular designs are one, two, or three modules deep and one or two modules high. A few companies build homes three modules high, although they require special approvals from the customer's home state to ensure that they comply with the building code. Most homes are two modules wide, with typical widths of approximately 24 feet, 26 feet, 27 feet 6 inches, and 31 feet 6 inches. The usual practice is to place modules side by side, with the long sides parallel to the road. Some designs, especially when built on narrow lots, turn the modules perpendicular to the road. Modules

34 Typical Modular Arrangements

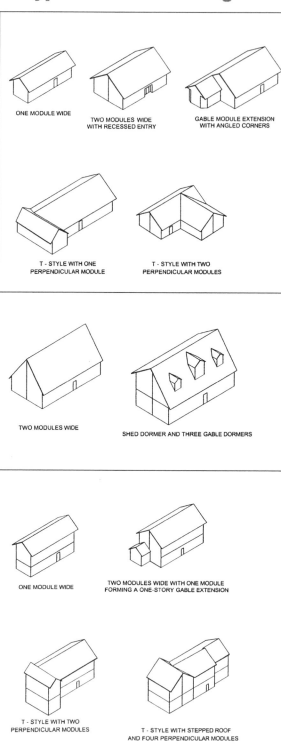

ONE MODULE WIDE

TWO MODULES WIDE WITH RECESSED ENTRY

GABLE MODULE EXTENSION WITH ANGLED CORNERS

T - STYLE WITH ONE PERPENDICULAR MODULE

T - STYLE WITH TWO PERPENDICULAR MODULES

TWO MODULES WIDE

SHED DORMER AND THREE GABLE DORMERS

ONE MODULE WIDE

TWO MODULES WIDE WITH ONE MODULE FORMING A ONE-STORY GABLE EXTENSION

T - STYLE WITH TWO PERPENDICULAR MODULES

T - STYLE WITH STEPPED ROOF AND FOUR PERPENDICULAR MODULES

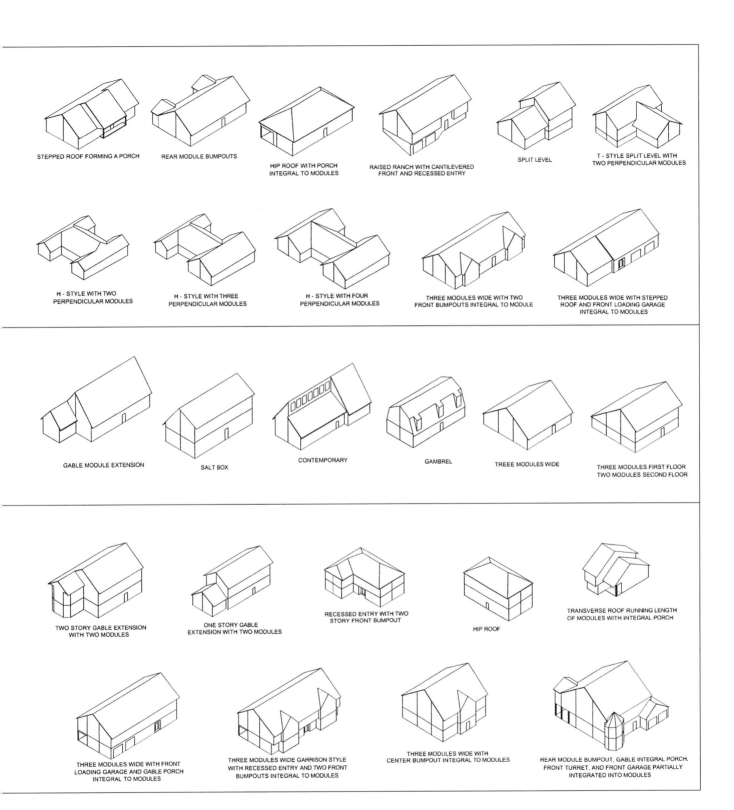

STEPPED ROOF FORMING A PORCH

REAR MODULE BUMPOUTS

HIP ROOF WITH PORCH
INTEGRAL TO MODULES

RAISED RANCH WITH CANTILEVERED
FRONT AND RECESSED ENTRY

SPLIT LEVEL

T - STYLE SPLIT LEVEL WITH
TWO PERPENDICULAR MODULES

H - STYLE WITH TWO
PERPENDICULAR MODULES

H - STYLE WITH THREE
PERPENDICULAR MODULES

H - STYLE WITH FOUR
PERPENDICULAR MODULES

THREE MODULES WIDE WITH TWO
FRONT BUMPOUTS INTEGRAL TO MODULE

THREE MODULES WIDE WITH STEPPED
ROOF AND FRONT LOADING GARAGE
INTEGRAL TO MODULES

GABLE MODULE EXTENSION

SALT BOX

CONTEMPORARY

GAMBREL

TREEE MODULES WIDE

THREE MODULES FIRST FLOOR
TWO MODULES SECOND FLOOR

TWO STORY GABLE EXTENSION
WITH TWO MODULES

ONE STORY GABLE
EXTENSION WITH TWO MODULES

RECESSED ENTRY WITH TWO
STORY FRONT BUMPOUT

HIP ROOF

TRANSVERSE ROOF RUNNING LENGTH
OF MODULES WITH INTEGRAL PORCH

THREE MODULES WIDE WITH FRONT
LOADING GARAGE AND GABLE PORCH
INTEGRAL TO MODULES

THREE MODULES WIDE GARRISON STYLE
WITH RECESSED ENTRY AND TWO FRONT
BUMPOUTS INTEGRAL TO MODULES

THREE MODULES WIDE WITH
CENTER BUMPOUT INTEGRAL TO MODULES

REAR MODULE BUMPOUT, GABLE INTEGRAL PORCH,
FRONT TURRET, AND FRONT GARAGE PARTIALLY
INTEGRATED INTO MODULES

There is a surprising number of ways that basic modules can be combined to create various exterior styles and interior space arrangements. Here are some of the more typical arrangements.

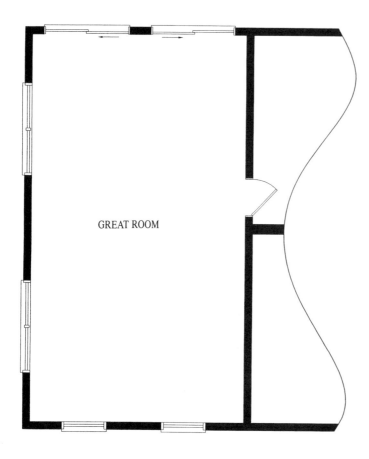

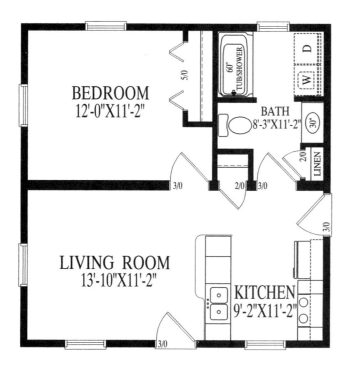

With little labor and materials required, the great room *(top)* is not an economical structure for a modular manufacturer to build. The smaller in-law apartment *(bottom)*, which requires more labor and materials, is.

can also be turned perpendicular to each other to create T-, L-, or H-shaped houses, which is one of many techniques the modular industry has employed to shed its image of making boring boxes.

Minimum Module Size

Modular manufacturers will build homes only if they can sell them for a competitive price and still make a profit. The minimum size they will build is a function of the amount of labor and materials required to build the home. Too little labor and materials makes a home uneconomical to build at the factory. This means that small additions usually won't work financially, and neither will larger additions that are essentially empty boxes. For example, a 16-foot by 27-foot 6-inch great-room addition is usually too small and devoid of value-added amenities to make economic sense. On the other hand, a 24-foot by 24-foot in-law addition with a kitchen and bathroom can work quite nicely.

Enlarging a Design

Adding length to a modular home is always easy to do from the manufacturer's point of view, as long as it stays within the maximum production and delivery dimensions. This is also true about adding bump-outs to the end of a module, such as a walk-out bay (see the Whately 1, page 56). Making a plan of a given size bigger by adding to the length of the modules is one of the best values in the entire construction industry.

Increasing the size of a particular plan by widening the modules is also a very good bargain. The cost per square foot is often a little more for adding width than length, because the floor system sometimes needs to be beefed up. For example, 2×8 floor joists for a 24-foot-wide home will need to be increased to 2×10s for a home with a width of 27 feet 6 inches. Widening a module, like lengthening a home, will always add more equity to a home than it costs.

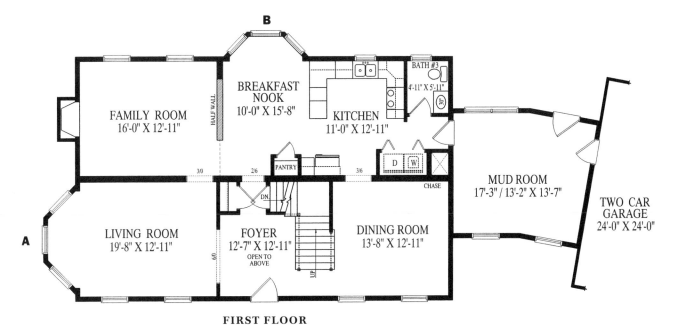

FIRST FLOOR

The walk-out bay in the living room (A) can be built by the manufacturer and delivered on a carrier as part of the front module. The walk-out bay in the back of the nook (B) cannot be delivered as part of the rear module, since it would make the module too wide to transport.

Adding a bump-out, such as the walk-out bay, to a long wall on a module that is already at its maximum width is more involved than adding one to the end of a module. The manufacturer must either build the bump-out as a separate miniature module or ship the necessary materials to the general contractor.

Attaching one or more additional modules perpendicular to the long side, as discussed above, can also enlarge a standard plan. This will require that a "saddle" or "cricket" be built to join the roofs.

Another way to enlarge a home is to attach a separate living unit, such as you might do to create an in-law addition or a two-family home. The second unit can be designed with either a custom or a standard plan.

When enlarging a given plan, make sure that any additional delivery fees are included in writing. Increasing the width up to 27 feet 6 inches should incur a relatively small fee. The fee for widening a home to 31 feet 6 inches, however, will be more substantial. Adding length to a home

One way to enlarge a modular home is to build a saddle that joins the roofs of perpendicular modules.

will sometimes require additional delivery carriers, which can add significantly to the cost. This happens when the original length of the house calls for delivering two modules on one carrier but the new, longer plan requires delivering the two modules on two carriers. Consequently, the new design requires an additional carrier for each pair of modules. For example, if a manufacturer's maximum length for shipping two modules on a carrier is 30 feet per module, it can deliver a two-module-wide, 28-foot-long in-law apartment on one carrier. However, if the apartment is lengthened by 4 feet, making each module 32 feet long, an additional carrier will be needed. When the original plan is a 28-foot-long two-story that is being lengthened to 32 feet, two additional carriers will be needed.

Additional carriers are also required whenever lengthening a home makes the modules too long for the manufacturer and general contractor to deliver to a site. For example, a narrow road to the site may make it impossible for a longer carrier to negotiate the turns. The only solution, other than keeping the home at its original length, might be to have the manufacturer divide each longer module into two shorter modules. The manufacturer will charge more to do this, but it may be the best way to get the home you want.

Enlarging a home increases the cost of any options that are affected by the expansion. For example, upgrading to a premium siding will cost more money when a home is made wider or longer, since more area needs to be covered. Other optional features will only increase in price if a particular room increases in size. For example, a dining room wood floor will cost more when a home is lengthened only if some of the additional length is put into that room. Increasing the size of a particular room can also force you to add windows or doors to meet the building code, which requires a minimum amount of "light and vent" for each room.

Exterior Dimensions

The marketing literature of most manufacturers typically lists the size and square footage of its house plans. The size refers to a plan's exterior length and width. The width is almost always listed in rounded-off dimensions. It is very common, for example, to list a plan that is 27 feet 6 inches wide as 28 feet wide. Many manufacturers also use rounded-off dimensions to calculate the square footage of the home, which will slightly inflate the size of a plan. However, all manufacturers give accurate dimensions when they complete the drawings they use to build a home. Make sure you know what the actual dimensions will be when discussing plans with a dealer.

Interior Dimensions

The interior dimensions, which define the usable space, are always less than the exterior because of the thickness of the walls. A room that is contained entirely within a module with a width of 13 feet 9 inches, an exterior wall of 6 inches, and a marriage wall of 3 inches will have an interior dimension of 13 feet. A great room that is created across two modules of the same size will have 26 feet 6 inches of interior space because the two 3-inch marriage walls will be eliminated.

Structural Walls

All homes must have a structure that can support its own weight. Each section of a modular home is usually designed so that two of the exterior walls, usually the long walls, carry more weight than the other two. Consider a typical ranch made up of two modules that have the long walls oriented parallel to the street. Since the front and rear of each module will bear more weight, there will be three weight-bearing areas: the front of the house, the rear of the house, and the middle of the house where the two modules come together. This middle section is generally referred to as the "marriage wall," or sometimes as the "mating wall." Because the marriage wall is really two walls joined together, it is 2 to 4 inches thicker than most interior walls in the remainder of the modular home and in site-built homes. This is one of the many reasons that modular homes are considerably stronger than typical site-built homes. This is noticeable only when there are passageways between the modules, such as a framed opening with or without a door.

Years ago, the only affordable way to eliminate this wall was to install a beam that dropped below the ceiling, sometimes known as a "dropped header," to carry the weight. While this opened up spaces between modules, it created a visible room divider in places where one was often unwelcome. Sometimes, however, customers featured the beam by covering it in stained wood. Today, thanks to the development of engineered wood beams made of laminated veneer lumber, sections of the marriage wall can be easily eliminated for a reasonable cost. Best of all, because the beams are not as tall, they can be installed flush with the bottom of the ceiling framing so that they do not protrude into the room. This creates a smooth ceiling, or clear span, across the two joined modules, making them into one large room. The master bedroom and great room of the Whately 2 are typical examples.

Laminated beams cannot always create the clear-span spaces you may desire. The more weight that bears down on the marriage wall from above, the larger the manufacturer must make the beams. The weight on the marriage wall increases as you remove more of it to increase the size of the opening. It also increases with certain roof

designs and when there are modules stacked above. Some of these structural situations require such tall beams that they will not fit in the ceiling without either protruding down, creating the dropped header effect, or jutting up, compromising the floor above of a two-story or any home with a usable attic, such as a Cape Cod. Since the structural designs of each manufacturer differ, you will need to rely on your dealer to determine how much of a clear-span opening you can create.

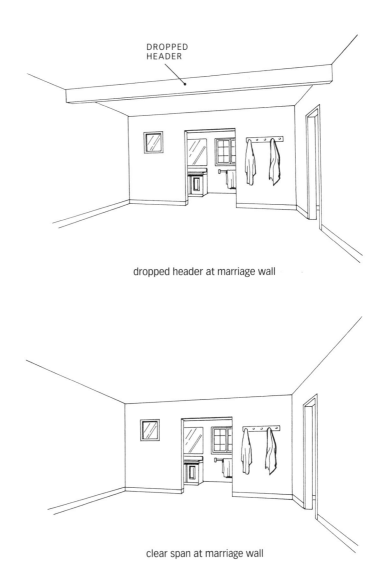

dropped header at marriage wall

clear span at marriage wall

A dropped header *(top)*, unlike a clear span *(bottom)*, hangs below the ceiling and divides the room.

Interior partitions

Most interior partitions are not structural walls, which makes it possible to move or eliminate them to improve the layout. Adding new interior partitions is also easy, and many manufacturers will do so for no additional charge.

A half wall is another way to create a good room divider while at the same time promoting a feeling of openness. Adding a painted or stained wood cap on top of the wall dresses it up while protecting it from damage. It also gives you a larger surface for plants, baskets, or other decorations.

Vaulted and Cathedral Ceilings

Modular homes can have cathedral and vaulted ceilings. Both are created in part by removing the ceiling of the first floor. A vaulted ceiling is open from the floor of the first story to the ceiling of the second story. Adding a vaulted foyer to a two-story home, such as in the Whately 1, creates an impressive entry. A vaulted space can also be created above another room, such as a dining room or family room. The disadvantage with all vaulted spaces is that they forfeit a potential room above.

A vaulted ceiling will create additional work for the general contractor, since he will have more drywall work to do where the modules meet between floors. When a vaulted foyer is involved, he will also need to install additional railings for the stairs and the balcony. A more affordable, but less impressive, alternative to a railing is to build a half wall with a wood cap.

cathedral to peak of roof

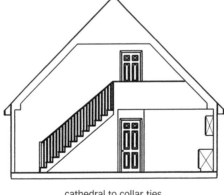

cathedral to collar ties

Cathedral ceilings can be built in very different ways. Some styles, such as the cathedral to peak, will cost more and take longer for the GC to finish. Be sure you know what you are getting.

cathedral boxed in to collar ties

A cathedral ceiling is open to the sloped roof above, as is done with the 12-in-12 roof of a chalet-style home. Making a flat ceiling into a cathedral ceiling creates a dramatic room. This feature will cost more than a vaulted foyer because it requires a significantly beefed up structural system to support the roof. It will also cost considerably more for the GC to complete his work, since the ceiling will likely be delivered without any insulation or drywall. Unless the GC completes the entire second floor, he will also need to build a partition that separates the unfinished space from the cathedral space. When the stairs to the second floor enter into the cathedral space, additional railings or a half wall to the balcony will need to be built.

Some manufacturers open their cathedral ceilings all the way to the peak of the roof, while others take them only partway, usually to where the slope flattens out at the collar ties that join the two sides of the roof. A cathedral ceiling to the peak is more striking, but also requires more finish work by the GC. If you prefer one or the other, make sure the dealer understands this.

Modular Home Styles

Modular homes can be built in a variety of styles, including one-stories, raised ranches, split levels, Cape Cods, and two-stories. As long as you take into account the design constraints mentioned above, each of these basic styles can be built in multiple sizes and configurations, and each can be rendered in classic designs, such as Colonial, Victorian, and Tudor, or take on more contemporary forms.

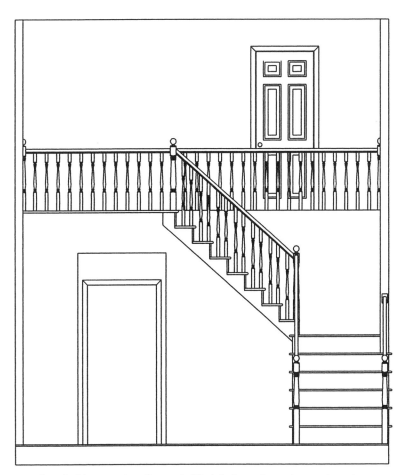

A vaulted foyer, such as the one in Whately 1, creates an impressive entry.

If you want three future bedrooms on the second floor of an unfinished Cape, make sure you see a drawing that shows if it works without a shed dormer.

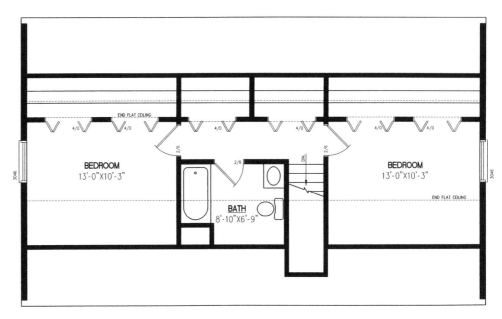

Three bedrooms do not fit in this 24 x 40 Cape Cod.

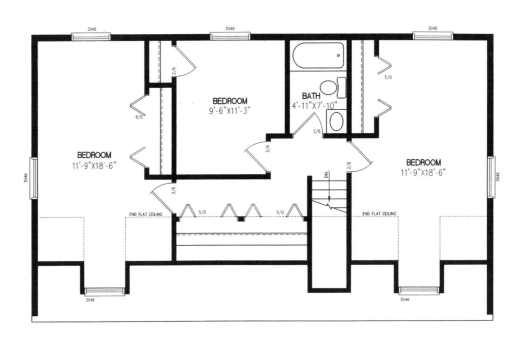

Three bedrooms fit in this 24 X 40 Cape Cod when a full shed dormer is added.

Designs with an Unfinished Story

A Cape Cod plan with an unfinished second floor is a very popular design, especially with young couples who hope to expand their families. You can create this unfinished but usable space in a ranch by adding a 12-in-12 roof pitch and a set of stairs to the attic. You can do the same thing with the attic of a two-story, although some states will require you to change other specifications as well to meet the fire codes for three-story homes. A few manufacturers will also build two-story homes with an unfinished second story so you or a future owner can build out the space.

If you purchase a home with an unfinished story, tell the dealer what the proposed floor plan will be so he can have it drawn to scale. Drawing the plan is the only way to ensure that it will work as you imagine. For example, if you hope to build three large bedrooms and a full bath on the second floor of a 24-foot by 40-foot Cape Cod without any dormers, you may be disappointed when you see it drawn. But you will be crushed if you do not see it until it is built, since the space will not be big enough to give you what you want. To take another example, consider how you will feel if you are building a larger Cape Cod with plenty of space on the second floor to do what you want, but the stair location prevents you from getting the big master bedroom suite you had hoped for. If you draw the proposed plan before the home is built, there will be time to add a full shed dormer to the smaller home and relocate the stairs in the larger home.

Another reason to draw your proposed floor plan is that the manufacturer should provide some rough plumbing, electrical, and HVAC work to the unfinished space, usually from the basement up through the first-floor walls. With these mechanical connections in place, it will be easier for future subcontractors to complete their work. Having a floor plan of the unfinished space will allow you and the manufacturer to locate these mechanicals in the most useful place.

Dormers

You can add substantial space with full headroom to the unfinished attic of a Cape Cod or any home with a 12-in-12 roof by building a shed dormer across the entire back of the roof. If you want to complete the second story yourself, now or in the future, you can purchase a panelized package from the manufacturer that will be assembled by the set crew. It is very important to have the second-story floor plan drawn with the dormer included so you can correctly locate the windows on the back and side of the home. The floor plan will also show you where to add a beam to make a large front-to-back room, which you may want to do with your master bedroom.

If you need the space finished immediately and would like the manufacturer to complete as much of it as possible, you can order an additional module for the rear of the second story that will serve as the dormer. When you purchase an additional module, the space under the sloped ceiling in the front half of the second story will still come unfinished. The general contractor will need to finish this area in order to maximize the second-story space. Also, the floor of the second-story module will be 8 to 12 inches higher than the floor of the unfinished space because of the height of the additional module's floor system. The GC will need to build up the floor area in the unfinished space to make the floors the same level. If you decide to have the manufacturer build the shed dormer as a module, you can have the manufacturer also build the front of the Cape, with its sloped roof, as an additional "wedge" module. The second story of the house would then be built completely by the manufacturer.

Shed Dormer Options

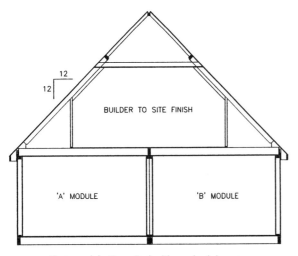

Two-module Cape Cod with no shed dormer

Here are three ways to add a shed dormer to a standard two-module Cape. A shed dormer will add significant space to an unfinished attic.

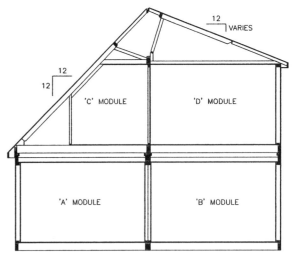

Two-module Cape Cod with panelized shed dormer

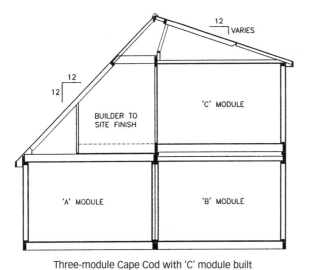

Three-module Cape Cod with 'C' module built as shed dormer

Four-module Cape Cod with 'C' module built as wedge box and 'D' module built as shed dormer

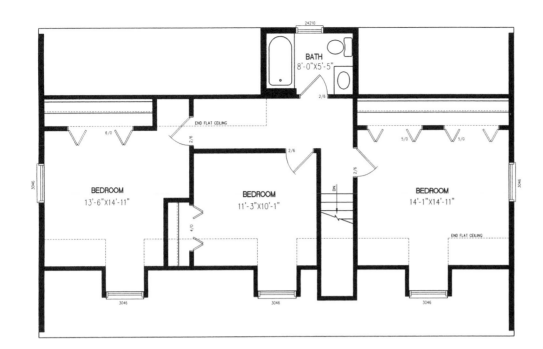

A small shed dormer large enough for a bathroom is an economic way to add space to the second floor of a Cape Cod.

Another way to make use of a shed dormer when you do not need all of the space or cannot afford to build it across the entire back of your home is to make a smaller shed dormer that is just big enough for a bathroom; a gable dormer can accomplish the same thing. The reason this is worth considering is that you are likely to want to put a bathroom across the hall at the top of the stairs. In a home with a width of 27 feet 6 inches, there will be only about 9 feet 6 inches across the back to build the bathroom. Without a dormer, the ceiling will begin to slope down within the first 2 feet of the bathroom, and the knee wall that supports the roof will likely provide only about 5 feet of headroom. This is not enough room to build a comfortable bathroom. A small dormer will solve this problem. Such a dormer can be especially helpful with a 24-foot-wide Cape Cod. If there is no room in the budget to build a dormer right away, it can be built on-site at a later date.

Yet another option is to add a shed dormer on the front of the house to create a faux saltbox look. You can do this with either a panelized dormer or an additional module.

Gable dormers (also called A-dormers and dog-house dormers) are another very popular option with Cape Cod homes, since they add space inside and character outside. They come in small and large sizes, with one or several windows. They can be dressed up with shutters and with circle-top windows or decorative moldings. Choose the size carefully, since it will affect how the home looks and how the second floor can be used.

Depending on delivery restrictions, the manufacturer may need to ship dormers on a separate carrier, which will add to the cost. In some cases it may make more sense for the general contractor to build them on-site instead. This is also a good option if the dealer's selection does not meet your needs. To assist the GC, the manufacturer should frame the opening for each dormer in the roof. The GC will need to help the set crew weatherproof the roof immediately after the set. He should then build the dormers as soon as he can.

Raised Ranches and Bilevels

A raised ranch is created when a ranch is built with a split-level entry on top of a raised foundation. Typically, the foundation is elevated at least 4 to 5 feet above the finished grade of the property. The exposed foundation is usually framed in wood and contains several medium- to large-sized windows. The entry to a raised ranch is split (that is, built halfway between the first floor and the basement), making the home a bilevel design. A platform is built at the front door that connects to two sets of stairs, one going up to the first floor and one going down into the basement.

Raised ranches are built for several reasons. The larger basement windows allow you to finish the basement affordably with good light. Raising the foundation out of the ground can solve a problem with a high water table. It is often easier to minimize excavation costs on a sloped property by building a raised ranch. In addition, if the property has sufficient slope, one side of the basement can be used for a drive-under garage, which is considerably less expensive to build than an attached or detached garage.

In designing a raised ranch, you will need to decide whether you want the front of the house flush with the front of the foundation or cantilevered over the top of the foundation. A cantilevered home, which is often preferred for its look, will have a foundation that is smaller than the main floor, which means you will get less usable space in the basement. You will also have to decide if you

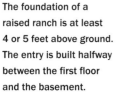

The foundation of a raised ranch is at least 4 or 5 feet above ground. The entry is built halfway between the first floor and the basement.

want the front entry to be flush with the front of the house or recessed. An advantage to a recessed entry, in addition to its appearance, is that it provides some overhead protection from the weather for anyone entering the front door.

If you plan to have any bedrooms in the basement, each one will have to have a window that meets the code requirement for egress. When thinking about the basement floor plan, pay attention to where the split-level stairs will be located. This is particularly important if you are building a drive-under garage, since the stairs should not enter into the garage.

Split-Levels

T-shaped ranches that are composed of a ranch on one leg of the T and a raised ranch on the other leg create a trilevel design. These split-levels, as they're often called, have some of the advantages of a raised ranch, although they do not work well on a flat lot with a high water table unless the ranch wing of the house is built on a crawl space. As with a raised ranch, these houses can also be built with either a flush or a cantilevered front and a flush or a recessed entry. Pay close attention to size of the basement windows and the location of the split-level stairs in relation to the basement space.

Hybrid Construction

Some manufacturers, especially those that serve the high end of the market, are now designing homes that combine modular and site construction. These hybrid plans usually have more complicated roof designs and multiple bump-outs. To create floor plans and elevations that have more eye-catching appeal, they are following the lead of the Japanese, who are committed to modular construction for their high-end homes, by using smaller modules when necessary. Using smaller modules allows for more flexible configurations. When

A split-level comprises a ranch on one leg of the T and a raised ranch on the other leg.

This creative hybrid modular plan, designed by Avis America, of Avis, Pennsylvania, requires more on-site construction than your typical modular home.

they cannot build a part of the plan in the factory with modules, the manufacturers use panels or sticks to complete the home on-site.

Although hybrid designs are usually more expensive to purchase from the manufacturer and complete on-site than the typical standard modular designs, they still represent a good value. That is because the types of plans that require hybrid construction are intrinsically more complex and expensive for everyone to build, including stick builders. And building them with modular technology considerably speeds up their construction.

The Exterior Look

Today's custom modular manufacturers, dealers, and general contractors apply every imaginable exterior finish and flourish to their homes. They side their homes in wood or fiber-cement board, when requested, and they trim out the siding with wider corner boards. In a neighborhood of stucco finishes, they apply stucco. In a neighborhood of cedar shakes, they provide the same. Depending on their customers' preferences, they accent the windows and doors with lineals, shutters, or mantels, and they dress up the eaves with frieze boards and dentil molding. They enlarge the roof overhangs and build gable returns to give the home a more custom look. They lend their homes more character by adding stepped roofs, dormers, decorative gables, and Victorian turrets. Once a modular home's exterior has been dressed up, it is usually impossible to tell whether the home was built at a factory or on-site.

As an example of how a simple house style can be turned into a classic statement, consider what can be done with a Cape Cod shed dormer, whether panelized or made

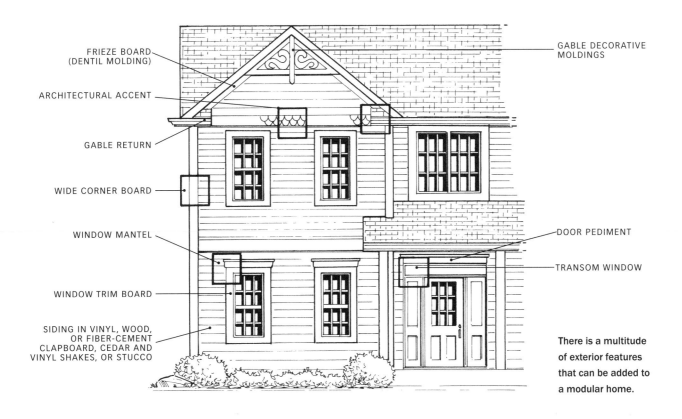

FRIEZE BOARD (DENTIL MOLDING)

ARCHITECTURAL ACCENT

GABLE RETURN

WIDE CORNER BOARD

WINDOW MANTEL

WINDOW TRIM BOARD

SIDING IN VINYL, WOOD, OR FIBER-CEMENT CLAPBOARD, CEDAR AND VINYL SHAKES, OR STUCCO

GABLE DECORATIVE MOLDINGS

DOOR PEDIMENT

TRANSOM WINDOW

There is a multitude of exterior features that can be added to a modular home.

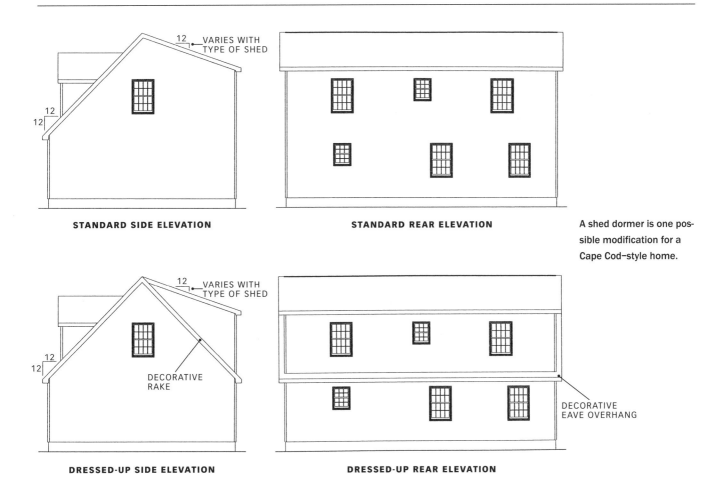

STANDARD SIDE ELEVATION

12 VARIES WITH TYPE OF SHED

12

12

STANDARD REAR ELEVATION

A shed dormer is one possible modification for a Cape Cod–style home.

DRESSED-UP SIDE ELEVATION

12 VARIES WITH TYPE OF SHED

12

12

DECORATIVE RAKE

DRESSED-UP REAR ELEVATION

DECORATIVE EAVE OVERHANG

with a third module. The most important detail is to create a 12-in-12 roofline in addition to the shed-dormer overhang. This can be accomplished by stepping the dormer in 2 feet on each gable end or by adding a decorative rake where the 12-in-12 gable roofline would be. The classic shed-dormer look can be enhanced by adding a decorative eave overhang on the rear of the dormer where the 12-in-12 roof would be.

The only limitations to what a home can look like are your personal vision and budget. Even the manufacturer does not have much say over how a home's exterior appearance is dressed up. If the manufacturer does not offer the finishes you want, you can have the GC apply the materials on-site.

Collecting photographs or drawings of what appeals to you most will help your dealer and GC understand your vision. The dealer will need to determine what contri-

bution will be made by the manufacturer and GC. The best way to ensure that everyone knows what you want is to have complete elevation drawings done of your home.

Site-Built Structures

Garages, porches, and decks are almost always built on-site by the GC rather than at the factory. It is usually impractical to make them part of the modular assembly, since they are not true boxes. An exception is when the structure is integral to the module, such as when a porch or garage is built into the box.

Many customers mistakenly think that a garage is a good candidate to make out of modules. A garage does not have a floor, it often does not have a true ceiling system, and it usually has one exterior wall almost entirely open for the overhead doors. In addition, since a typical garage is much

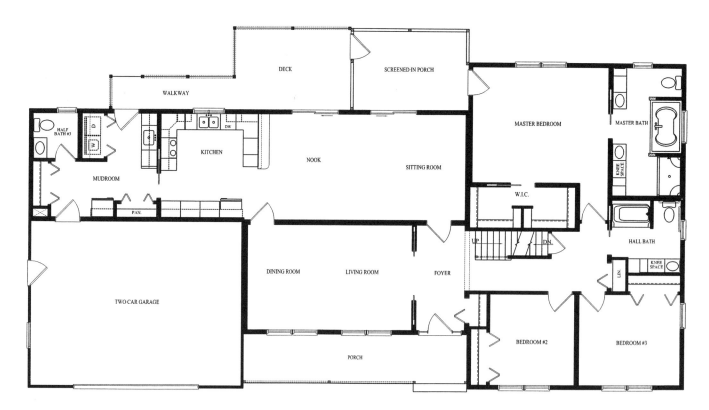

Modular homes are often built with additional structures, such as a garage, porch, and deck. Because these structures are usually not true boxes and are difficult to deliver in one piece, they are nearly always built on-site.

too wide to deliver in one piece, the manufacturer must divide it in half, which eliminates yet another wall, rendering it too weak to transport. Manufacturers who do provide garages, porches, decks, and other such structures with their homes usually offer them as panels or precut kits. Many GCs, however, prefer to build them on-site. Other rooms that are not designed to standard modular dimensions, such as some family rooms and mudrooms, can also be built on-site. The angled mudroom of the Whately 1 (see page 56) is a good example.

Universal-Design Criteria

Houses that incorporate the principals of universal design constitute a growing trend in residential construction. Though it is often considered synonymous with "handicapped-accessible" construction, universal design is actually a much broader concept, intended to create houses that are usable by and accessible to all people, regardless of their age, size, abilities, or disabilities. By incorporating features such as level (that is, stepless) entrances, wider hallways, and larger doors, a universal-design home becomes easier to navigate when pushing a child in a stroller, moving furniture in or out, or operating a wheelchair, temporarily or permanently. A home that features universal design will meet a family's needs now and into the future, allowing homeowners to age in place without having to consider moving when their abilities start to slip.

Modular homes can readily accommodate the principles of universal design. Most standard plans can be modified to have the doorways, hallways, and bathrooms enlarged. This may require that some rooms change shape and size, but this usually will not present problems if the

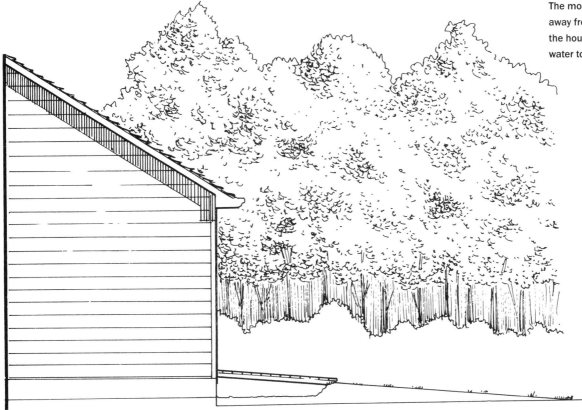

Adding a bridge can create a level entry without having to build a ramp. The moat keeps the dirt away from the front of the house and allows the water to drain away.

floor plan is large enough. For example, universal-design principles were applied to the Sugarloaf 3 to create the Sugarloaf 5. This entailed reconfiguring the bedrooms and bathrooms and then adding 2 feet to the length of that wing. In addition to changes in the floor plan, universal design incorporates user-friendly items like lever door handles and faucets, which are discussed in more detail in chapter 4.

Building a house that can be entered without having to climb steps, that allows someone in a wheelchair or with a walker to circulate easily, and that has a bathroom on the main floor allows visitors with a wide range of physical abilities to feel welcomed.

Younger people often fail to build a house that is accessible for elderly parents or disabled friends. Designing a house with a full bathroom on the first floor adjacent to a room that could be used as a bedroom would make it possible for an aging parent to move in some day. Given the astronomical costs of nursing homes, this is often the only viable option for families. If you are building a Cape Cod or two-story home that you intend to retire in, consider including room for a future chairlift or elevator, since space for these items cannot be easily added after the home is built.

Here are a few examples of universal-designed kitchens and bathrooms created by the Center for Universal Design at North Carolina State University for The Home Store and Excel Homes of Liverpool, Pennsylvania. Universal design makes your home accessible to everyone.

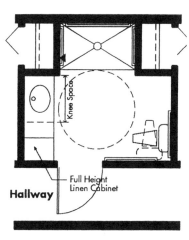

STANDARD TUB

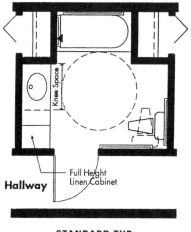

3X5 ROLL-IN SHOWER

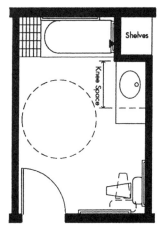

TUB WITH TRANSFER SEAT

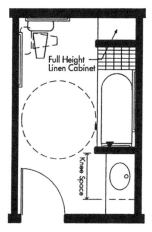

TUB WITH TRANSFER SEAT

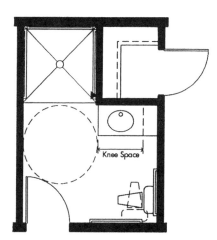

5X5 ROLL-IN SHOWER

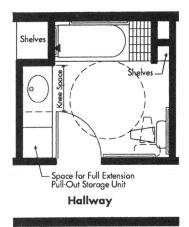

Shelves

Knee Space

Shelves

Space for Full Extension
Pull-Out Storage Unit

Hallway

TUB WITH TRANSFER SEAT

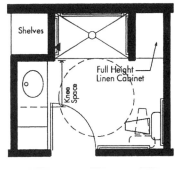

Shelves

Knee Space

Full Height
Linen Cabinet

Hallway

3X5 ROLL-IN SHOWER

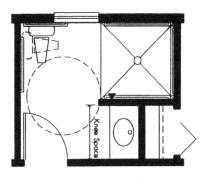

Knee Space

Hallway

5X5 ROLL-IN SHOWER

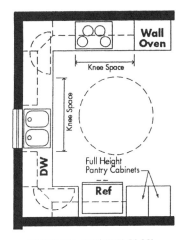

Wall Oven

Knee Space

Knee Space

DW

Full Height
Pantry Cabinets

Ref

OPTIONAL KITCHEN PLAN

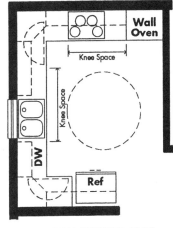

Wall Oven

Knee Space

Knee Space

DW

Ref

OPTIONAL KITCHEN PLAN

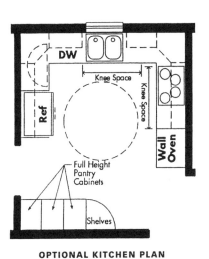

DW

Knee Space

Knee Space

Ref

Full Height
Pantry Cabinets

Wall Oven

Shelves

OPTIONAL KITCHEN PLAN

DW

Wall Oven

Knee Space

Ref

Full Height
Pantry Cabinets

W   D

OPTIONAL KITCHEN PLAN

The standard hallway width in modular homes is 36 inches. Adding 6 inches can make a big difference for someone in a wheelchair or moving furniture. There should be no extra material or labor charge for making the space bigger. However, you might be charged an engineering fee to redesign the plan, since you will need to take some space from the adjoining rooms.

It is not necessary to find a dealer with expertise on universal design. As long as a dealer knows what you need, he should be able to obtain the necessary help from his manufacturer. If you wish to create a fully accessible bathroom or kitchen, show your dealer the layouts on pages 92–93 designed by Ron Mace for The Home Store and Excel Homes, a modular manufacturer in Liverpool, Pennsylvania.

Construction Drawings

Before your home can be built, the manufacturer will need to draw a complete set of plans. Even if the dealer has already provided you with some good drawings, the manufacturer will need to create its own set, which it will complete with a computer-aided design (CAD) program. The drawings should be done to scale and include a floor plan, exterior elevation plan, electrical and HVAC plan, and foundation plan. Each of these plans will tell you something important about what the manufacturer will be building. In most cases, the dealer will request one or more revisions to the drawings to include changes initiated by you, the GC, or the dealer.

When you finally authorize the dealer to build your home, he will build it according to the drawings you have approved. This makes it very important that you carefully review each page, as well as any accompanying notes. Only in this way can you be sure that you are getting what you think you are getting. If you do your homework now, you will be able to catch and correct any mistakes or misunderstandings before your house is built.

When reviewing the drawings, keep in mind that however hard the dealer and manufacturer worked to get every detail just right, they might have made a mistake. This is especially true if you design a custom plan, or if the original plans or revisions submitted to the manufacturer contained many handwritten changes and notes. Sometimes the plans and changes are easy to interpret and complete, and the resulting drawings will look just as you anticipated. Other times the plans and changes are difficult to interpret and impossible to complete exactly as you imagined. This will be especially true if the initial plans were drawn by hand or with a drawing program rather than with a professional CAD program. In that case, several of the plan's features, such as room dimensions, closet sizes, and window spacing, may change noticeably when they are drawn to scale. This makes it critical that you look at these types of details closely and review the drawings with your GC.

Foundation Plans

The modular manufacturer will draw a foundation plan for your home. This plan may show the particular custom features you have selected, such as a walk-out bay and fireplace, or it may be a generic plan that your GC needs to redraw with the specifics added. If you are getting a full basement, one of the critical details is the manufacturer's location of the columns that support the marriage wall in the basement. Check whether the columns are located in areas you intend to use, such as in the middle of a room. This would be a particular problem, for example, if a column is placed where a pool table is planned.

Most manufacturers will prepare foundation plans only for the structures they are building. For site-built structures, the GC normally must prepare his own plans.

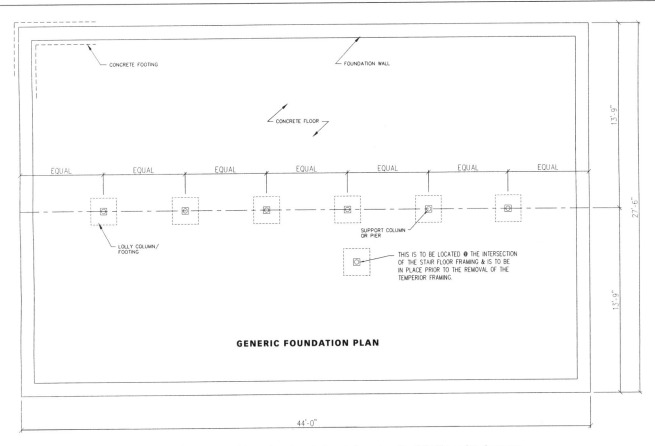

GENERIC FOUNDATION PLAN

If a manufacturer provides only a generic foundation plan, the GC will need to draw one of his own that shows the details of your plan.

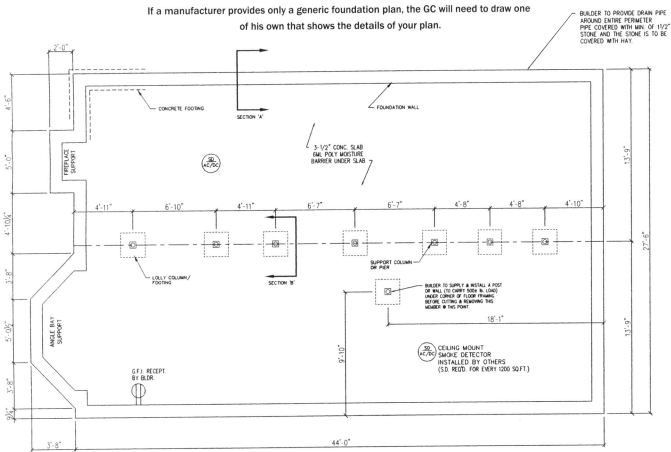

DETAILED FOUNDATION PLAN SHOWING SPECIFICS OF YOUR DESIGN, INCLUDING WALK-OUT BAY, FIREPLACE, AND EXACT LALLY COLUMN LOCATIONS

Electrical Plans

The manufacturer will prepare a drawing of the electrical layout. This will show the locations of all receptacles, switches, lights, circuits, TV and phone jacks, smoke detectors, door chimes, and thermostat wiring. Specialty wiring, such as for a stereo or security system, should also be shown. The locations of these items can have a big impact on how you enjoy your home, so it is smart to study the drawings carefully.

One detail you should verify is the location of the electrical panel box. The decision of where to locate the box should be made by the GC and the electrician. If the electrical power is coming from a telephone pole

Sample electrical layout drawing for a T-ranch. You should study your electrical plans carefully.

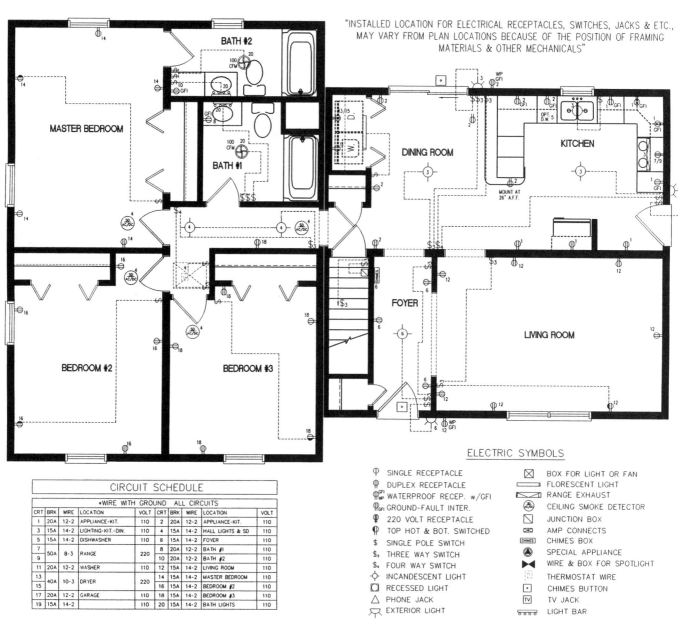

"INSTALLED LOCATION FOR ELECTRICAL RECEPTACLES, SWITCHES, JACKS & ETC., MAY VARY FROM PLAN LOCATIONS BECAUSE OF THE POSITION OF FRAMING MATERIALS & OTHER MECHANICALS"

CIRCUIT SCHEDULE

*WIRE WITH GROUND ALL CIRCUITS

CRT	BRK	WIRE	LOCATION	VOLT	CRT	BRK	WIRE	LOCATION	VOLT
1	20A	12-2	APPLIANCE-KIT.	110	2	20A	12-2	APPLIANCE-KIT.	110
3	15A	14-2	LIGHTING-KIT.-DIN.	110	4	15A	14-2	HALL LIGHTS & SD	110
5	15A	14-2	DISHWASHER	110	6	15A	14-2	FOYER	110
7	50A	8-3	RANGE	220	8	20A	12-2	BATH #1	110
9					10	20A	12-2	BATH #2	110
11	20A	12-2	WASHER	110	12	15A	14-2	LIVING ROOM	110
13	40A	10-3	DRYER	220	14	15A	14-2	MASTER BEDROOM	110
15					16	15A	14-2	BEDROOM #2	110
17	20A	12-2	GARAGE	110	18	15A	14-2	BEDROOM #3	110
19	15A	14-2		110	20	15A	14-2	BATH LIGHTS	110

ELECTRIC SYMBOLS

Symbol	Description	Symbol	Description
⏚	SINGLE RECEPTACLE	⊠	BOX FOR LIGHT OR FAN
	DUPLEX RECEPTACLE		FLORESCENT LIGHT
WP	WATERPROOF RECEP. w/GFI		RANGE EXHAUST
GFI	GROUND-FAULT INTER.		CEILING SMOKE DETECTOR
	220 VOLT RECEPTACLE		JUNCTION BOX
	TOP HOT & BOT. SWITCHED		AMP CONNECTS
$	SINGLE POLE SWITCH	CHIMES	CHIMES BOX
$₃	THREE WAY SWITCH		SPECIAL APPLIANCE
$₄	FOUR WAY SWITCH		WIRE & BOX FOR SPOTLIGHT
	INCANDESCENT LIGHT		THERMOSTAT WIRE
	RECESSED LIGHT		CHIMES BUTTON
△	PHONE JACK	TV	TV JACK
	EXTERIOR LIGHT		LIGHT BAR

NOTE: AN ARC-FAULT BREAKER IS REQUIRED (BUT NOT LIMITED TO) ALL CIRCUITS WHICH HAVE AN OUTLET LOCATED WITHIN A BEDROOM, INCLUDING THE SMOKE ALARMS, LIGHTS, AND FAN LIGHT. PER ARTICLE 210.12B OF THE 2002 N.E.C. TWO OF THE THREE CIRCUITS THAT ARE STANDARD FOR THE SECOND FLOOR OF A CAPE WILL BE ON A 20 AMP ARC FAULT BREAKER.

on the street, the electrician should first consult with the utility company, which will decide which pole to use. That decision, in turn, will influence where the meter is located, which will then determine the best location for the panel box. The manufacturer will take the electrician's suggestions and build the house with enough wire for the electrician to locate the panel box as requested. If you or the electrician gets this information wrong, the fix can add up to several hundred dollars.

HVAC Plans

If you are purchasing any part of your HVAC system from the manufacturer, you will receive a drawing of the system. Depending on the types of HVAC system, the drawing should show the locations of hot-water and electric baseboard units and forced-air registers.

Manufacturers sometimes cannot place baseboard units where the drawings indicate due to framing conflicts. If you need to precisely locate a baseboard unit in a particular room, note that information on the drawing. The manufacturer may be unwilling to comply with your request, however, without charging for the work involved in changing the framing.

When installing a conventional forced-air system in a two-story or Cape Cod design, an important design consideration is to create space for a "chase," which is an open box from floor to ceiling, to carry the supply and return ducts from the basement

Sample hot-water baseboard heating layout drawing for a T-ranch.

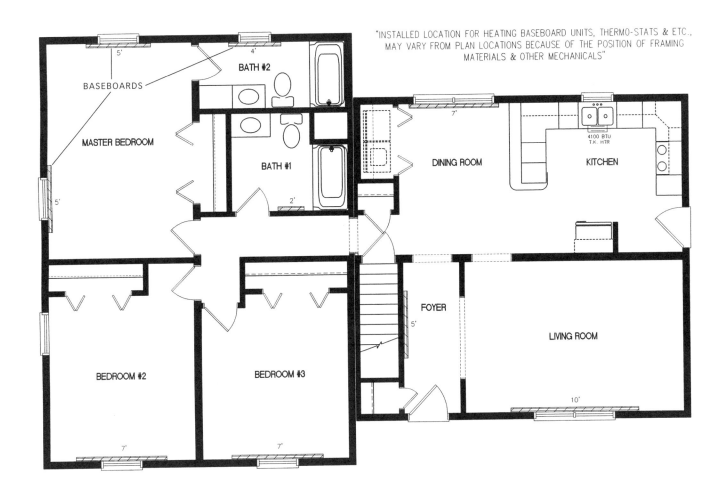

"INSTALLED LOCATION FOR HEATING BASEBOARD UNITS, THERMO-STATS & ETC., MAY VARY FROM PLAN LOCATIONS BECAUSE OF THE POSITION OF FRAMING MATERIALS & OTHER MECHANICALS"

up to the attic. Planning the chase's location is sometimes difficult because you do not want it to disrupt the floor plan. An option is to build an insulated exterior chase, especially when you can hide it in a garage.

Heating and cooling systems function more efficiently and provide more comfort when they are zoned, which means that separate thermostats control different areas of the house. You will want your HVAC drawing to show which areas of your home have their own zone and where the thermostats are located that control each zone. This is particularly important for hot-water baseboard systems on the second floor of a two-story, since the pipes will need to be correctly connected in the factory when the modules are constructed.

A modular manufacturer's drawing will contain a wealth of detail that will determine exactly what you will get. Study it closely, and ask questions of your dealer if you are unsure what a notation means.

- 2X10 SPF #2 FLOOR JOIST
- ROOF SYSTEM TO BE 16" O.C.
- EXT WALL 16" O.C./2X4 MARRIAGE WALLS 1ST FLOOR/2X3 MARRIAGE WALL 2ND FLOOR 16" O.C.
- 9' CEILING 1ST FLOOR/8' CEILING 2ND FLOOR
- MASS CODE REQ.
- WINDOWS: 7D

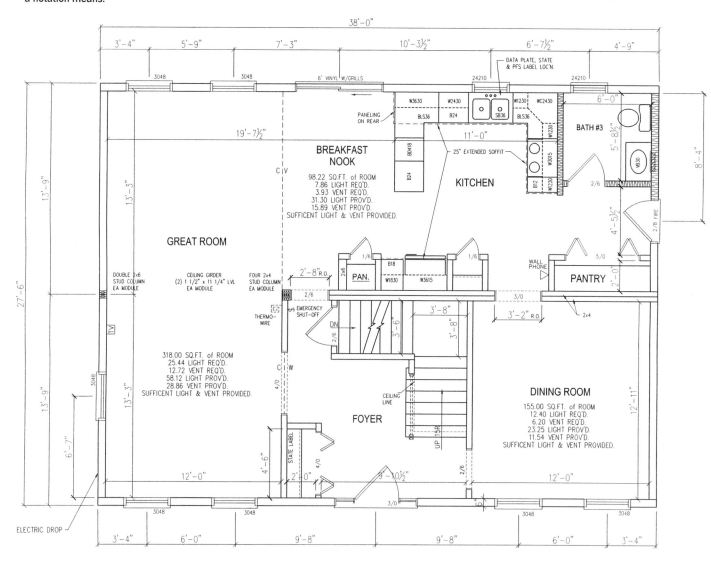

FIRST FLOOR

Floor Plans

A floor plan is a view of the house from above. Reading a floor plan is not difficult when it comes to identifying the rooms, but understanding what the symbols, dimensions, and notes mean can be difficult for the uninitiated. Many customers tend to skip over these details, but this can be a mistake. For example, a dotted line might indicate where one type of flooring ends and another begins, but it might also represent a landing at the bottom of the stairs, something you were not expecting and do not want. Or it could mean that the marriage wall in the 26-foot-wide great room

will have a painted drywall beam, which will make the space feel like two rooms, exactly what you were trying to avoid. The manufacturer should include notes on the plan indicating what the symbols mean, but if it does not, you definitely need to ask the dealer to explain them to you.

Begin by looking at the size of the home. This is especially important if you have a custom plan that was first drawn by hand, since the draftsperson could misunderstand your intentions. Next, carefully check the size of each room. Ideally, your copy of the plans will be drawn to scale, since this will allow you to make your own measure-

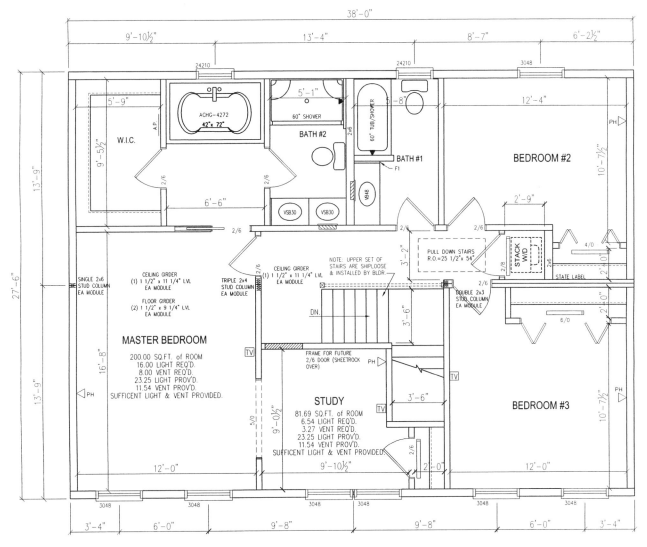

SECOND FLOOR

ments of each room. The size of a room will be meaningful only if you have a reference point. Sometimes you can create one by determining the room sizes in your current home or your modular dealer's models. Then compare the size of these rooms to those in your own plan.

When a room has a jog in it, which can be caused by a closet in the room or in an adjoining room, note whether the room measurements were taken between the widest or narrowest points of the room. For example, consider bedroom 3 in the Sugarloaf 7 (see page 70). The width of the room is listed as 9 feet 8 inches, front to back, because the width is measured from the closet wall to the front wall. If you were to measure the width from the entry door to the front wall, which would show you the room at its widest, you would get a 12-foot-

wide room. This would be misleading, because you cannot effectively use the extra 2 feet 4 inches in front of the door. One of the best ways to determine if a room size will work is to visualize all of your furniture in the room. It will help if you know the size of your furniture, so get out your tape measure. Making miniature pieces of furniture to scale and locating them on the scaled drawings provided by the dealer can also be helpful. Some lumberyards sell kits for this purpose.

Kitchen Layout

It is important that you understand the cabinet designations listed on the floor plan. For example, with most manufacturers a W1830 indicates a wall cabinet that is 18 inches wide and 30 inches tall, while a BD18 indicates a base cabinet 18 inches

Wall cabinets and accessories will have their own designations on your floor plan. This floor plan shows the typical space created for a refrigerator, range, dishwasher, and washer and dryer.

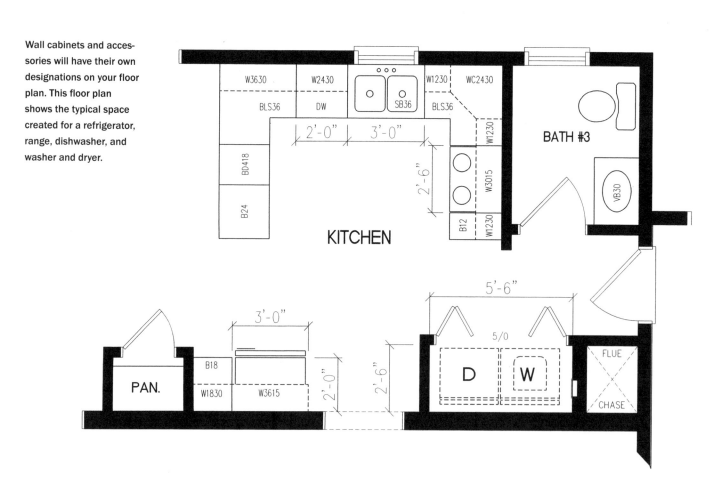

A Floor-Plan Checklist

Here is a list of items worth looking at when you review floor plans. Obviously, not every house will include everything that follows. As you study the plans, check off those details that are accurate or not applicable. Also, carefully read the information contained in each note written on each page of the floor plan and review the plans with your GC.

- ☐ Layout
- ☐ The location of sunlight in the morning, afternoon, and evening
- ☐ The best views
- ☐ Traffic patterns (for example, do you have to cross a large room to get to another room?)
- ☐ Shape and size of each room
- ☐ Location and length of each wall
- ☐ Furniture placement in each room (for example, think about your L-shaped sofa and television table, your queen-size bed and two large dressers, or your dining room table and hutch)
- ☐ Location and size of each closet
- ☐ Size and configuration of stairs to the second floor
- ☐ Location and direction of all flooring boundaries (where carpet, vinyl, wood, or tile surfaces meet)
- ☐ Location and size of each passageway, including those in the marriage wall
- ☐ Location, size, and style of a cathedral ceiling
- ☐ Location of each HVAC chase
- ☐ Location of a walk-out bay and other bump-outs
- ☐ Location of each fireplace (and possible interference with furniture placement)
- ☐ Doors and windows; refer to the manufacturer's door and window schedules that come with your plans
- ☐ Location, size, and swing of each door (for example, make sure that the swing of the door makes it easy to enter the basement from the kitchen)
- ☐ Location and size of each window
- ☐ Height of windows above the floor and height of your furniture

- ☐ Number of front-door sidelights
- ☐ Which side of a double door or slider opens
- ☐ Size of each wall, base, and specialty kitchen cabinet
- ☐ Location and swing of each cabinet door
- ☐ Sufficient space for the appliances
- ☐ Size of each vanity
- ☐ Location of each vanity drawer
- ☐ Location, type, and size of special vanity cabinets
- ☐ Location of towel bars and toilet-paper holder
- ☐ Location and size of mirror
- ☐ Location and size of medicine cabinet
- ☐ Size of each shower and number of seats
- ☐ Size and style of soaker or whirlpool tub
- ☐ Space around toilet and vanity
- ☐ Location of plumbing and electrical runs to the second floor of unfinished Capes or other usable attics
- ☐ Location of each phone and cable jack
- ☐ Location of each electrical switch
- ☐ Location of each electrical receptacle (be sure to have a receptacle within reach of each end table with a lamp)
- ☐ Location of each switched receptacle
- ☐ Location and type of each light
- ☐ Closet lights
- ☐ Location of special wiring, such as ceiling fans
- ☐ Location of each thermostat
- ☐ Location of each doorbell and transformer
- ☐ Location of the electrical meter and panel box, as required by the electrician
- ☐ Location of smoke detectors
- ☐ Location of plumbing access panels
- ☐ Location of each heating baseboard unit
- ☐ Orientation of plan
- ☐ Dimensions of plan
- ☐ Location of site-built structures, including mudrooms and fireplaces

wide with all drawers. There will be special designations for each specialty cabinet, such as a wall oven, cooktop, pantry, and microwave cabinet. Accessories, such as roll-out shelves and lazy Susans, will also have their own designations. If you do not understand the designations on your plan, ask your dealer to explain them.

The location of each cabinet can be as important as its size. For example, having the right cabinets next to the dishwasher and range can make a big difference in how easy it is to use your kitchen. Also pay close attention to whether there is enough countertop space next to the cooktop, refrigerator, and dishwasher. Reading a few articles on kitchen design in magazines or on the Web can help you make better decisions.

Unless you purchase kitchen appliances from the modular manufacturer, the location and size of each appliance is also very important. The manufacturer will use standard appliance sizes, as shown in the illustration on page 100, when creating space for each of the appliances. Appliance manufacturers, however, increasingly are making models that differ from standard sizes. If you purchase appliances that require a nonstandard space, inform your dealer and make sure it is recorded on the plans. Since there is no established standard size for cooktops, wall ovens, and microwave ovens, specify the size of these appliances to your dealer. Before ordering appliances, be sure to consider which way the appliance door should swing if it does not come with a reversible door.

Elevations

Elevations are the drawings of the vertical or side view of a house, inside or out. A complete elevation plan will show what your home will look like, taking into account the property's slopes and contours after the GC completes his button-up work and site-built structures. To provide this plan, a draftsperson must integrate three types of detail. The first shows how the home will look after the GC completes his button-up work, and assumes that the GC is building no other structures and your property is perfectly flat. The second type of detail adds all of the GC's site-built structures along with any finishes he is applying to the home. The third level of detail depicts how the property's grades and landscaping will impact what is actually built.

All modular dealers will provide an elevation plan of your home, but not all of them will draw the exact home you are building. They may instead give you a generic plan showing the style of home you have selected. For example, if you are building a 1,700-square-foot, two-story home with the front door to the right side of the house and a walk-out bay on the opposite gable, they may show you a 2,200-square-foot two-story with a center-entry front door and no bay on the end of the house.

You should not settle for a generic elevation. Ask your modular dealer to provide you with an elevation plan of what you intend to build before you authorize him to build it. If his manufacturer will not provide the plan and he is unable to create it himself, ask him to have someone else draw it. Even if you have to pay extra for an accurate plan, you need to see it. Little things like the spacing of the windows can matter a lot, especially when you have taken a standard plan and added length or windows or you have moved some windows within a room to accommodate furniture. In addition, if you are dressing up a plan's no-frills standard look, you will want to see if your vision holds up to your own scrutiny.

FRONT ELEVATION

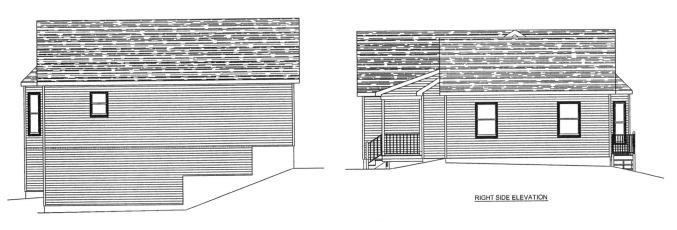

REAR ELEVATION

LEFT SIDE ELEVATION

RIGHT SIDE ELEVATION

A complete set of elevations should show the modular home, the site-built structures, and the finished grading. Generic elevations often do not show how a specific house will look on a specific site. In this example, the drawings show that the grade slopes from left to right and front to rear. This creates a walk-out in the rear.

Asking the dealer for plan-specific elevation plans will not be sufficient when your GC is building additional structures. Unless your elevation plans include everything the GC is building, you will not be able to review what your home will look like. However, you might not get the needed assistance from your dealer if he is not serving as your GC. Your dealer may feel that he does not want responsibility for drawing what someone else is building. He may also point out that the regulations governing modular construction in your state do not allow the manufacturer to include any of these site-built structures in its permit plans. He may correctly insist that state and local officials only allow the manufacturer to draw what it is building. While this is true, some dealers and manufacturers will help you by creating a separate set of plans that show the GC's work. Before completing the drawings, they will ask you and your GC to provide the details, insist that you take responsibility for their accuracy, and charge you an additional fee for the assistance. You should make this investment.

An alternative to having the dealer complete the elevation plans is to ask your GC to complete his own set of plans. The advantage to this approach is that he will know exactly what he is building. The disadvantage is that you will now have two sets of incomplete plans, one of your modular home and one of your GC-built structures. The only way they will appear on the same page is if the GC re-creates the modular plans or another person integrates the details into a third set of plans, an impractical and wasteful step. This is why it is better to pay the dealer to provide a complete set of plans.

Few builders, modular or stick, provide the third set of details, those that capture the property's topography. This is not important if the land is perfectly flat. It is important, however, when the elevation plans depict the home on a flat lot when the property slopes front to back or side to side. For instance, if the finished grade varies more than a couple of feet around your home, more of the foundation will be exposed at the low points. Once you see an accurate plan showing a large section of the foundation above the finished grade on one side, you may want to consider replacing that section with wood-framed knee walls or walk-out walls. This may in turn lead you to relocate the furnace and water heater to maximize the benefits that the added windows will provide. To accomplish this, you may have to modify the house plan to move the chimney closer to the new furnace location. If you do not discover this situation until after the excavation work has begun, it may be too late to change the house plan and relocate the chimney, which could mean that the furnace is stuck in the middle of what could have been a very convenient and affordable basement family room or office.

An accurate elevation plan may also make you aware that the slope in your backyard is so steep that it will require additional steps to the rear porch. You may prefer to avoid a long set of stairs. Learning of this potential situation in advance will allow you to eliminate the problem by purchasing additional fill for the low spot. Since the fill will cost a bit of money, however, you may not be able to afford it unless you omit something from your modular contract, which you will be able to do only if you make the decision when you review the elevation plans. Waiting until the GC begins the excavation will be too late, since you will have already signed off on the plans and specifications.

An elevation plan with topographical detail can also reveal when there needs to be a step down or up between parts of your home. For example, you might need three steps to enter the home from the garage because of a gentle slope across the front of the property. One way to avoid the steps is to build a retaining wall on the side of the garage so that additional fill can raise the garage floor without the threat of erosion. You will surely want to know about this condition while you are still in the planning stages so you can budget the additional funds required.

As these examples illustrate, the natural contours of your land can significantly affect how you build your home. The more you know before construction begins, the more options you can consider and factor into your design and budget. Therefore, ask the dealer to show the property's topography when he draws your home and site-built structures. Your GC or a surveyor can provide the dealer with this information.

The most accurate topographical detail comes from using a transit or its equivalent. The GC may suggest that he can come

Reviewing Elevation Plans

Here is a checklist to help you review exterior elevation drawings. Make sure the features that apply to your house are shown on the drawings and look as you expect them to look. Some of these items are discussed in greater detail in chapter 4.

ROOF
- ☐ Pitch
- ☐ Shingle type
- ☐ Eave and gable overhang size
- ☐ Gable-end returns: size and style
- ☐ Decorative gables: size and pitch
- ☐ Skylight location

SHED DORMERS
- ☐ Size
- ☐ Style
- ☐ "Fake rakes" along gable ends and rear

GABLE DORMERS
- ☐ Size
- ☐ Style
- ☐ Symmetry

WINDOWS
- ☐ Quantity
- ☐ Size
- ☐ Spacing and symmetry
- ☐ Grills
- ☐ Transoms
- ☐ Specialty styles

DOORS
- ☐ Quantity
- ☐ Size
- ☐ Spacing and symmetry
- ☐ Sidelights
- ☐ Grills
- ☐ Transoms
- ☐ Door lights

SIDING
- ☐ Type
- ☐ Corner-post style and width

DECORATIVE FINISH
- ☐ Dentil moldings and frieze boards
- ☐ Window mantels, lintels, and shutters
- ☐ Door pediments

RAISED RANCH AND SPLIT-LEVEL FINISH
- ☐ Front elevation: cantilevered or flush with foundation
- ☐ Entry: recessed or flush
- ☐ Knee-wall windows and doors: quantity, location, and size

EXTERIOR STAIRS
- ☐ Tread width
- ☐ Landing size
- ☐ Number of steps
- ☐ Railings: location and style

CHIMNEY
- ☐ Size
- ☐ Style
- ☐ Location

close enough by walking the property, but line-of-site judgments made with the naked eye are often inaccurate, especially when a property is heavily wooded or covered with thick vegetation. The only way to accurately determine the topography is for someone to take detailed site measurements with the appropriate equipment. You will have to pay for this service, but unless your property is perfectly flat, it will be worth the expense.

Keep in mind that even if you receive elevation plans that conform to your property's grades, they may not represent exactly how your home will sit on the lot after it is built. That's because the actual finished grade will depend on how deep the foundation is installed. This in turn will partly depend on the soil and groundwater conditions discovered after the basement hole is dug. Groundwater and ledge can require the house to be raised substantially higher than what was drawn on the elevation plan. The finished grade will also depend on how much fill, if any, is brought to or removed from the site to compensate for these conditions. Unless you dig some test holes on the property before finalizing the decisions on your home (see pages 189–190), you will not be able to anticipate and plan for these elevation changes. Only after your property is finish-graded and landscaped will you truly see what it is going to look like.

4 Specifications and Features

ANUFACTURERS OF MODU-
LAR homes are offering
their customers more and
more customization. The
market demands it, and
their competitors are providing it. This
chapter discusses hundreds of options that
are usually available and gives advice on
which offer the most value and which the
least. It suggests what things are best left to
the manufacturer and what to the general
contractor. And it discusses those features
that should be taken care of right away and
those that can be handled later.

This chapter will not, however, provide a
detailed survey of each and every option. A
knowledgeable dealer is a good source for
additional and up-to-date information. But
when it comes to the many specific products that go into a finished home, customers
should seek out a variety of informational
sources, such as specialized books, consumer magazines, and Web sites for residential construction.

The number and variety of options listed
in this chapter may be surprising to customers. Some dealers and manufacturers
will be surprised as well, since the typical

manufacturer does not offer all of these
options. In fact, there are only a few manufacturers that offer everything suggested in
this book. Manufacturers can afford this
range of customization only if they raise
their prices compared to their competition.
Not surprisingly, those companies that
present the most choices are among the
most expensive.

But even those manufacturers that offer
the fewest choices are embracing customization more than many site builders,
especially production builders who build in
their own subdivisions. The philosophy of
these builders is to limit the customer's
choices to color and finish selections and a
couple of plans with a few predetermined
alterations. Modular manufacturers would
like to conduct their business in this manner, but customer demands make it difficult
to do so.

Choice, however, comes at a cost. It is
easy to overspend when you let your wants
get ahead of your budget. To protect yourself, generate a wish list. Rank the items in
terms of your family's priorities. Figure out
a realistic budget to determine how many
items on the list you can afford. Start by

asking each family member to pick the five most important items he or she would want in a new home. Ask them to consider which features of their current home they like and which they would change. You should revise the priority list after learning of the selections available from your modular dealer. When finalizing your priorities, keep in mind the discussion that follows about which items add the most value and which can be delayed to a later date.

Adding Value

The best value in modular construction is the cost of square footage or space. When you enlarge a home, you get additional framing, insulation, electrical, drywall, roofing, siding, flooring, moldings, and so forth for a very good additional cost. That is because modular assembly-line systems are very efficient. So when a customer faces a choice of more usable space or an option that could easily be added later, it is smart to choose the additional square footage. The cost of installing an additional bathroom is another great value; it can be as little as half the price a local plumber would charge a stick builder.

On the other hand, most amenities that make a home more attractive do not necessarily add appreciably to its value. They do not increase the lender's appraisal or the future resale price by as much as they cost. For example, selecting wood windows and doors over vinyl, cherry kitchen and bathroom cabinetry over oak, cedar siding over vinyl, and 16-inch-on-center framing over 24 inches in the walls and roof can add thousands of dollars to the cost of a home without adding value in the eyes of most lenders. Relatively inexpensive upgrades such as lever locksets, rocker electrical switches, larger windows, and wider stairs are even less likely to add measurable value. Yet all of these features are worth considering because they do add quality to a home.

In the eyes of some customers, that quality adds to the resale value.

Omitting Materials from the Manufacturer

Customers often ask which materials from the manufacturer it makes sense to omit so they can buy their own and have their GC install them. Some materials are easy for the manufacturer to omit and the general contractor to install. This includes both required materials, such as a faucet, and optional ones, such as a zero-clearance fireplace. Other materials can create significant additional work for the GC when not done by the manufacturer, although it is still reasonable for a customer to exclude them. Interior moldings and doors fall into this category. Still other materials, such as electrical wiring, make no practical or economic sense to omit. The lists (facing page) give typical examples of materials that fall into each of these three categories.

When deciding to omit the manufacturer's materials, make sure that the GC understands what additional work he is expected to perform; this is particularly important when the GC is not also the dealer. Also, consult with the GC and dealer to determine what the manufacturer should and should not do to prepare the house for the GC's on-site installation.

Delaying Optional Amenities

Most customer budgets do not have room for every optional item on their wish list. Customers sometimes get discouraged because they assume they will never get all of their hoped-for features unless they obtain them when they first build their home. But patient customers can often get some of their wish-list items after their home is built. As long as it is realistic for customers to afford some amenities in the future, they can reorder their priorities in terms of which options must be built into their home now and which can be affordably and simply added at a later date.

The strategy of postponing optional amenities to maintain a budget works best when the items are expensive. The more money saved with each deferred item, the more that is available to spend on other amenities. For example, delaying a masonry fireplace helps overcome budget constraints a lot more than postponing crown molding in the dining room.

There are several factors that determine how economical and easy it is to add an optional item after a home is built. One factor is the work involved to install the feature. If the installation can be done without remodeling, it is a good candidate to postpone until a future date. Decorative finishes, such as crown moldings, chair rail, and wainscoting, for example, can be added easily at a later date, since they do not require removing or rebuilding anything in the house.

Another factor is the cost of replacing the manufacturer's standard materials. It is practicable to postpone adding an amenity when the manufacturer's standard materials are so inexpensive that the customer is not seriously penalized for later replacing and discarding them. For example, if the manufacturer offers carpet in the living room as part of the standard package, it may be smart for a customer to accept the carpet even though she knows she wants a wood floor. That is because the bulk-purchased carpet may be a very inexpensive item for the manufacturer. In this case, the manufacturer might offer a wood floor for $1,800 but provide a credit of only $200 for the unused carpet. If that additional $1,600 is a strain on the budget, it might be wise to take the carpet, use it for a couple of years, and then install the wood floor. The used carpet could then be put in another room.

Omitting Materials: Dos and Don'ts

The following materials can be omitted when ordering a modular home with little consequence as long as the customer and the general contractor plan properly. It may cost more to buy and install them locally, but not much more:

- Gable dormers installed in unfinished attics
- Appliances
- Bathroom medicine cabinets and mirrors
- Closet shelving
- Door handles
- Exterior doors
- Faucets
- Flooring
- Kitchen cabinets
- Kitchen countertops
- Light fixtures
- Decorative gables
- Siding
- Sinks
- Skylights
- Stair railings
- Toilets
- Zero-clearance fireplaces

The following materials can be omitted from the manufacturer, but doing so may create a significant amount of work:

- Electrical switches and receptacles
- Interior doors
- Interior moldings
- Roof system
- Shingles
- Stairs to upper floors
- Tubs and showers

The following materials do not make economic sense to omit:

- Drywall
- Electrical wiring
- Floor, roof, or wall sheathing
- Second-floor plumbing
- Windows

Other products that can be easily delayed include solid-surface or granite kitchen countertops. If a customer purchases one of these countertops, the manufacturer might give a $250 credit for omitting the standard plastic-laminate countertop and then charge an additional $4,000 for the replacement. Taking the standard item now leaves $3,750 in the budget that could be applied to another upgrade that may not be replaced so easily in the future. The cost for using the standard countertop for a few years would be only $250.

Another opportunity for deferring an optional feature is when the customer has the skills to install it but cannot do the work right away. For example, if the customer is able to install a tile floor but won't have time for two years, he should take the standard flooring rather than have the GC do the work now. Keeping the standard flooring takes some pressure off the customer in case two years becomes three. Delaying the expense helps him hit his budget, and completing the work saves him money.

Sometimes it makes good financial sense to have the manufacturer prepare the house for an expensive item, which can then be purchased and installed at a later date. A fireplace is a good example of this strategy.

Universal-Design Features

If you want to build a home that is usable by all of your family and friends, regardless of whether they're young or old, tall or short, strong or weak, able-bodied or with a disability, consider including some of the following options in your home. You may find that your modular dealer is unaware of the universal-design concept and its benefits. Even so, he will almost certainly understand each of the specific features listed below, and he should be able to provide those you ask for.

☐ Lever-handled locksets for exterior doors

☐ 36-inch-wide exterior doors

☐ Low exterior-door thresholds

☐ 36-inch-wide interior doors

☐ Lever-handled interior-door handles

☐ Extra-wide 42-inch hallways

☐ E-Z-Fold hinges for bifold closet doors

☐ Raised electrical outlets

☐ Rocker-style electrical switches

☐ Top position ground plug at all outlets

☐ Easy-to-reach thermostat location

☐ Thermostats with large numerals and an intuitive operation

☐ Task lighting for specific activities, such as cooking, reading, and shaving

☐ Varying height kitchen work surfaces

☐ Knee space at a kitchen countertop for a chair

☐ Pull-out shelf for oven food transfer

☐ D-shaped cabinet handles

☐ Extra maneuvering space in the kitchen and bathrooms

☐ Knee space at the bathroom vanity for a chair

☐ Offset tub/shower anti-scald controls

☐ Curbless shower with a flexible water dam

☐ Handheld showerhead

☐ Blocking for grab bars at toilets, tubs, and showers

☐ Bathroom grab bars

Some people lower the electrical switches to make it easier for young children to turn the lights on and off. Others raise the electric outlets to help people with arthritis plug and unplug fixtures. Many people install lever door handles to make it easier for everyone to open the doors. Lever handles do not need to be grasped. In fact, they can be opened with an elbow by someone carrying a bag of groceries. More and more people are adding grab bars in their bathrooms, since anyone can slip on a wet surface. Everyone appreciates the additional support when bending down, rising from a lowered position, or stretching to reach something. For more on universal design, see pages 91–94.

When a customer knows that she wants a fireplace but cannot afford to pay for it right away, she should have the manufacturer frame the openings in the walls and ceiling and have the general contractor install the necessary foundation pad. Completing this work while the home is under construction minimizes the installation costs for a fireplace later on. Some customers also apply this strategy to whirlpool tubs. They have the manufacturer install the rough plumbing and frame the platform and access panels while the house is being built, making it much easier for the plumber and carpenter to install the tub at a later date.

Features That Should Not Be Deferred

There are many optional features that incur a substantial expense if not installed when the home is built. Extra square footage, a steeper roof, 9-foot-high ceilings, and architectural shingles cost thousands of dollars extra if they are added at a later date. This also goes for wood windows, upgraded kitchen and bath cabinetry, wood siding, stained-wood interior doors with matching moldings, and oak stair treads. In general, if adding a feature after a home is constructed requires extensive remodeling and additional engineering, do it when you build your home or reconcile yourself to living without it.

There are two groups of options that are relatively affordable as long as they are included when the house is built but very expensive when installed at a later date: energy efficiency and accessibility, or universal design. Although both options provide enduring benefits, many customers do not consider them until after they have built their home.

Some of the things that make a home more energy efficient can be postponed. For example, increasing the ceiling or basement insulation can be done at any time. But other items, such as energy-efficient windows and sealing air leaks, should be done when the home is constructed. As an investment, energy-saving options pay for themselves relatively quickly and ultimately save money. They offer protection against future leaps in the cost of fossil fuels. And energy-efficient homes are a lot more comfortable, because they maintain their temperature without drafts.

As discussed on pages 91–94, a home that incorporates universal-design features is more accessible for both the homeowners and their visiting friends and family

Modular Mansions

Modular builders in Greenwich, Connecticut, have built several multimillion dollar homes. Certainly part of what makes these homes so expensive is the cost of the land. But it is the things these builders are doing to make their homes into mansions that makes them sell quickly to very happy customers.

Some of the builders start by working closely with an architect. All of them select manufacturers who have creative engineering departments and flexible purchasing and production staffs. They complete many of their homes' exterior embellishments and interior and exterior flourishes on-site with local subcontractors. Some builders use top-of-the-line production cabinets, while others install custom cabinetry made by a local woodworking shop. The same goes for interior doors and moldings, with custom crown moldings, chair rail, and wainscoting complementing impeccably detailed window and door trim. It is not uncommon for the builders to skim-coat all of the drywall with plaster before repainting the entire home. They finish it with custom hardwood, tile, and laminate flooring. And they landscape the property to perfection.

Why do these builders use modular homes to construct multi-million-dollar mansions? Cost is one reason, of course, but reliability, speed, and quality are the principle incentives, since the local subcontractors are too busy and their quality is too inconsistent. With such success, these cutting-edge, luxury builders have no intention of going back to stick construction.

members. The most important of these features are best included when the house is being built. Wider doorways, larger bathrooms, and blocking in the bathroom walls for future grab bars are only modest investments when the house is built, but they are considerably more expensive to add later on, since these investments will most likely require extensive remodeling of your home.

Standard and Optional Selections

The following sections discuss many of the standard and optional specifications and amenities offered by modular manufacturers. Although most of the optional selections will incur an additional cost, no estimated prices are included. This is because prices vary considerably from dealer to dealer within and across regions. In addition, due to a variety of reasons, prices in the building industry often change considerably year to year, especially products made of wood.

Some items described below as an upgrade actually are included in some manufacturers' standard specifications. When a feature has merit, it is recommended that you consider it. It is not expected, however, that you need or want every recommended feature. In fact, your dealer should discourage you from selecting options you do not need. One of his responsibilities is to help you build a home within your budget, which means he should help you stick to your priorities. Ultimately, however, the decisions are yours; you need to carefully weigh which items are most important to you.

Foundation and Sill Plate

Before building a modular home, the manufacturer needs to know the type of foundation the GC will install and the thickness of the plates he will use to secure the home to the foundation.

Building to Code

Modular dealers and manufacturers are responsible for building homes that meet the state's building codes. However, the customer or general contractor should check with the local building department to learn if it enforces any codes that differ from those of the state. If any governmental institutions are financing the home, they may also have specific code requirements. If any adjustments to code requirements are necessary, the dealer should be told soon enough to incorporate them into his order with the manufacturer. Any additional costs are the responsibility of the customer.

In theory, this should not be a concern in any state with a preemptive building code, since all building inspectors in that state must comply regardless of their personal preferences. In practice, however, some local inspectors choose to enforce more stringent codes, or a few codes more stringently, sometimes without recognizing that they are doing so. Knowing this in advance can prevent conflicts down the road.

Foundation

It is not practical to build a modular home on a slab, since the slab cannot be used as the first floor of the home. Also, modular homes require access to the plumbing below the first floor, which requires either a full basement or crawl space. The customer needs to specify which type of foundation he wants. For a full basement, the manufacturer frames a stairwell opening. A crawl space does not require stairs, although you may want the manufacturer to create access through the floor. With a crawl space, the electrical panel box and heating system are usually placed on the main floor.

Sill Plate

The GC must fasten the home to the foundation. If the foundation is a poured or cement-block wall, he affixes a sill plate, usually single or double 2×6s, on top of the foundation. The set crew then secures the house to the foundation by fastening the bottom of the house to the plate. The dealer needs to be informed if the GC will use a single or double plate, since this determines where the manufacturer installs the bottom row of siding. If the manufacturer assumes that the GC is installing a single plate when he is using a double plate, or vice versa, the siding will not cover the sill plate correctly. To correct the problem, the GC will need to remove and reinstall all of the siding.

Insulation and Housewrap

Customers may increase the amount of insulation in the walls and ceiling. They may also choose to install housewrap under their siding as a way to reduce air infiltration and protect the sheathing.

Insulation

Some manufacturers allow you to beef up the amount of insulation in the exterior walls by either replacing the standard materials with higher-density fiberglass insulation or adding rigid-foam insulation to the exterior. The latter option requires the manufacturer to extend the window and door jambs to accommodate the additional wall thickness.

The ceiling or attic insulation can be increased easily by the manufacturer, usually up to R-38. An alternative is to have the GC blanket the attic with blown-in insulation, which has the added advantage of reducing air infiltration in the ceiling. It is also a very effective way to insulate any HVAC ducts installed in the attic.

Adding insulation to the interior walls is a reasonably effective way to reduce noise transfer between rooms or around plumbing pipes, although it does not completely soundproof them.

If the basement ceiling is going to be insulated, plan also to insulate the stairwell to the basement and add a door sweep on the bottom of the door to the basement.

Housewrap

Most manufacturers do not automatically install housewrap, such as Tyvek, under their vinyl siding unless required to do so by the building code. The manufacturer's logic is that this material was designed to cover all of the cracks and crevices found in normal site-built homes. Since modular manufacturers build their homes to prevent air infiltration, the material is redundant when applied to a modular home. The Energy Star testing has found that a well-assembled modular home does not benefit significantly from an air-infiltration barrier. This logic applies, however, only if the set crew and the GC properly seal all of the cracks and gaps that are formed when the modules are joined together.

Many industry professionals argue that housewrap also helps keep water away from the sheathing. This is particularly useful if the GC is not able to side the house immediately after the set. It is also good to use under wood and fiber-cement sidings.

Energy Star Homes

One of the best parts about building a modular home is that you get superior energy efficiency as part of the standard package. You can even improve on that by having your home built to the Energy Star specifications. The Energy Star initiative is a federal program administered by the Department of Energy and the Environmental Protection Agency. Conventional builders need to take many extra steps at greater cost to their customers to meet these energy standards. For modular customers, the steps require very little effort and no appreciable expense.

A typical Energy Star home increases energy efficiency, lowers utility bills by hundreds of dollars per year, and improves a home's comfort and safety. The higher levels of insulation and lower levels of air infiltration make the interior surfaces of the home quieter and warmer.

The mechanical ventilation included in an Energy Star home improves indoor air quality. This is important because some of the enhanced energy efficiency is created by reducing air infiltration, which can allow air pollutants to build up indoors. Household cleaners, carpet adhesives, carbon monoxide from heating systems, dust, and pet dander can accumulate in a well-insulated home. In addition, showering, cooking, and breathing produce humidity that can build up if not vented to the outside.

In fact, due to today's building codes, even new site-built homes that are not Energy Star rated build up too many indoor air pollutants. You might think the answer is to abandon the new building codes, or at least to avoid going beyond what the code requires. But tighter homes are not only more energy efficient, which saves money; they are also much less drafty, which makes them more comfortable. The solution is to build tight homes, but to let them breathe with controlled continuous ventilation.

Increasing numbers of modular manufacturers are following the Energy Star program by offering supplemental ventilation systems. The most popular system is an appropriately rated bathroom fan connected to a variable-speed timer. Costing less than $200 to install, this addition continually refreshes indoor air for only one tenth the cost of heating the air in a leaky house.

If you would like to participate in this program, look for a modular dealer who has experience in building energy-efficient homes. Ask a dealer who lacks that experience if he is willing to learn the Energy Star procedures, which are simple to master. He will need the manufacturer's assistance to be successful. If the manufacturer does not have experience building Energy Star homes, it will need to consult a local representative from the program.

The most important extra step the manufacturer might need to take is to apply extra air-sealing techniques to the exterior shell. This also requires the manufacturer to use recessed lights rated for low air infiltration. The manufacturer might need to insulate the basement stairwell, increase the ceiling insulation, upgrade to low-e windows, switch to a properly rated bathroom fan, and use a vented range hood for a gas range. Many of these features, however, are standard with most manufacturers, and they are routine options for the other companies.

The GC needs to install an energy-efficient heating system and water heater, insulate the basement, and take additional air-sealing steps. A whole-house fan is also worth considering.

The Energy Star program assists the dealer with designing your home for energy efficiency when you submit your house plans along with a list of energy-related specifications. The program compensates the dealer, the GC, and you for some of the additional costs. Most local gas companies give the GC a rebate if you select gas heat. And you'll receive a rebate for Energy Star–approved appliances and light fixtures, both of which use less energy.

The Energy Star program helps you build a better home. It costs you less money to own and gives you more comfort to enjoy. Building a modular home makes this easy and affordable to achieve.

Exterior-Wall Construction

If the manufacturer's standard specification is to frame exterior walls with 2×4s spaced 16 inches on center, upgrade to 2×6s spaced 24 inches on center. The modules handle the delivery and set better with thicker walls. Also, 6-inch walls allow an extra 2 inches of insulation. For modules over 54 feet in length, it is best to build them with 2×6s spaced 16 inches on center, since this helps the modules during the delivery and set.

To further increase the strength of each module, it is smart to frame the marriage walls, which are the walls that join two modules, 16 inches on center. This strategy can be used for all interior walls, but as a general rule it is unnecessary.

Most manufacturers install either ⅞- or ½-inch sheathing on the exterior walls, and either does fine.

Floor Framing

Most manufacturers build their floors with conventional floor joists. For narrow homes, they use 2×8 floor joists and for wide homes they use 2×10s. Some companies, however, use 2×8 joists for the second story of wide homes when this is all that is required by the building code. In this case, upgrading to 2×10 joists produces a sturdier floor.

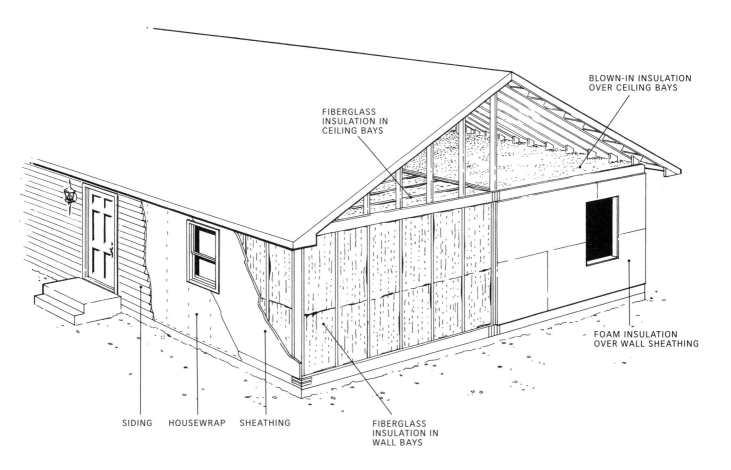

BLOWN-IN INSULATION OVER CEILING BAYS

FIBERGLASS INSULATION IN CEILING BAYS

FOAM INSULATION OVER WALL SHEATHING

SIDING   HOUSEWRAP   SHEATHING

FIBERGLASS INSULATION IN WALL BAYS

The walls and ceilings of modular homes are usually insulated with fiberglass batt insulation. Foam insulation is sometimes offered to increase the R-value of the walls. Some manufacturers use blown-in insulation in the attic. When housewrap is used, it is placed over the sheathing and under the siding for increased energy efficiency.

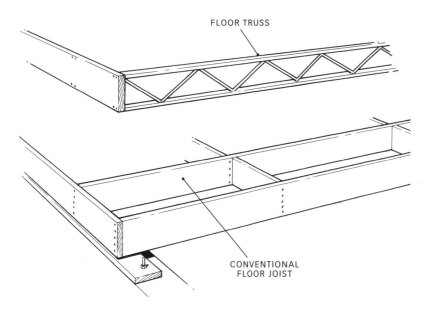

FLOOR TRUSS

CONVENTIONAL
FLOOR JOIST

Compared to conventional floor joists, engineered floor trusses have the benefit of allowing the GC to run pipes through the floor system.

Floor trusses, which have a web design, are worth considering if the manufacturer offers them. Trusses allow the GC to run at least some of the plumbing pipes, heating pipes, and HVAC ducts within the floor system, which produces more headroom in the basement.

It is also a good idea to use ¾-inch-thick sheathing for the floor rather than ⅝-inch sheathing. Sturdy floors are almost always worthwhile upgrades.

Types of Sheathing

Manufacturers use two types of sheathing to build their roofs, walls, and floors: oriented strand board (OSB) and plywood. In the past, many builders and customers misidentified OSB as "particleboard" and considered it to be an inferior product. Today, OSB is used widely in homes in every price range. Unless you have a strong preference for plywood, and your budget can comfortably afford the additional expense, opt for OSB if you have a choice. OSB is made with hardwood wafers bonded together with waterproof resins under high pressure and temperature. Each type of

OSB sheathing is engineered for structural applications. Unlike plywood, there are no knotholes, no core voids, and no layers or plies to come apart. The surface is smooth and uniform, and tongue-and-groove edges fit snugly for extra rigidity. OSB panels hold nails securely, do not puncture or tear, and resist racking. Particleboard, which is basically made of sawdust, is made with an interior-grade glue that is prone to fall apart when in contact with water. OSB is a very different product.

Making plywood requires peeling layers from trees, which creates a lot of wood waste. OSB, on the other hand, uses all of the tree material. In addition, because plywood is made from large trees, more and more of the older forests must be destroyed to create it. OSB, on the other hand, is made from small, fast-growing trees, and is a lot more environmentally friendly.

Roof System

The pitch of the roof plays an important role in how a home looks and what can be done with the attic. Framing and sheathing options can affect the strength of the roof. There are several choices that need to be made to control moisture and ensure good attic ventilation. Finally, a roof can be dressed up with some available upgrades.

Roof Pitch

Roof pitch plays a major role in the appearance of a house. Modular manufacturers understand this and offer several different pitches. At the very least they offer 5-in-12, 7-in-12, 9-in-12, and 12-in-12 roofs. On a 24-foot-wide home, a 5-in-12 roof is 5 feet tall, a 7-in-12 roof is 7 feet tall, and so on. Wider homes are proportionally taller for all roof pitches.

Most manufacturers build their ranches and two-stories with a standard 5-in-12 roof and their Cape Cods with a 12-in-12 roof. For a home that comes with a 5-in-12 roof,

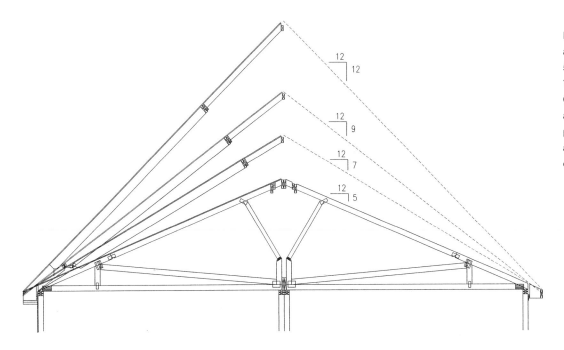

12 | 12

12 | 9

12 | 7

12 | 5

Ranches and two-stories are typically built with a 5-in-12 roof pitch or a 7-in-12 roof pitch; Cape Cods are usually built with a 12-in-12 roof pitch. The pitch of your roof can affect the attractiveness of your home.

increasing the pitch to 7-in-12 provides more character.

Upgrading a ranch or two-story to a 9-in-12 roof with storage trusses creates an even more attractive home, along with great storage space. You can maximize the usefulness of the space by adding full-sized or pull-down stairs to the attic and by having the manufacturer provide sheathing for the attic floor. However, the headroom will probably be too low to build rooms that meet the building code for living space.

Upgrading a ranch or two-story to a 12-in-12 roof with storage trusses, floor sheathing, and a full set of stairs to the attic creates enough headroom to allow the attic to be finished, although in some states the additional space does not meet building-code requirements. Even if the attic is not finished, it can serve as a huge storage area.

Hip Roof

Some manufacturers offer hip roofs. This classic look is quite attractive, although it is also more expensive. A flat hip roof can be unattractive, however, so it is important to increase the roof pitch to complement the roof style.

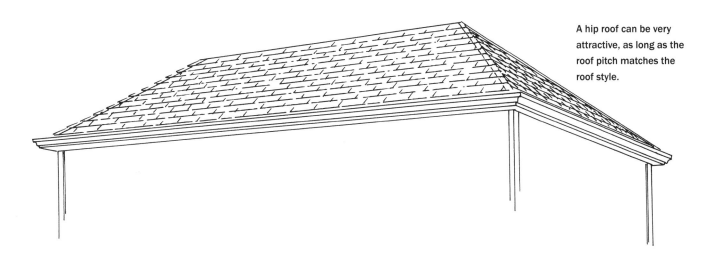

A hip roof can be very attractive, as long as the roof pitch matches the roof style.

Roof Framing

In some areas, the building code allows roof trusses or rafters to be 24 inches on center. In other areas, especially in the snowbelt, spacing must be 16 or even 12 inches. When 24 inches is the code minimum, it is worth considering an upgrade to 16 inches. This strengthens the roof, and the reduced spacing lessens the chance that unsightly dips will be visible between the roof members. In areas with high winds, such as along the shore, the roof needs to be strengthened with special fastening systems.

When adding stairs to the attic, note that pull-down stairs require no additional space from the floor plan, but they are very steep and provide a narrow opening into the attic. They should be located where there is room in the attic to stand up and where there is room below to unfold them. A full set of stairs is easier to climb safely and makes it possible to bring larger items to and from the attic. However, they take more space and cost more money. A full set of stairs to an unfinished and uninsulated attic cannot be left open. The easiest way to seal the attic from the heated or cooled space below is to put a door at the bottom of the stairs. An alternative is to close off the stairwell with walls and a door at the top of the stairs.

Attic Storage

More and more builders of all types are using engineered trusses rather than rafters to build their roofs. Storage trusses allow for easy storage, and web trusses do not. Storage trusses function more like traditional rafters in that they create an open area in the middle of the attic. In addition, they come with a bottom section that creates a true weight-bearing floor system after sheathing has been secured to the bottom chord of the truss. All true Cape Cod designs come with a 12-in-12 storage truss. By contrast, the structural members of a web truss crisscross the attic, rendering the space unusable for serious storage. In addition, web trusses are not meant to serve as a strong floor system.

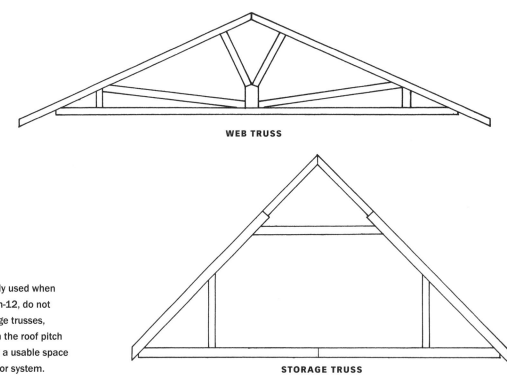

WEB TRUSS

STORAGE TRUSS

Web trusses, which are usually used when the roof pitch is 5-in-12 or 7-in-12, do not allow room for storage. Storage trusses, which are typically used when the roof pitch is 9-in-12 or 12-in-12, provide a usable space and a structurally stronger floor system.

Attic Bath Floor Trusses

If you are planning on building a bathroom in a home with an unfinished attic, as you would do in a Cape Cod design, ask your dealer to install roof trusses with the bottom built as a floor truss in the area where you expect to build the bathroom. This makes it much easier and less expensive for the plumber to run pipes to the fixtures.

Roof Sheathing

Most roof systems come with either $\frac{7}{8}$- or $\frac{1}{2}$-inch sheathing. There is often an option to upgrade to $\frac{5}{8}$-inch, but the best way to strengthen the roof is to upgrade to closer spacing of the trusses or rafters rather than thicker sheathing.

Roof Overhangs

All manufactures include a standard eave overhang, but only some of them include a true overhang on the gable ends of the house. For some customers, a modest gable end overhang is very important. Oversized eave and gable end overhangs may also be available.

Decorative Gables

Decorative gables, also known as "reverse" or "turned" gables, are the triangular peaks that often get added to the front of a house. With most manufacturers, they are only decorative. However, they can be made into a dormer that adds space to an attic made with storage trusses. Pay close attention to the width and pitch of a reverse gable, since it can look out of place if done to the wrong proportions. Even flatter roofs look better with steeper-pitched reverse gables.

Ice and Water Shield

In areas where snow and ice are likely to build up on the roof, the dealer should replace the standard felt paper that covers the sheathing on the eaves and in the valleys with an ice-and-water-shield product. This reduces the chance of leaks due to ice dams.

Make sure a decorative gable is done to the right proportions so that it blends with and enhances the house style.

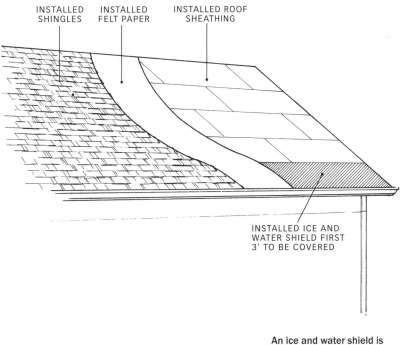

An ice and water shield is placed between the sheathing and shingles.

Attic Ventilation

Gable-end vents are sometimes used to remove excess heat and moisture from the attic. But they are not nearly as effective as a ridge-vent system that induces a continuous airflow from the soffits under the eave to the peak of the roof. A "shingle-over" ridge-vent system is more attractive than the typical exposed aluminum system, since it blends in with the roof.

Exterior Finishes

There are many options for dressing up the exterior of a modular home. The manufacturer provides selections of siding, decorative moldings, and shingles. If these do not meet a customer's needs, the customer can omit the standard offerings and have his or her GC supply and install the materials locally.

Siding

The standard siding offered by modular manufacturers is vinyl. The typical profile is 4/4, which means it gives the appearance of overlapping clapboards with a 4-inch exposure. It can come in either a textured or a smooth finish. Most dealers offer other choices, such as 3/3, German, and Dutch styles. Some siding companies offer a vinyl product that looks like stained cedar, in either clapboards or shakes. This product provides some of the class of expensive cedar siding with none of the costly maintenance. Some manufacturers also offer wood and fiber-cement siding, although most companies leave it to the GC to install these products. Cultured stone, which is a cement product that looks convincingly like real stone, is another option being offered by some manufacturers, especially to accent a section of the house.

Siding can also be bought locally and installed by the general contractor. This option provides for more choice, but the customer usually does not receive much of a credit for the omitted siding. The modular manufacturer buys it in such volume that it gets a great price break.

Exterior Trim

Some companies always install white J-channel and undersill with their vinyl siding, while others match the window exteriors. Since most windows are white, this usually does not matter. But if you select another color for the windows, you may want to specify a different color J-channel and undersill.

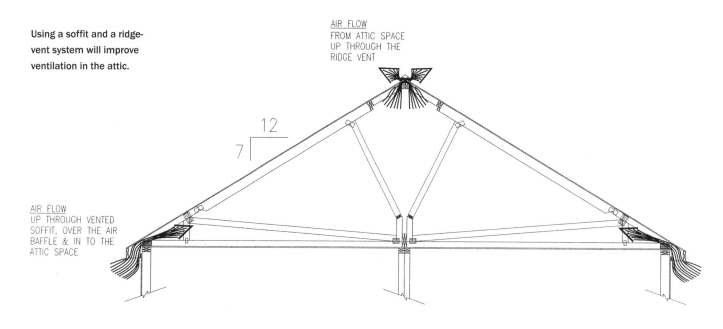

Using a soffit and a ridge-vent system will improve ventilation in the attic.

AIR FLOW
FROM ATTIC SPACE
UP THROUGH THE
RIDGE VENT

12
7

AIR FLOW
UP THROUGH VENTED
SOFFIT, OVER THE AIR
BAFFLE & IN TO THE
ATTIC SPACE

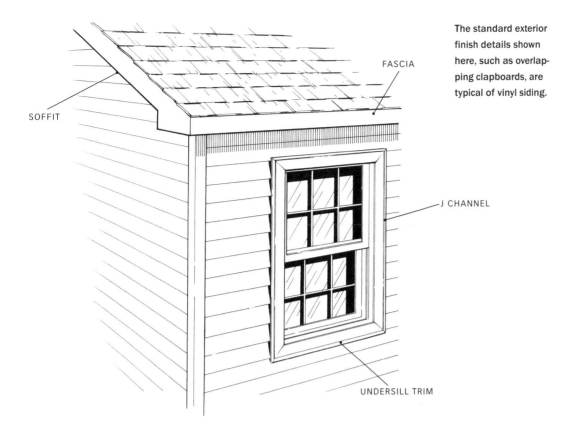

SOFFIT

FASCIA

J CHANNEL

UNDERSILL TRIM

Vinyl siding usually comes with 4-inch corner boards. An easy and affordable way to dress up the exterior of a home is to upgrade to 6-inch corners.

The standard color for soffit and fascia is white. For dark color siding, a darker soffit and fascia might be advisable.

When installing wood or fiber-cement siding, consider installing an upgraded soffit and fascia on-site. If you omit the standard soffit and fascia, tell the manufacturer how you want the overhangs framed and what kind of adjustment you want for the drip edge. The GC should know how to handle this work.

Shutters

Most manufacturers offer shutters and trim boards to decorate window exteriors. Shutters are usually available in louvered or raised-panel styles, with the latter often costing a little more. Some neighborhoods tend to favor one or the other. One frustrating fact is that shutters come in limited sizes, since they are only decorative, so they are not always the exact same length as the windows. Trim boards, which are placed around the perimeter of a window, often come in both 4- and 6-inch widths.

Decorative Exterior Moldings

One way to dress up the front of a home is to put mantels over the front windows and a pediment over the front door. Both of these come in a few different styles. You can also deck out the windows and doors with trim boards. Outlining the roof in frieze boards and dentil moldings can give a home a custom look. If the dealer does not offer moldings that suit you, the GC can buy and install them. Keep in mind that some window mantels do not fit over second-story windows because they hit the underside of the roof overhang.

Roofing

Architectural shingles are a very popular upgrade for customers wanting to boost the curb appeal of their home. Some manufacturers make these their standard shingle. Not only do they look more attractive to most people, they also tend to hide minor flaws in the roof. This is significant because the hinge points in a modular roof can sometimes be seen as a faint line if you are in just the right light and have the standard three-tab shingles. Although this problem is merely cosmetic, making it invisible is a good idea. If you choose standard three-tab shingles, make sure they have at least a 25-year warranty.

Metal or slate roofing must be installed on-site. When the roofing materials are being installed, it is critical that the GC help the set crew protect the house against weather damage, since the manufacturer will have finished most of the interior, and water that finds its way past an unfinished roof could cause serious problems.

Windows

There are many good window manufacturers, so it is usually safe to choose by features rather than brand. Most manufacturers offer a choice of a standard vinyl window and an upgraded wood window. For many customers, the former is the better value because of price, while for others the opportunity to stain the wood window is paramount. Energy-efficient glass and tilt-in sashes are important considerations. Some

Architectural shingles are often the more popular choice because they are more attractive and hide the hinge points in a modular roof.

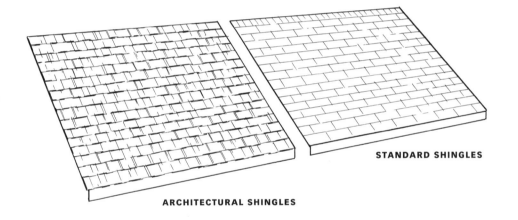

ARCHITECTURAL SHINGLES

STANDARD SHINGLES

Three gable-end overhang options.

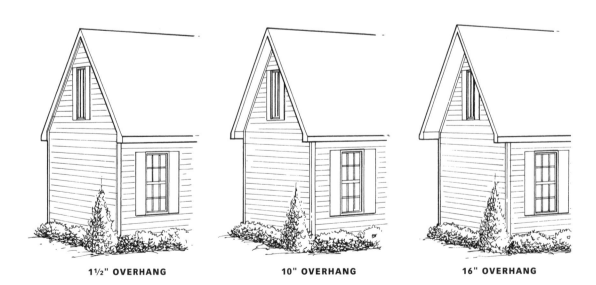

1½" OVERHANG   10" OVERHANG   16" OVERHANG

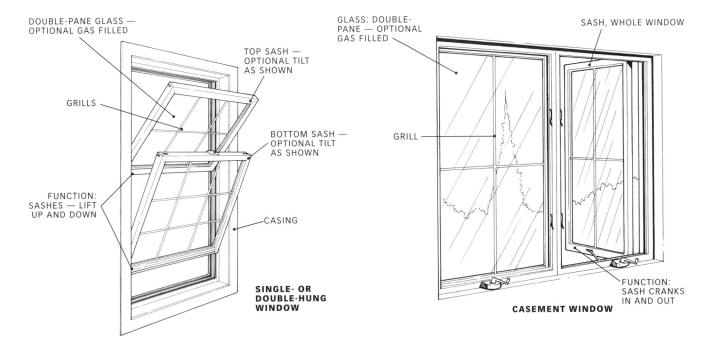

DOUBLE-PANE GLASS — OPTIONAL GAS FILLED

GRILLS

FUNCTION: SASHES — LIFT UP AND DOWN

TOP SASH — OPTIONAL TILT AS SHOWN

BOTTOM SASH — OPTIONAL TILT AS SHOWN

CASING

SINGLE- OR DOUBLE-HUNG WINDOW

GLASS: DOUBLE-PANE — OPTIONAL GAS FILLED

GRILL

SASH, WHOLE WINDOW

FUNCTION: SASH CRANKS IN AND OUT

CASEMENT WINDOW

people prefer double-hung windows so they can open both sashes. Regardless of the type of window selected, the size of each window can make a big difference in how it looks from both the exterior and the interior of the house.

When a room does not have any exterior walls or privacy is important, a skylight or suntube is a good alternative to a window.

Vinyl or Wood Frame

The most basic choice is between wood- and vinyl-framed windows. Wood windows need more maintenance, although they may be clad in vinyl or aluminum on the exterior. On the other hand, they can be painted or stained, at least on the inside. Vinyl windows are almost carefree, the frames have better insulation value, and they are much less expensive. But most vinyl windows do not have as strong a warranty and are unlikely to last as long as good wood windows.

Energy Efficiency

The best windows are double pane with low-e-rated glass, which has an invisible metallic coating. They also have an inert gas, such as argon, in the space between the double panes. The metallic coating and gas restrict heat flow, keeping the house warmer in winter and cooler in summer. Low-e glass also reduces the fading of interior finishes and fabric caused by ultraviolet rays. Some companies offer "insulated" glass windows that have two panes of glass but lack the metallic coating or gas.

Single-Hung, Double-Hung, or Casement

Another fundamental choice is between single- or double-hung and casement windows. Casements, which open and close with a crank, are usually optional, and some modular manufacturers do not offer them. For most customers, the choice is about appearance, since the two types of windows look very different. However, there is also a significant difference in function. While single- and double-hung windows lift up and down, casement windows swing out, which means they intrude into the outside space, such as onto a deck. On the other hand, reaching over a sink to open a single- or double-hung window can be more difficult than turning a crank. Many customers choose casements for these hard-to-reach locations and single- or double-hung windows for the rest of the house.

Two typical window styles. Casement windows may be best for hard-to-reach locations. The bottom sash is fixed with a single-hung window.

Tilt-In Function

Some single- and double-hung windows tilt in for easy cleaning. How they tilt depends on whether both the top and bottom sash (double-hung) or only the bottom sash (single-hung) opens. For many people, a single-hung window with a fixed top sash is not much of a problem, since they are not inclined to open it, but it does mean that only the bottom sash tilts in for easy cleaning. It is considerably easier to clean the top sash of a single-hung window when the bottom sash tilts in than a double-hung window that does not tilt at all. Another difference is that a single-hung window comes with a half screen. The choice between single- and double-hung windows often comes down to budget, since double-hung windows cost more. One compromise is to upgrade to double-hung windows only for the second story of the house, where window cleaning is more of a problem.

Window Color

The color of the standard window is likely to be white, unless it is an unfinished wood sash. Other colors may be available for an additional charge. Another specification that affects the appearance of windows is whether or not they come with grills, which are the dividers that make a window appear as if it is made of individual panes of glass. Sometimes the grills are installed between the panes of glass, which makes for easy cleaning but impossible removal, and sometimes they are attached to the inside of the window. As a general rule, vinyl windows have the grills between the panes of glass, while wood windows have them on the interior surface of the window. Some manufacturers of wood windows also offer wood grills that can be painted or stained for a traditional look; others offer less-expensive plastic grills that can be stained with a special product.

Window Size

The size of windows is important, and not just for the amount of interior light they let in. Taller windows have a different exterior look that is preferred by some customers, but they also complicate furniture placement in front of the window. Some tall windows come with even-sized sashes while others come with the bottom sash 50 percent bigger than the top sash, creating what is known as a "cottage" style. These windows are sometimes installed on the first story of two-stories.

Specialty Windows

Most manufacturers offer a selection of specialty windows, such as circle-top, Palladian, and quarter-round designs. They also offer bay and bow windows. After looking at all of the windows available from the manufacturer, some customers decide they want a specialty window that they can get only from a local supplier. In this case the manufacturer should be asked to frame a rough opening in the wall, which will make it easy for the GC to install the window. The modular dealer must be informed of the window's exact dimensions and location so that the manufacturer can add the specifications to the plans. If a customer does not get this information to the manufacturer, the GC will have to reframe the opening. Although this is typically manageable, installed electrical wires, plumbing pipes, and heating ducts may need to be relocated, which wastes time and money. If the window location is on one of the sides that will be sided at the factory, the manufacturer needs to hold back the siding around the window opening. After the window is installed, the GC can complete the siding.

Skylights and Suntubes

Skylights offer an effective and attractive way to add light to a room when a window will not work. They also help when privacy is an issue. Modular manufacturers are unable to complete the shaft of a skylight at the factory, however, since the roof is delivered folded down. The GC can build the shaft after the set crew has lifted the roof into place. Since a skylight is sometimes located so that it cuts through one of the roof's structural trusses or rafters, the manufacturer must be informed of any skylights that will be installed on-site so it can frame the openings in the correct location.

Suntubes are another way to get outside light into a room. They are less expensive and easier to install than skylights, because they do not require a shaft. Their light is more indirect than the light from a skylight, however, and not everyone likes them as much.

Exterior Doors

Steel swing doors are the industry standard for the front and rear entries. Most manufacturers also offer a fiberglass alternative. When selecting a door, consider the size of the opening as well as the type of door. Many doors can be dressed up with a variety of accessories.

Main Entry Doors

Exterior doors come in many styles and sizes. Most swing-entry doors are steel clad with an insulated core. Some manufacturers offer these doors clad in vinyl, which offers the advantage of never needing to be painted but has the disadvantage of always being the same color. Fiberglass doors have better insulating properties, will not warp, and are less likely to dent. Some versions can even be stained to look convincingly like expensive wood doors. Homes with an attached garage require a fire-rated door.

All door manufacturers offer several glass inserts, made of clear, leaded, crystalline, or beveled glass in a variety of styles. In addition to dressing up a door with decorative glass panels, you can install a transom window above the door or sidelights along one or both sides. A brass kick plate that is installed on the exterior also enhances the door's appearance.

The front door should be 3 feet wide, and when there is room in the floor plan, 3 feet is a smart choice over the more standard 2 feet 8 inches on side and rear doors. Larger doors make it much easier to move objects in and out of the house. Screens are not usually included with swing doors, but the general contractor can purchase a screen kit from a local supplier. Adjustable door thresholds are recommended for all exterior swing doors to help seal out the elements.

Accessible Entries

You can make the exterior entrances of a home more accessible for people with impaired mobility by installing a low threshold with each door. Lever handles are easier to operate than round handles for people with limited hand dexterity or strength. A thumb latch is an attractive option for the front door, but it is hard to use for people with a weak grip. The manufacturer should be expected to install interior-style door handles on the inside of any exterior swing door that comes with a thumb latch. If deadbolts are an option, select one for every swing door. Request that all entry doors with keyed locksets be keyed alike; some manufacturers give a different key for each door.

French Doors and Sliders

The same swing doors used for a single entry also come in pairs in a French or atrium style. They are available in widths of 5 feet and 6 feet, with two hinge styles. The two doors can be hinged in the center, which creates one active door and one fixed

French Door Styles

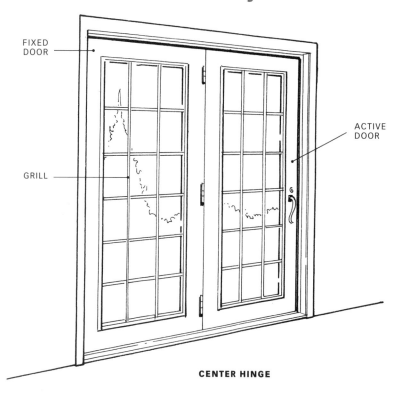

FIXED DOOR

GRILL

ACTIVE DOOR

CENTER HINGE

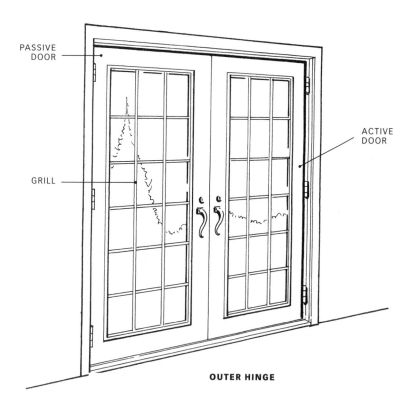

PASSIVE DOOR

GRILL

ACTIVE DOOR

OUTER HINGE

Know which type of French door you are receiving. A center hinge may not be the best choice if you plan to move furniture through the doors.

door that cannot be opened. Or each door can be hinged separately on its outside edge, which creates one active door and one door that can be opened by unlocking a clip. The latter style in a 6-foot width is the best choice. The 6-foot opening is great for moving furniture or hosting a party. Some manufacturers also offer an upgraded wood French-style door, which can be stained. These type doors can be very attractive, but also quite expensive.

Sliders are likely available in both vinyl and wood. Many sliders are not as energy efficient or secure as French doors, and they provide less of a passageway when fully opened. In fact, a 6-foot slider does not provide a 3-foot opening even with one of the panels fully opened. On the other hand, sliders have the advantage of requiring no additional space to open within a room. If you want to use a slider in an opening that must be accessible to someone with mobility limitations, upgrade to an 8-foot model. Ask the dealer to upgrade the glass in a slider or French door to low-e if this is not standard. This can be especially helpful when the door is right next to a dining room table.

Both sliders and French doors have an option of grills. As with the windows, some doors come with the grills installed between the panes of glass and others with the grills installed on the inside surface of the doors. The interior grills may be made of vinyl or wood. Select a keyed lockset for a slider, if available, and a deadbolt for a French door. Consider selecting bright brass hardware for wood sliders or French doors.

Specialty Doors

If you would like to purchase a special door that you cannot get from the manufacturer, have the manufacturer frame a rough opening for it. If the door location is on a side of the home that will be sided at the factory, the manufacturer needs to hold back the siding around the opening. After the door is installed, the GC can complete the siding.

Fireplaces

Masonry fireplaces are slowly giving way to zero-clearance systems, which have the advantages of price, speed of installation, energy efficiency, and design flexibility. With the advent of cultured-stone products that look amazingly realistic, you can give a zero-clearance fireplace a great exterior look. Gas fireplaces have become especially popular since they are almost maintenance-free and provide flames that are hard to distinguish from those in a wood fire.

When selecting a zero-clearance fireplace, note the size of the firebox opening, whether it comes with a combustion-air exchange kit, and if the glass doors are included. Also note whether the interior of the firebox is finished black or in a brick pattern and whether the facade is trimmed out in brass or black. Well-functioning fans are very helpful, since they carry the heat into the room. Remote controls are also available and convenient.

All wood-burning fireplaces require a flue, which is usually a metal pipe, although you can connect a zero-clearance fireplace to a tile flue within a masonry chimney. Gas fireplaces come in three types. The most popular is a direct-vent system, which vents directly out the rear or top of the firebox without the need of a flue pipe. Ventless systems offer the most flexibility, since they do not have to be installed in an exterior wall. Some states have banned these, however, because they generate a lot of carbon dioxide and moisture that is retained in the room. The third type of gas fireplace system requires a flue.

A flue pipe for a wood or gas fireplace adds to the cost of a fireplace system in two ways. The flue is expensive and is usually enclosed in a chimney chase to hide the metal pipe. For homes that require a flue pipe for another purpose, such as an oil-burning heating system, the same chimney can be used for the fireplace flue.

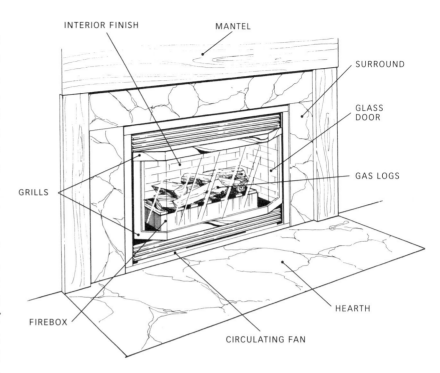

Gas fireplace features, such as glass doors and a patterned brick interior firebox, make these zero-clearance systems very attractive.

Zero-clearance fireplaces can be embellished with surrounds, hearths, and mantels in many styles, materials, and colors. Slate, real stone, cultured stone, brick, tile, marble, and granite can all be used, as can stained and painted wood mantels.

HVAC

All manufacturers supply the hot-water or electric baseboard units to each room when customers select these heating systems. But only a few companies supply the ducts needed for forced-air systems, and few supply the boiler or furnace. Most manufacturers expect the GC to supply and install the HVAC equipment on-site.

Manufacturers that supply the complete heating system can save the GC some time completing a home. This may make the purchase of the system a good value even if it is slightly more expensive than the GC's estimate for comparable equipment. For Cape Cods or two-story homes with forced-air systems that will be installed on-site, the dealer should locate a "chase" in the plan. A chase is an open box from floor to ceiling that can carry the supply and return ducts from the basement furnace to the second floor and attic.

Some manufacturers have started installing high-velocity forced-air systems, which work with small-diameter ducts that can be run within wall and floor cavities more easily than standard ducts. This keeps the ducts in the conditioned (heated and cooled) space of the home, which is very energy efficient. High-velocity systems are more expensive, however.

Customers should create heating and cooling zones when their home is larger than 1,200 square feet. This will enable them to control different areas of the house independently. Typically, one zone is created for the living area and one for the bedrooms. Customers should discuss with their dealer how many zones they want, since each one requires a separate thermostat. If the manufacturer is installing hot-water baseboards, he needs to know which rooms will be included in each zone on the second story of a two-story so the necessary connections can be made behind the walls and under the floors.

For an oil-fired heating system, the dealer should install an emergency shut-off switch in order to be in compliance with the building code.

Running Utilities to Unfinished Attics

Manufacturers supply rough electrical and plumbing from the basement to the second floor of unfinished Cape Cods. They usually include two or three electrical circuits, hot- and cold-water supply lines, and a drain line. For hot-water baseboard heat, they should provide a supply and return line. You may want to ask the manufacturer to supply additional electrical circuits and another supply and return for a second zone to the heating system.

To better handle future electrical needs, consider having the manufacturer install one or more 3-inch PVC conduits from the basement to the attic in a centrally located wall. The conduits allow an electrician to easily run additional circuits to the second floor, as well as any lines needed for phones, TVs, and computers.

For future plumbing needs, have the manufacturer enlarge an interior partition to a 2×6 wall. A larger cavity that can serve as a chase will likely be needed for the HVAC system; sometimes they must be as much as 2 feet by 2 feet.

For ranches or two-story homes that have been upgraded to a 12-in-12 roof pitch, consider having the dealer run rough electrical, plumbing, and HVAC to the attic, or at least include a chase from the basement all the way to the attic.

Rough Electrical

An important first step in getting the electrical system right is specifying where the electrical panel box should go. You need to choose the type of electrical switches, the number of phone and cable jacks you need, and where you want thermostat wires. If you have any special wiring needs, such as for computers or a security system, spell those out to the dealer.

Electrical Panel Box

If a 200-amp electrical panel box with 20 circuit-breaker slots is standard, upgrade to at least 40 slots, which allows for plenty of new circuits in the future. Even more slots or a subpanel is useful if the house has appliances that need dedicated circuits and if the GC is constructing several structures on-site. Review the appliance literature to see what is required, and discuss your options with the general contractor.

If the GC knows on which side of the house he wants to install the electric meter and panel box but is uncertain exactly where he will install them, ask the dealer to provide additional wiring from each of the modules so the panel box can be moved several feet in either direction.

Electrical Switches

Some manufacturers offer conventional toggle switches as their standard; others offer rocker switches, which are easier to use. Make sure there are light switches at all entry points to each room. Dimmer switches are an option worth considering for the dining room and other rooms.

Phone and Cable Jacks

It is better to have the manufacturer install all of the jacks rather than to pay the phone or cable company to do so. The manufacturer can run the wires and cables within the structure. If given a choice, select the best phone wire available for Internet access. At this time, it is category-5. Tell the manufacturer if you need something other than RG6 coaxial wire for either the cable or the satellite system.

Thermostat Wires

Each HVAC zone needs a thermostat, which the manufacturer supplies only when it is providing the complete system. Otherwise, the GC must supply them. The manufacturer, however, should run the wire for the thermostats. For a forced-air system, request 18-8 wire so the GC can make all of the connections for heating and air-conditioning.

Height of Electrical Controls

The modular manufacturer installs all switches, receptacles, phone jacks, TV jacks, thermostats, and chimes at a standard height. Some manufacturers allow the customer to adjust the height, usually for an additional fee. When specifying the height,

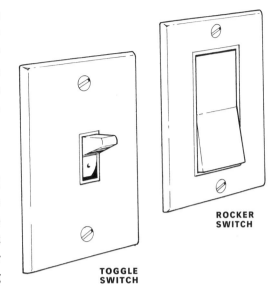

ROCKER SWITCH

TOGGLE SWITCH

Manufacturers use one of these two electical switch options as their standard.

measure from the height of the finished floor to the center of the device.

Door Chimes

Not every company includes in its standard package door chimes for both the front and back doors. If you want them, make sure to order them.

Special Wiring

Manufacturers can often provide wiring for special needs, such as connecting all rooms with an intercom system, piping music throughout the house, networking computers, or installing a security system. Specify the wire you need and where it should go. Ask the manufacturer to run the wire but not install the connectors or adapters.

Not all manufacturers are equally flexible and accommodating, so you might be constrained by a manufacturer's assembly line. If a manufacturer will not do what you request, such as install security or computer network wiring, see if it can do anything else that might facilitate the on-site installation. For example, the manufacturer might install conduits or pull-strings to assist the on-site subcontractors.

Lighting and Other Electrical Fixtures

Although you will get a standard light package that covers most of the house, you might not get lights for the bedrooms, living room, family room, and closets.

Interior Light Fixtures

Every manufacturer offers a standard light package that includes overhead light fixtures for the foyer, kitchen, dining room, nook, halls, and bathrooms. Some manufacturers do a better job of selecting attractive light fixtures than others. Some even offer a few optional upscale selections.

If you are not excited about the standard and optional fixtures offered, consider keeping them anyway. The credit for omitting fixtures is minimal. By ordering standard fixtures, each room comes fully wired, allowing for fixtures to be easily replaced at a later date.

Light fixtures will probably not be provided for the living room, family room, and bedrooms. Instead, each room will get one switched receptacle, which allows you to turn a lamp on and off at the wall switch. As a general rule, only the bottom half of these receptacles is connected to the switch; the top half is permanently hot so that a TV or alarm clock can be connected. If you order a ceiling light for the room, the manufacturer will eliminate the switched receptacle unless told otherwise.

Closet lights should be included for walk-in closets, but they are likely to be an option with other closets. Plan to order lights for a washer and dryer closet as well as a pantry closet.

To add a light in a room that does not already have one, you can either purchase a light fixture from the dealer or prep the room for one by having the manufacturer install a switch, wire, and box. If you have the room prepped, buy a fixture locally and ask the electrician to install it.

Recessed Lights

All manufacturers offer recessed ceiling lights as an option. They are very popular, but their cost can add up quickly since a large room needs several. Also, standard recessed lights are very energy inefficient, allowing warm air to pass into the ceiling or attic. Avoid this problem by asking the dealer to install the more energy-efficient Insulation Contact Air Tight (ICAT) recessed lights that are now available.

Energy-Efficient Lights

Consider selecting compact-fluorescent light fixtures, if the manufacturer offers them, since these fixtures use one-third the energy and last much longer than incandescent bulbs.

Location of Interior Lights

Most people prefer to locate ceiling lights in the center of a room or in some other specific location. Ceiling framing, however, often does not allow this unless structural adjustments are made. To avoid this costly step, most manufactures leave the framing alone and try to locate ceiling lights as close as possible to the desired location. If this solution is not good enough for you, be sure to tell the dealer, and expect to pay extra. If you are installing a heavy light fixture, such as an antique chandelier, have the dealer order a special brace box in the ceiling to support the weight.

Basement Light

A basement that can be accessed from inside the house should have a three-way switch at the top and bottom of the stairwell. That way, the GC can connect the stairwell light to other lights in the basement, all of which can be turned on and off at the top of the stairs. If you do not ask for this, you will likely get a two-way switch, which means you will have to turn on each light one at a time or pay the GC to rewire the switch at the top of the stairs.

Exterior Lights

Every exterior door comes with at least one standard light fixture. You may want to add a second light at the front door, especially if it is a double door or has two sidelights.

An option to consider when building an attached garage is to install two three-way switches at the door to the garage, one with 25 feet of wire for an interior garage light and one with 50 feet of wire for an exterior garage light. The electrician then connects the two switches to two additional switches inside the garage. This way, the two lights can be controlled from inside the house or from the garage.

Floodlights, which are optional, do a good job of lighting up a deck, backyard, or driveway. The manufacturer can provide the switch, wire, and box for each floodlight location, but the electrician needs to supply and install the fixtures.

Ceiling Fans

If you want a ceiling fan, you may be able to purchase one from your dealer. If you can't, have him install a special brace box in the ceiling to support the fan's weight. Order two switches, one for the fan and one for the light, if the fixture has both.

Whole-House Fan

If you do not intend to install an air-conditioning system, consider installing a whole-house fan. These fans are powerful enough to bring in enough fresh air to cool an entire home. Ask the manufacturer to provide the appropriate wiring and to frame a rough opening for the fan.

Smoke Detectors

Some building codes require a smoke detector in every bedroom and one outside every sleeping area, usually in the hall. Strongly consider adding them to all bedrooms even if they are not required. The cost is trivial when compared with the protection they provide.

Carbon Monoxide Detectors

If a carbon monoxide detector is not included with the home, order one.

Exterior Receptacles

Two exterior-rated receptacles are standard. Putting in another one or two receptacles is often worthwhile.

Rough Plumbing

The manufacturer is likely to give you only a couple of choices for rough plumbing. You should select a kitchen-sink cleanout and shut-off valves for all water lines. But you are unlikely to have a choice of the type of material used for the plumbing lines.

Drain, Waste, and Vent Material

Some manufacturers use PVC pipes for their drain, waste, and vent (DWV) lines, while others use ABS. Although both are good materials, plumbers accustomed to working with one type of material are often unhappy using the other. In addition, some local plumbing inspectors insist on one or the other. If your GC has this problem, see if the manufacturer is willing to switch materials. But keep in mind that most manufacturers are unwilling to do so since it disrupts their assembly-line efficiencies and quality control.

Kitchen-Sink Cleanout

You will want a kitchen-sink cleanout in case you drop something down the drain that you need to recover. If soap scum, hair, or food gets caught in the trap, the cleanout makes it easier to unclog the drain without having to cut the pipes. Almost all plumbing today is connected by glued fittings, so they cannot be unscrewed.

Shut-Off Valves

If the manufacturer does not install shut-off valves on its hot- and cold-water lines, ask for them. Shut-off valves make it much easier to carry out plumbing maintenance in the future.

Laundry Tub

If the laundry room is large enough, the manufacturer can install a large utility laundry sink.

Radon Vent

Radon is a radioactive gas found in some homes that in sufficient concentrations can cause health problems. Adding a PVC vent pipe from the basement to the attic allows you to more easily install a radon mitigation system, should you need one.

Appliances and Hookups

Manufacturers usually offer competitive prices for the limited selection of appliances they offer. If you do not purchase your appliances from the manufacturer, instruct it to complete the preparatory work so that the GC can easily install them.

Appliance Choices

Each manufacturer offers at least the essential kitchen appliances along with a washer and dryer. Usually the selection is limited, but the pricing is competitive. Buying from the manufacturer saves you time shopping, and it saves the GC time, since the appliances arrive already hooked up or ready to be connected. If you are building an Energy Star home, make sure the appliances are appropriately rated. When comparing the prices of the manufacturer's appliances with those offered locally, be sure to consider the manufacturer's prep fees and the GC's hook-up fees.

Appliance Hookups

If you purchase your own appliances, have the modular manufacturer complete the electrical and plumbing prep required to hook them up. The GC completes the hookups after the appliances are delivered.

Both the manufacturer and GC will charge for this work.

The plumbing that must be done by the manufacturer to prep the home for an appliance is straightforward, but the electrical work varies considerably among appliances. A manufacturer will not necessarily know the electrical needs of each appliance unless it supplies them. This means you must either purchase appliances that work with the manufacturer's standard circuit schedule or select the appliances before the home is built and provide the manufacturer with the power requirements for each appliance.

The manufacturer needs to know if you will have any gas appliances, such as a stove, wall oven, cooktop, and clothes dryer. Gas appliances require a 110-volt line while some electric appliances use 220 volts. Some appliances need larger circuits than others, and some require a separate, dedicated circuit.

Completing the necessary prep for a dishwasher, icemaker, and washing machine is routine. You will want a pan and drain installed under the washer if it is on the second floor, and maybe even if it is on the first floor. Request that the washer shut-off valve be a single-lever control.

Consider a dedicated circuit for a window-style air conditioner if it requires 220 volts. For a central vacuum system, have the manufacturer install the boots and bring the pipes to the cleaner's location. This work can be done more easily while the home is being constructed than afterward.

Range Hood

If a recirculating range hood is standard, upgrade to a vented hood, which prevents the odors, smoke, and grease from being recirculated back into the kitchen. If you do a lot of cooking, especially frying, you can omit the standard range hood and purchase a heavy-duty unit on your own. You should also omit it if you are installing a downdraft cooktop or if you want a nonstandard size to match a nonstandard-size range. When omitting the standard range hood, ask the manufacturer to provide the electrical work and venting to the outside. Another option is to replace the hood with a space-saver microwave oven that has a built-in fan.

Flooring

You have more choices when selecting flooring than with any other item in your home. Modular manufacturers typically offer a few different materials, each in several colors. These choices can be supplemented with many other materials available locally. Cost and speed are the primary advantages of selecting one of the manufacturer's options.

Carpet

When selecting carpet, look at the stain protection and warranty. Spending a little extra for better carpet may be a bargain in the long run. It is especially worth the money to get a good pad. A rebond pad is superior to a low-density foam pad.

Flooring Installation

Not all flooring supplied by the manufacturer can or should be installed at the factory. The GC needs to finish any flooring that crosses the marriage wall, such as a great room covering the width of two modules. Carpet is the easiest flooring for these situations, since it can be installed by the manufacturer in both modules and then seamed together on-site. Vinyl flooring cannot be installed this way, nor can tile, wood, or laminate flooring, because it is impossible to build two separate modules and consistently have the patterns and pieces of flooring join together exactly as they should. The problem can be eliminated by installing all of the flooring in one module and shipping the remainder to the GC. The floor installer can then start at the marriage wall with the correct size and pattern.

The GC runs into a similar situation when he attempts to install the same flooring in two separate but adjoining rooms, such as vinyl floor in a foyer and kitchen that meet at a 3-foot passageway on the marriage wall. To use the exact same flooring in both rooms without creating a mismatched pattern at the passageway, you need to follow the procedure described above. An easier solution is to select a different pattern of vinyl flooring for the two rooms. The manufacturer can install both pieces, and the GC can install a wood or marble threshold between them in the passageway. Since the two pieces have different patterns, it will not appear as if the two rooms are misaligned.

Sometimes it is better for the manufacturer to omit the installation of its flooring, especially carpet. Building in the winter in climates with a lot of sleet and snow can create a muddy mess on-site. This mud is likely to get tracked into the home during construction. Most manufacturers put some protective covering on their flooring, and the GC should take some responsibility for ensuring that the covering is doing its job. But the reality of a construction site is that it is very difficult to adequately protect carpet in the winter. For peace of mind, consider paying the GC to have the flooring installed at the end of the project.

Locally Purchased Flooring

When purchasing flooring from a local supplier, tell the dealer what you are installing and how the manufacturer should prepare each room. For a wood floor, the GC needs to decide how he wants the manufacturer to install the baseboard: as the manufacturer normally would, raised to some specified height, tacked in place, or shipped uncut and uninstalled. Tacked in place is the best compromise, since the GC can easily remove the baseboard before installing the flooring and then reinstall it without having to spend time cutting every piece to fit. If you are putting a wood floor in the kitchen or bathrooms, ask the dealer to raise the cabinets and vanities to accommodate the thickness of the floor.

It is critical to tell the dealer the exact thickness of the flooring at the top or bottom of the stairs. The building code requires that the height of each step be consistent, which can be done only when the thickness of the floor is taken into account. If you are building a Cape Cod with an unfinished second floor, tell the manufacturer what you intend to install on the second floor in the future.

When the manufacturer lays ceramic tile, vinyl, or glued-down carpet, it installs the appropriate subfloor according to the product guidelines. If you are purchasing the flooring through local suppliers, you should still have the manufacturer provide the proper underlayment. It is better for the manufacturer to do the underlayment, since it can place the material under the cabinetry in the kitchen and bathrooms as well as under the stairs to the second floor in the foyer. Cement board or plywood should be used under ceramic tile. Plywood or lauan should be installed under glued-down carpet, and lauan should be installed under vinyl flooring.

Stairs and Railings

Stairs are required for Cape Cod, two-story, raised-ranch, and split-level designs. They come in different configurations, widths, and finishes, and are usually fitted with a stained or painted railing system.

Stairs on Cape Cod and Two-Story Designs

A full set of stairs to a second or third story can come straight, L-shaped, or U-shaped. Straight stairs take up considerably less space on both floors. They are also less expensive. On the other hand, L-style stairs

create a more dramatic look and more spacious foyer. They also have the advantage of being slightly safer if someone falls, since the fall will be broken earlier.

Each side of a set of stairs can be enclosed by a full wall, half wall, or railing system. A full wall separates the stairs from the adjoining room, while a half wall and railings opens up the stairs to that space. The advantage of a half wall is that it does a better job of defining the space, and if it is anchored properly to the stairs, it is also a stronger structure. In addition, railings are more expensive in materials and labor. Most people consider railings more attractive, however, and they make the space feel more open.

Each side of the stairwell does not need to be built the same. In the Whately 1 (see page 56), for example, you can close off the stairs on the dining room side but leave them open with railings on the foyer side. If the stairs create a balcony, with the second story overlooking the first, you also have a choice of railings or a half wall.

When installing railings on one side of the stairs, decide how many steps should have railings. Some manufacturers leave four steps open, while others leave more. If given a choice, ask the dealer to build the stairs with railings on the first five and a half steps. This gives an open feeling without bringing the railings too close to the ceiling.

The typical set of stairs is 36 inches wide. If you are willing to give up 6 inches of space from the adjoining rooms on both floors, you should consider widening them to 42 inches. Wider stairs are more distinctive. They are also more practical when you want to move furniture to and from the second floor. If you are building an unfinished Cape Cod, you will have an easier time bringing construction materials to the second floor when you are ready to finish it.

One way to dress up the stairs is to have one or both sides of the bottom step rounded into a bullnose. If the dealer offers

STRAIGHT STAIRS

2ND STORY

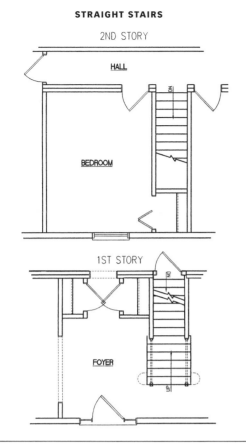

HALL

BEDROOM

DN

1ST STORY

FOYER

UP

L-STAIRS

2ND STORY

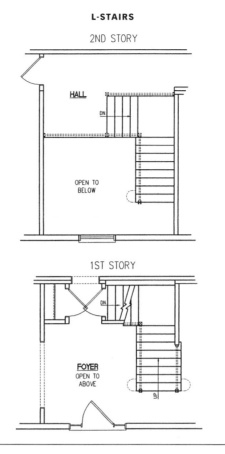

HALL

OPEN TO BELOW

DN

1ST STORY

FOYER
OPEN TO ABOVE

DN

UP

There are several stair options for Cape Cod and two-story homes. You can enclose each side of the stairwell with a full wall, a half wall, or a railing system.

WRAPAROUND STAIRS

2ND STORY

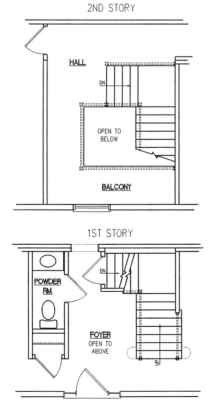

HALL

DN

OPEN TO BELOW

BALCONY

1ST STORY

POWDER RM.

DN

FOYER
OPEN TO ABOVE

UP

U-STAIRS

2ND STORY

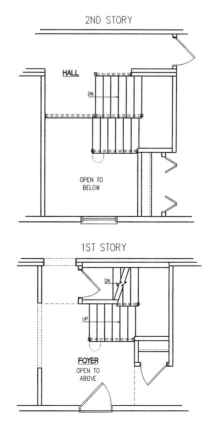

HALL

DN

OPEN TO BELOW

1ST STORY

DN

UP

FOYER
OPEN TO ABOVE

this option, each bullnose requires a special railing piece, usually a volute or turnout, so that the railing follows the curve of the bullnose. The installation of the railing system adds to the cost of this option, since it requires more skill and time.

Stairs on Raised-Ranch and Split-Level Designs

Raised-ranch entries and split-level stairs can be quite attractive, even a primary feature of the home. One of the ways to make them more attractive is to widen them. Modular manufacturers typically frame the opening so the GC can build two sets of 36-inch-wide stairs, which are usually U-shaped with one going up to the first floor and one going down into the basement. You might prefer 42-inch-wide stairs, although this costs you 1 foot of space in the adjoining rooms (6 inches for each set of steps).

Manufacturers sometimes provide the materials for the GC to build the raised-ranch or split-level stairs. For this to happen, the GC needs to tell the manufacturer the height to the basement floor. Since this is not always possible with enough lead time, some GCs prefer to supply their own materials.

In most raised-ranch and split-level plans, the living room overlooks the stairs. Tell your dealer whether you want a half wall or railings on the balcony. Expect to pay more for the railings.

Stair Finish

The standard stairs are usually carpeted, covering a "carpet-grade" material such as plywood or pine boards. Some carpet-grade stairs do not have a true skirtboard along the side of the treads; the framing lumber used for the skirtboard is hidden by drywall. Even when a carpet-grade tread looks like finished woodwork, it can have knots and cracks and other imperfections

When entering a raised ranch, you step into a landing that connects to two sets of stairs. One leads up to the first floor and the other leads down to the basement.

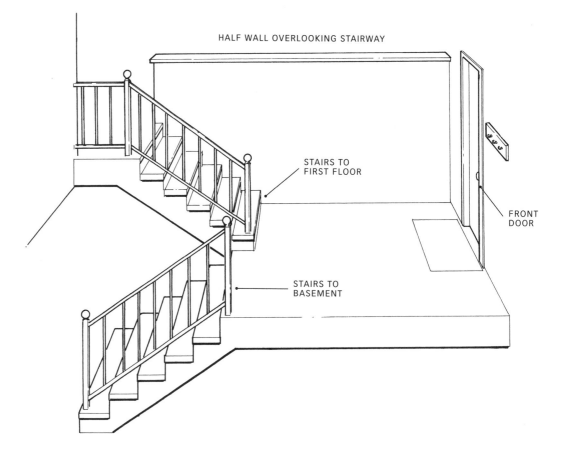

HALF WALL OVERLOOKING STAIRWAY

STAIRS TO FIRST FLOOR

FRONT DOOR

STAIRS TO BASEMENT

because it was not designed to be visible. If you want true stain-grade wood stairs, you need to pay for an upgrade from the manufacturer.

The first level of upgrade provides unfinished, stain-grade stair treads and paint-grade risers and skirtboards. If you want to stain the risers and skirtboards along with the treads, order the next level of upgrade. Some manufacturers, however, do not offer an all-hardwood stair system. The second level is likely to cost quite a bit more than the first.

An L- or U-shaped set of stairs has a landing. If you upgrade to stain-grade wood stairs, order the same material for the landing. Otherwise, the manufacturer may design it for carpet.

Stair Railings

Most manufacturers supply a standard railing system in one type of wood with one or two finishes, but they usually do not offer an upgrade. Some manufacturers also install as much of the railings as possible; others leave the entire installation for the GC. Those manufacturers that install some of the railings almost always use a partially preassembled system, which may not have the custom look of a traditional railing but is much easier to install. The GC might need to stain or paint the railings, although some manufacturers provide railings that are prefinished to match their moldings. For a truly custom look, have the GC purchase the materials from a local woodworking shop.

Interior Trim and Doors

The manufacturer is likely to have a limited selection of interior moldings and doors. Expect white to be the standard, with one or two stain colors optional. If your budget allows, you could have the GC purchase and install custom moldings and doors. Pay close attention to how the manufacturer finishes its window jambs and sills when the moldings are a different color from the sash.

Interior Moldings

The standard interior package of most manufacturers includes painted-wood moldings. Some manufacturers offer stained-wood moldings as a standard feature, and those that do not tend to offer it as an option. If you select unfinished moldings, the GC can choose among having them installed, tacked in place, or shipped loose and uncut. It is usually better to have them tacked in place, which enables the GC to remove the moldings from the walls before they are stained and then to reinstall them without having to measure and cut each one. If you intend to do the staining yourself, ask for the unfinished moldings to be tacked in place.

The GC can buy and install custom moldings bought locally, but do not expect to receive a substantial credit for omitting the manufacturer's standard moldings. You will have to decide whether you want the manufacturer to install the window and door jambs. If you do not want the door-jambs installed, the GC needs to install the doors as well, which adds to his labor cost.

Interior Doors

The manufacturer matches the finish of its standard interior doors with its standard interior moldings, which are usually painted white. It also offers some optional doors that match its optional moldings, usually in one or more color stains. If you want doors and moldings with different finishes, one stained and the other painted, think about how the doorjamb will be finished. Some manufacturers finish the jamb to match the moldings; others finish it to match the door.

Triple hinges on interior doors is an option worth paying for, if you have a choice. They reduce warping and ensure greater stability, allowing the doors to operate easier and last longer.

The standard door handles are likely to be round or bell-shaped. Most manufacturers offer lever handles as an option. In addition to being attractive, children and adults with arthritis or other disabilities often find lever handles easier to use. You might prefer to omit the standard handles and install a style purchased locally. In this case, decide whether you want the manufacturer to bore its standard hole in the door and jamb. Not all handles work the same. If you drill the wrong holes, you will not be able use the door handles or you will need new doors. On the other hand, choosing handles that will work with the manufacturer's standard holes will save the GC time installing them.

Pocket doors slide in and out of the wall, eliminating the problem of a door that swings inconveniently into another door or hallway. If you want a pocket door installed in the marriage wall, the manufacturer should frame the opening in the wall and the GC can complete the installation on-site. Request a lock for a pocket door in a room that needs privacy, such as a bathroom.

A French door between rooms can add an element of formality and class. Since the door is likely to be delivered to the manu-facturer as unfinished wood, verify if the manufacturer finishes it or delivers it unfinished to the site.

Passageways

The traditional way to finish a passageway between rooms is to encase it in the same wood moldings and jambs used to trim out the interiors of the windows and doors. A contemporary alternative is to finish the opening in drywall. The manufacturer is likely to finish the installation of its bifold and bipass doors in the same way that it finishes its passageways.

Unless you prefer the drywall look, see if wood openings are available at an affordable price. Wood does a much better job of protecting the edges of the openings from nicks and dings, which are inevitable in small passageways.

Interior Windows

The main parts of a window are the sash, which holds the glass; the jamb, which is the inside of the frame that surrounds the window; the sill, which is the bottom of the jamb; and the moldings, which frame the

Two ways to finish a passageway. A cased passageway will hold up better against scratches and dings, while a drywalled passageway gives a contemporary look to the home.

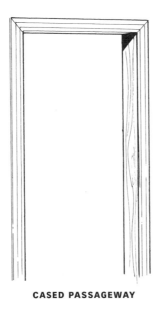

CASED PASSAGEWAY

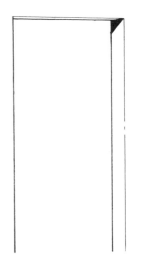

DRYWALLED PASSAGEWAY

window on the wall. These parts can all match or they can have different finishes. Different modular manufacturers follow different rules about the finish of the jamb and sill when the moldings are a different color from the sash. Some manufacturers finish the jamb and sill to match the sash, others to match the moldings. This creates two very different looks. You will not always be able to get a complete match among the four components even when the manufacturer is flexible. For example, if you purchase vinyl windows, the sash is always the same color, usually white, even when you select stained moldings. To take another example, you cannot stain the jambs if they are finished in drywall. Ideally you should see a model home or photograph that shows how your selections will look.

Some manufacturers install extended wood windowsills, while others leave out the extended sill and carry the jamb around the bottom to create a true picture-frame look. Extended sills create nice places to put plants or odds and ends, and can serve as a perfect sunning spot for the family cat. Most manufacturers that install an extended sill finish it the same as the moldings. Other companies use this as an accent point and install a different type of wood with a different finish. For example, with painted white moldings they might install an oak sill stained to match the cabinetry. Be sure you know what you are getting.

Few manufacturers offer a stained window sash on wood windows because they cannot buy the windows prestained, which is how they buy their doors and moldings. Instead they deliver the windows unfinished and they must be stained on-site. Do not underestimate the work involved.

Pay attention to how the manufacturer installs two windows side by side, or mulled. Some manufacturers place a stud between the windows for structural support. This is a particularly good idea for

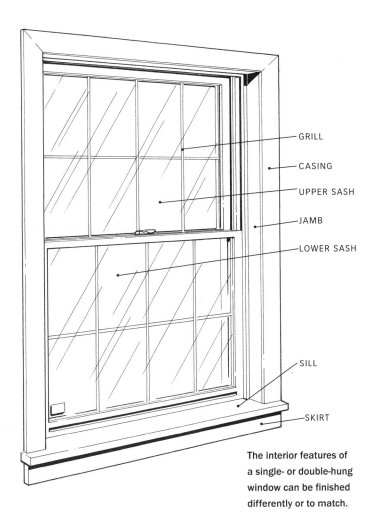

The interior features of a single- or double-hung window can be finished differently or to match.

vinyl windows, since their frames are not as rigid as those of wood windows. The additional stud changes the appearance of the windows, since it requires an extra molding to cover the space that is created. This can be attractive, but it is always good to know in advance what you are getting.

Ceilings

Most manufacturers give you a choice of ceiling heights and a few decorative features. You can add character to a room by having the manufacturer slope the ceiling about 18 inches from the outside wall to the marriage wall. Adding a tray in a room also dresses it up. For the most dramatic effect, design a room with a vaulted or a cathedral ceiling.

Ceiling Height

The standard ceiling height for most manufacturers is 8 feet, but a few build their homes with 7-foot 6-inch ceilings. With some manufacturers, the second floor of an unfinished Cape Cod has a 7-foot 6-inch ceiling even when the first floor is 8 feet. Upgrading a one-story or two-story to a 12-in-12 roof pitch with these companies usually gives you the same 7-foot 6-inch ceiling in the attic. A home with a stepped roof, which is created when you have a plan that is narrower on one end than the other, might have a 7-foot 6-inch ceiling in the narrower section even when it has an 8-foot ceiling in the remainder of the house.

Manufacturers also offer 9-foot ceilings, an increasingly popular option. Because of the shipping height limitations, however, it is more difficult to ship modules with 9-foot ceilings that have a roof over them. Ranches, Cape Cods, and the second floor of two-story designs all push right up against the maximum height regulation. Sometimes the only way to make this work is to change the roof trusses, which adds to the cost of the taller ceilings. Manufacturers also make this work by delivering the modules with the roof overhangs detached from the roof, since this sometimes reduces the height just enough to meet the regulations. Either the set crew attaches the overhangs or the GC does so immediately after the set. This does not work for all manufacturers and is especially difficult to do with taller roofs, since they require a roof design that folds twice, with the second flip on top of the first. The additional flip makes it harder to comply with the height restriction.

Tall windows are a nice complement to 9-foot ceilings. The tall windows can be installed at the same sill height as standard windows, providing room for furniture underneath the window.

With 9-foot ceilings, you have to decide how to handle the additional space above the wall cabinets in the kitchen.

Sloped Ceilings

Modular manufacturers can build a sloped ceiling that adds 12 to 18 inches of pitch from the outside wall to the center of the house. The roof load is carried by a beam, usually finished in drywall, that drops below the ceiling at the marriage wall. Sloped ceilings often require a 7-foot 6-inch ceiling height where the bottom of the slope begins as well as on the remainder of the house. And they work only with a 5-in-12 or 7-in-12 pitched roof.

Tray Ceilings

The manufacturer can create a tray ceiling in a room of your choice by building a soffit around the perimeter. You can embellish the tray with crown moldings both inside the tray and where the soffit joins the wall.

If you build a two-story home, you have another way to build the tray on the first floor. Because a modular two-story has a ceiling on the first story that is independent of the floor of the second story, it is possible to remove the first-story ceiling from everywhere within a room except around the border. Keeping the ceiling along the border creates the soffit needed for a tray ceiling. A nice benefit of this tray is that you raise the ceiling height in the room; a traditional dropped soffit creates a lowered perimeter rather than a raised ceiling.

Vaulted and Cathedral Ceilings

When designing a home with vaulted or cathedral ceilings (see pages 80–81), plan the lighting carefully. For example, consider a chandelier in a two-story vaulted foyer, as a small foyer light looks out of place. You may also want a ceiling fan to help recirculate the air.

Drywall Finish

The walls and ceilings of all manufacturers are painted a shade of white with a flat latex paint. The paint is sprayed on and then

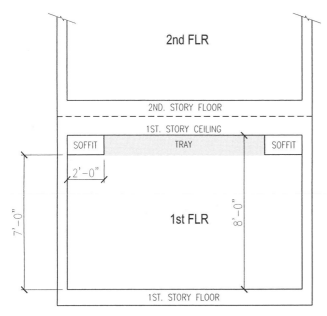

2nd FLR

2ND. STORY FLOOR

1ST. STORY CEILING

SOFFIT · TRAY · SOFFIT

2'-0"

7'-0"

1st FLR

8'-0"

1ST. STORY FLOOR

TRAY CREATED BY BUILDING A PERIMETER SOFFIT

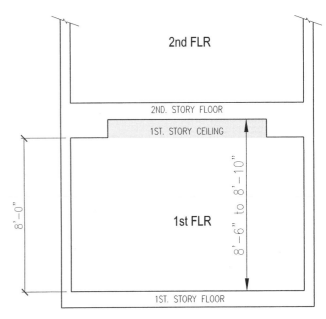

2nd FLR

2ND. STORY FLOOR

1ST. STORY CEILING

8'-0"

8'-6" to 8'-10"

1st FLR

1ST. STORY FLOOR

TRAY CREATED BY REMOVING CENTER OF CEILING FRAMING

rolled before it dries. Since another coat is not applied after the paint dries, most modular manufacturers consider their paint to be a primer. If you want a different color, or would prefer a semigloss finish, which is easier to clean, you or the GC can do the painting before you move in. If you intend to leave the paint as is for a while, order a couple of extra gallons of matching paint to help you make future touch-ups. If you select painted interior moldings and doors, you want the finish to be a semigloss, if the manufacturer gives you a choice. Flat paint on interior moldings and doors looks inferior to semigloss, and also wears poorly.

Many manufacturers do not give their customers a choice of ceiling finish. Some offer only a smooth finish; others provide only a textured finish. Textured ceilings hide imperfections, but they are harder to maintain and keep clean. It is messy and expensive to change the ceiling finish, especially from textured to smooth, so be sure you are comfortable with what you are receiving. Before insisting that a manufacturer provide a smooth ceiling when its preference is actually textured, consider that smooth ceilings

require more skill and training, which the manufacturer's production team may lack.

Some manufacturers automatically drywall the ceiling and walls in the stairwell; others do not. If you care about the look and the price is right, you might want to have this done.

Closet Shelving

Coated-wire closet shelving should be standard in all closets. Although a few manufacturers offer custom shelving, you might find a better and less expensive selection locally. Closet shelving is easy to install.

Kitchens

Manufacturers usually offer a good selection of cabinets. Expect to have a choice of wood type and finish as well as some specialty cabinets. Make sure you have enough of the right type of cabinets located in the most useful places. The space over the wall cabinets can be left open, trimmed out with a galley rail, closed in with a soffit, or filled in with taller cabinets. Laminate countertops

Because the first and second stories of a modular home are independent of one another, you have an additional way to create a tray ceiling in a two-story home. Above are two options.

with either a square or beveled edge are usually offered. If you prefer, you can select a solid-surface countertop, from either the manufacturer or a local supplier. Most manufacturers give you an option of a cast-iron sink to replace their standard stainless steel. If the manufacturer does not offer the sink or faucet you want, consider purchasing it from a local supplier and asking the GC to install it.

Kitchen Cabinets

Oak, maple, and white plastic-laminate kitchen cabinets are typical offerings, and cherry and hickory are sometimes available. Doors usually come in a couple of choices, such as recessed and raised panel. The wall cabinets might come in a few choices, such as square, arch, and cathedral, but the base cabinets will likely be square. Some of these choices are included with the standard kitchen package while others cost a little more. Ask if each base corner cabinet comes with a lazy Susan in the standard package. If not, you will receive a blind-corner cabinet with a lot of inaccessible space. Also ask if you get cabinet hardware for the doors and drawers.

Having the right amount of cabinets installed in the right places requires some careful thinking about how you use the kitchen. If you need additional kitchen cabinets, order them from your dealer. Although you could get them on your own at a later date, including them right away forces you to design the kitchen so that everything fits the way it should.

Cabinet companies offer several specialty cabinets. Some of the more popular selections include appliance garages, pantry cabinets, wood range hoods, and glass doors. Roll-out shelves are an excellent option, although you should note whether they are for the bottom or the middle of the cabinet. For 9-foot-high ceilings, consider 42-inch-high wall cabinets. Anyone not using a conventional oven is likely to need a cooktop and oven cabinet. When ordering an oven cabinet, be sure to specify either a single or a double oven; for the latter, clarify if it is one or two appliances. If you are purchasing a single oven, tell the dealer if you intend to put a microwave in the other section of the cabinet. Another way to keep the microwave off the countertop is to put it in a specially made cabinet.

Combining a 24-inch-deep wall cabinet above the refrigerator with flanking end panels that enclose the refrigerator is a nice touch. Installing a second set of doors on the back side of island or peninsula cabinets makes the kitchen more convenient to use. So does adding a desk. Instead of purchasing a desk cabinet, create one by combining a base cabinet that has drawers with a knee space under an extended countertop. This enables you to size the cabinet to your liking, and might cost even less than a kitchen desk cabinet. If the desk countertop is too high with a standard kitchen cabinet, switch to a vanity cabinet.

There are many other cabinet options that you can learn about only if you do some exploration on your own or ask your dealer to show you the cabinet company's product book. Some specialty cabinets are quite inexpensive, such as a tilt-out soap tray for the kitchen sink. Many, however, are expensive, and most manufacturers limit the available options.

An affordable way to create additional kitchen storage is to build a pantry closet. It does not need to be a walk-in pantry, which takes up a bit of space. It can be a 3-foot by 8-foot closet with one or two doors and multiple shelves. And it does not need to be 3 feet deep to be useful. Extra depth in pantry closets is not always advantageous, as it can be hard to see items in the back.

Taller people sometimes like to raise the height of base cabinets, which can be done by having the manufacturer build a platform under the cabinets. The platform can then be hidden with moldings that match the cabinetry.

If you decide to omit the manufacturer's cabinets because you want to buy your own, make sure the manufacturer understands how to prepare the kitchen for the installation. Give the manufacturer an accurate drawing of your kitchen layout. This helps it know where to place switches and receptacles, including those needed for appliances and the rough plumbing. Do not expect to receive a sizable credit for omitting the cabinets, since the manufacturer's cost for its standard cabinets is less than you would pay.

Soffits above Wall Cabinets

The typical 12-inch space above the wall cabinets can be left open for baskets, cookware, or decorations, or it can be closed off with a drywalled soffit. If you keep the space open, you can add crown molding on top of the wall cabinets to provide an attractive border. The drywalled soffit can be extended in front of the wall cabinets, and recessed lights can be installed in the soffit. The extended soffit brings the lights in front of the wall cabinets and closer to the work area. If you have a soffit, ask for a recessed light over the sink rather than a surface-mounted fixture.

Nine-foot ceilings have 2 feet of space above standard wall cabinets. You can leave the space open or you can create a stepped 2-foot soffit with recessed lights above the cabinets. Another alternative is to replace the standard 30-inch wall cabinets with 42-inch cabinets, which provide more storage, although at a higher reach.

Kitchen Countertops

The standard countertop will almost certainly be plastic laminate in a choice of several colors. Some manufacturers offer a wide range of optional colors for an additional fee. You usually have a choice of a

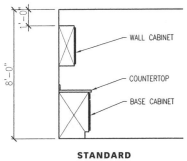

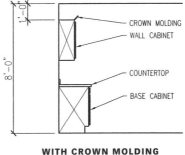

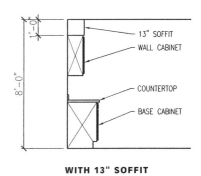

The space above the kitchen wall cabinet can be left open for storage or you can build an extended soffit.

square or beveled front edge, although you may need to pay a little extra for the bevel. Some manufacturers offer only the beveled look in plastic laminate, while others also offer the look in wood. The wood edge may or may not be an exact match with the kitchen cabinets.

If a solid-surface countertop captures your fancy, you can purchase it from the dealer or from a local countertop specialist. Very few manufacturers offer granite or other natural-stone countertops, but that should not stop you from getting one locally. A solid-surface countertop is expensive no matter who you buy it from. Cement, wood, and stainless-steel countertops are also available locally.

Kitchen Sinks and Faucets
Expect to receive a double-bowl stainless-steel sink with the standard laminate coun-

tertop. Make sure the sink is wide and deep enough for your needs. Most manufacturers offer a few other sizes, styles, and configurations. An enameled cast-iron sink is an affordable, attractive alternative to stainless steel that also goes nicely with laminate countertops. If you select a solid-surface countertop, you may want a solid-surface sink.

The faucet is likely to be a single-lever design, although some manufacturers use dual handles. Most manufacturers supply a sprayer. Optional faucets are often available for an additional fee. Sometimes you are better off omitting the faucet and having the GC install one you like.

Bathrooms

Bathrooms require a lot of decisions. Most modular manufacturers offer a selection of vanities, countertops, and fixtures. If a customer is not happy with the selections, she can omit the manufacturer's offering and buy locally.

Bath Vanity
You have the same basic choices for bathroom vanities as for kitchen cabinets. Be sure to ask if hardware for the doors and drawers is included.

Most modular manufacturers do not include a bank of drawers with their standard vanities. And some bathroom layouts are too small to add a vanity drawer base next to the sink. But if you have room, you can usually make much better use of a vanity with three drawers than one with a large door and one drawer.

Vanities are usually shorter than kitchen base cabinets, with a typical vanity 33 inches tall and a typical kitchen cabinet 36 inches tall. But some modular manufacturers now offer taller vanities as their standard specification.

There are four typical ways to finish the edge of a kitchen or bath countertop, though the standard countertop is plastic laminate with either a square or a beveled front edge.

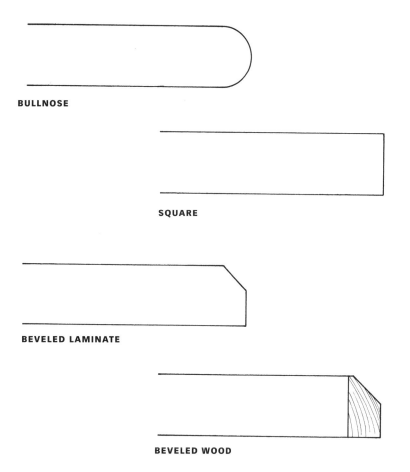

BULLNOSE

SQUARE

BEVELED LAMINATE

BEVELED WOOD

Bathroom Countertops, Sinks, and Faucets

Most manufacturers provide either a cultured-marble countertop with an integral sink or a plastic-laminate countertop with a china-bowl sink. Both are likely to come in several colors, although you will pay extra if the color you want is not on the standard list. Some cultured-marble patterns nicely mimic granite tops. You can trim a bathroom plastic-laminate countertop with the same edge choices as in the kitchen. The manufacturer may also offer an optional solid-surface countertop and sink, and you can always omit the factory's standard materials and purchase something locally.

A pedestal sink is a popular option. One disadvantage of a pedestal sink is that it does not provide any storage space. If you install a pedestal sink in a bathroom that does not have a linen closet, make sure there is other storage space within easy reach.

A second sink in the master bathroom is another option worth considering. It can be installed next to the other sink or in its own space. The bathroom-sink faucet is usually a single-lever or single-knob design, but dual-handle options and other upgrades are often available. The standard finish for faucets and bathroom accessories is chrome. Some manufacturers also offer brass finish as an option.

If you are concerned about the height of a vanity, keep in mind that the type of countertop affects the finished height. Plastic-laminate countertops are usually thicker than cultured-marble or solid-surface countertops.

Medicine Cabinets and Mirrors

Modular manufacturers offer various combinations of medicine cabinets and plate mirrors as standard and optional fare. You will probably want one or the other on the wall above the sink. A medicine cabinet can also be mounted on the wall beside the sink. Since most manufacturers glue their plate mirrors to the wall to make them as secure as possible, make sure they are located where you want them. If the vanity contains two sinks, consider adding separate mirrors and medicine cabinets.

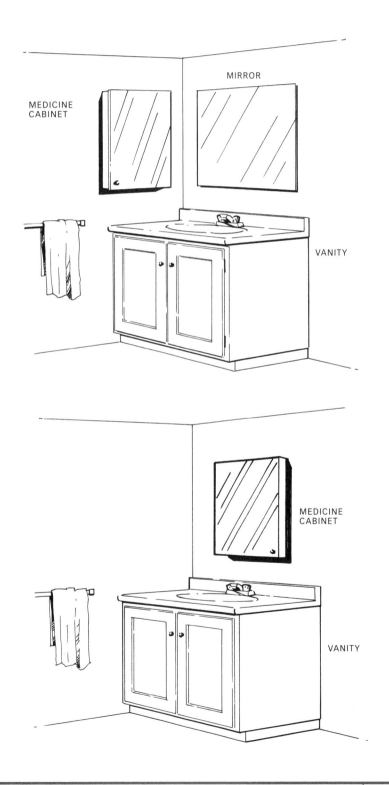

Two popular options for mounting a bathroom medicine cabinet and mirror.

Pay attention to the size of the mirrors and medicine cabinets to be sure they meet your expectations. An 18-inch medicine cabinet or a 24-inch plate mirror may look disproportionately small when situated above a 60-inch vanity. Also note the medicine cabinet's style, such as whether the glass is trimmed in wood or beveled glass. You may have an option to select a medicine cabinet that matches the vanity.

Bathroom Lights and Fans

You need a light above the sink. It could be part of the medicine cabinet, mounted on the wall, or attached to the ceiling. Some people find wall-mounted lights too bright and others find ceiling-mounted lights too dim. If you opt for ceiling lights, a soffit with recessed lights is a nice touch. Make sure you have enough recessed lights, since they tend to be of low wattage in this application. If you are not happy with any of the choices available, have the manufacturer install a switch and wire so the GC can mount a fixture you buy locally.

All bathrooms should have mechanical ventilation. This can be accomplished by installing a ventilating fan and light combination in the ceiling. In a large bathroom or one with separate compartments, you may want two fans. There are three options you should ask about. The first is a quieter fan. Fans play an important role in removing moisture generated by the shower and bath, but noisy fans often do not get used. Fan noise is rated in sones: The lower the sone rating, the quieter the fan. The second option is a timer that allows you to control when the fan is on and for how long. You may want to follow the lead of the Energy Star program and use the timer to run the fan for a period of time every day to bring fresh air into your home. The third option is a heat lamp that warms the bathroom quickly without having to warm the rest of the home. You may need to purchase a sep-arate heat lamp if you also upgrade to a quieter fan with a timer.

Toilets

Low-flush toilets are a federal requirement. Fortunately, most of today's toilets actually work the way they are supposed to. Ask your dealer if he would recommend an upgrade. Elongated toilets make life easier than round ones for little boys and grown men. A raised toilet for anyone with special needs is available as an option.

Grab Bars

Virtually everyone appreciates the support provided by properly anchored grab bars. When installed in a tub or shower, everyone can bathe more easily and safely. Installing grab bars alongside a toilet helps anyone with a mobility problem, weak knees, or poor balance. Grab bars come in white and other colors. Many people use the larger sizes for towel holders. If you are not ready to install them, you should still consider having the manufacturer put solid blocking in the bathroom walls so you can properly mount grab bars in the future without having to remove the drywall.

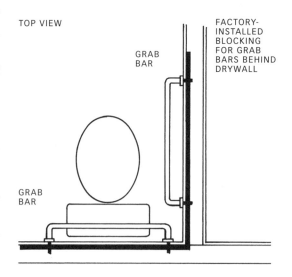

To attach grab bars securely to a wall, they should be mounted on a plywood backer installed behind the drywall.

Tubs and Showers

The standard tub/shower combination is 5 feet long and may come with a shower door. Some standard tub/shower units are quite narrow and shallow, so make sure the standard size works for you. If not, you should be able to order a larger size. You may want to add grab bars to the standard tub/shower, or consider an optional tub/shower that has more universal-design features, including plenty of size, built-in grab bars, and a seat. A dome top might be available, but it would be better if it came as one piece with the tub. When the dome is attached as a second piece, it creates a visible seam that becomes a home for mildew.

The showerhead will be mounted at a standard height unless you specify otherwise. You may want to consider a handheld showerhead with an adjustable pole, which makes it easier for kids and anyone else who wants to sit down to take a shower. It also makes it easier to clean the tub.

The typical shower faucet and diverter can control only the temperature, not the pressure. Some manufacturers offer faucets that control both, but with the advent of temperature-control (anti-scald) valves, these faucets are no longer as common. If the manufacturer does not offer an optional shower faucet that is acceptable, you could install your own. The problem with this strategy is that the only way to get to the shower diverter is to remove the shower or to enter from the other side of the wall in the adjoining room. Removing the shower is not practical, and entering through the other wall means cutting a hole in the finished drywall. An alternative is to instruct the manufacturer to leave the shower uninstalled or to provide an access panel in the adjoining room.

Ship-Loose Materials

If you are building structures on-site and want materials to match the house, such as siding and shingles for a garage, the GC should order them from the modular manufacturer so that it can ship the materials with the home. These uninstalled materials are often called "ship-loose." For example, the GC will need all the siding accessories, such as corner posts, J-channel, and starter strip. He will need a drip edge and ridge vent for the roof, along with the soffit and fascia for the roof overhangs. Matching windows and shutters, as well as mantels and pediments, might be necessary so the GC can make his site-built structures blend in with your modular home. Ordering exterior door locksets and lights for the site-installed doors will allow him to match the site-installed doors with those installed with your modular home.

If you want the interior of your site-built structures to match the interior of your home, the GC should order all of the moldings, doors, paint, and stain to match. He can even order matching plumbing fixtures for site-built bathrooms.

Ask your GC for his list of ship-loose materials and their quantities as soon as possible, since GCs tend to procrastinate on creating this list. Ideally, the dealer has a list of potential items to help guide the GC. Should you want to order ship-loose materials that are different from materials being used on your house, be sure to tell the dealer.

Many of the same considerations that apply to a tub/shower unit also apply to a stand-alone shower. You may want a shower door, built-in grab bars, a faucet that controls pressure as well as temperature, and a raised or handheld showerhead. Size is also an important consideration. If you have a choice and the room, do not select a 36-inch shower, which is too small for most people. Larger sizes include 42, 48, and 60 inches. Another option is for the GC to build a tile shower, in which case the modular manufacturer can assist by installing a shower base and cement board on the walls.

Larger showers sometimes come with one or two seats. A 48-inch shower with one seat or a 60-inch shower with one or two seats is particularly nice for people with physical limitations, as well as for women looking to shave their legs comfortably. Have the dealer document how many seats are included. For showers with only one seat, you may also want to specify which side it is on.

A special handicapped-accessible shower benefits people with special needs. Proper installation of these showers requires them to be recessed into the floor to reduce the lip on the base of the shower. Ask the dealer to make sure the manufacturer follows the installation instructions carefully. Another option is to have the GC build an oversized tile shower.

A soaker or garden tub usually comes in only a couple of lengths, typically between 5 and 6 feet, but it may be available in several different shapes, widths, and depths. It can be an oval, rectangle, hourglass, or triangle. It can be wide and deep enough for a comfortable soak for two or feel crowded with just one person. Look closely at what you are getting.

Since a soaker tub usually holds a prominent place in the bathroom, you may care a little more about its faucet. The manufacturer may offer you an upgrade selection.

Most soaker tubs sit in a platform that can be tiled, usually by the GC on-site. If you intend to tile the tub, ask the manufacturer to install cement board around the platform.

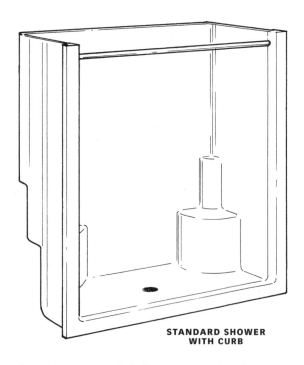

STANDARD SHOWER WITH CURB

A universal-design shower typically will have a curbless bottom.

CURBLESS SHOWER

A whirlpool tub is basically a soaker tub with a pump that recirculates the water through several jets. If you have the choice, select an in-line heater to help maintain the water temperature. Recirculating cold water takes some of the fun out of a whirlpool.

Wood Finishes

When selecting the color of kitchen cabinets, bathroom vanities, hardwood floors, and interior woodwork, you may have an option to select a finish that is either natural or almost natural, with only a hint of light-colored stain. Before you make this selection, you need to understand and accept what you will receive.

Unfinished wood has unique natural characteristics. Pieces cut from the same tree can have considerable color variation. There may be shades of white, red, gray, or even green in areas of the wood. Mineral streaks may be visible. Grain pattern and texture can vary considerably from piece to piece. This is the unique beauty of natural wood. (See the story at right.)

Clear finishes tend to bring out these natural characteristics. Because each piece of wood is unique, what you receive will be different from the sample the dealer may have shown you. In addition, one piece of wood may look entirely different from an adjoining piece within the same cabinet door or hardwood floor. Modular dealers cannot be held responsible for the type or amount of variation in wood products. If you want to minimize the visibility of these natural variations, consider selecting a dark stain or paint.

The Warranty

All modular manufacturers provide the minimum warranty required by state law, which is most often one year on all features of the home. Some manufacturers provide

Speaking from Experience

WHEN I BUILT ONE OF MY TWO-STORY MODEL HOMES, I decided to dress up the first floor with oak trim and doors, all finished in clear polyurethane, so that my customers could see what this option looked like. The first customer who ordered this upgrade called me very upset soon after we set his home. He said that some of his oak moldings had a much darker grain pattern than the others, which he felt was not the case in my model home. Without first looking at my model, I went to his house to see what he was unhappy with. When I saw the variation for myself, I ordered replacement moldings. Unfortunately, the new moldings came in with as much variation as those installed with his home. Finally, after ordering three sets of replacement moldings, we were able to match all of the moldings in his house almost perfectly. (In retrospect, I cannot believe that my manufacturer provided me with all of these moldings for no additional charge.)

A couple of months later, the manager of a custom woodworking shop visited my model home. I told her about the problem with the oak moldings and I showed her the rejected moldings. She then walked me through my model home and pointed out that I had the same "problem." Even more surprising, she said that all of her high-end custom stick-built customers had the same "problem" when she provided them with clear finished wood moldings. Most customers actually prefer this natural variation, she told me.

In addition to teaching me about the natural qualities of wood, this experience taught me how easy it is to miss the true appearance of a home's features. Over the years, I've noticed that while most customers do not look closely at model homes, they do put a microscope to their own home. And when they do, they see both real and imagined imperfections that they do not realize are typical of all homes, including their dealer's model homes.

an extended warranty that covers two years on the mechanical systems and 10 to 15 years on the structural systems. If you are offered the extended warranty, you should take it, even if the dealer charges for it. An extended warranty seldom costs more than a couple of hundred dollars. Although it is unlikely that you will ever need to use the extended coverage, it will give you peace of mind and might also help you sell your home, since the warranty is transferable to the next owner.

Verifying Your Order

When you finally authorize the dealer to build your home, he will construct it according to the specifications and features that have been written in your signed agreement. You will get these specifications and features even if they are different from what he told you. This makes it very important that you carefully verify that the specifications, features, pricing, and plans, as well as any accompanying notes, are accurate.

This advice might seem self-evident to you. But the truth is that many customers do not review their house details thoroughly, and many wind up with some surprises. After all, the customer and the dealer are both human, and communication, understanding, and memory are mistake-prone, even with professionals. The customer may ask for one thing when she means to ask for something else; the dealer may misunderstand a customer's request or write it down incorrectly. Think of it this way: If you and your dealer generate a list of 500 specifications and features and you get 99 percent of them right, there will still be five mistakes. They may be minor mistakes or major disappointments, but they can all be avoided by making sure that the specifications and features written in your signed agreement are what you want.

Change Orders

Change orders happen because people change their minds. Should you decide that you want to make some changes to your home after authorizing the dealer to build it, and it is not too late, expect to be charged for the change. Some items, such as special-order windows and materials that require long lead times, cannot be easily or affordably changed. The best way to avoid these costs is to work closely with your dealer to determine exactly what you want before you sign a contract.

A Modular Home Gallery

Cape Cods
PAGES 152–156

Modular Cape Cods are built with the same distinctive rooflines as their site-built siblings, and they come in both traditional and contemporary styles.

MANORWOOD HOMES

Ranches
PAGES 157–158

Modular ranches can be simple and straight, but they can also have decorative gables and porches and be built in other configurations to add character.

MANORWOOD HOMES

Two-Story Homes
PAGES 159–162

Modular manufacturers have learned how to design and build most sizes and styles of two-story homes, from modest Colonials to spacious contemporaries.

ALL AMERICAN HOMES, LLC

Mansions
PAGES 163–164

Custom one-of-a-kind modular mansions, often designed by architects, are appearing in wealthy communities. Some are built with more than 20 modules.

EPOCH HOMES

Interiors
PAGES 165–166

A modular home's interior can be as beautiful and luxurious as your heart desires, your imagination envisions, and your budget allows.

PENN LYON HOMES CORPORATION

cape cods

Cape Cods, which come in traditional and contemporary styles, can be made into chalets and configured in T-shapes. Cape Cods are often built with the second story unfinished. This makes them more affordable, which makes them very popular with young couples looking to add rooms in the future for their growing families. When fitted with gable and shed dormers, the second story of a Cape Cod becomes more attractive on the outside and brighter on the inside. As modular homes become more upscale, modular manufacturers are offering to complete some of the second story at their factory.

The stepped roof adds curb appeal to a simple design.

Dormers can add light to a Cape Cod's second story and interest to its roofline.

The mudroom and garage add character to the house.

Traditional New England styling with a country porch adds visual appeal.

Three gable dormers crown a classic porch.

COURTESY OF ROBERT COOLIDGE, ARCHITECT

A custom design created by architect Rob Coolidge and built by The Home Store (see page 22).

COURTESY OF ROBERT COOLIDGE, ARCHITECT

The back of Rob Coolidge's design has lots of glass to face the sun and lake.

The classic details of this house fit in both old and new neighborhoods.

The T-shape and front bump-out lend this Cape Cod a contemporary feel.

This modern design features a gable dormer that doubles as a porch roof.

A contemporary design is made striking with casement windows, skylights, and vertical wood siding.

A classic chalet with stained cedar siding creates a vacation retreat.

ranches

Most customers build straight ranches, usually because they are more affordable. But T-, L-, and H-shaped ranches are increasing in popularity. These configurations add a lot of character, especially when combined with taller roofs, decorative gables, and porches. Raised ranches are sometimes selected for the opportunity they provide to finish the basement with reasonable sunlight. Narrow building lots often require gable-entry designs.

THE HOME STORE

The Home Store's Sugarloaf 5 model home with a front porch is a popular choice (see page 68).

GENESIS HOMES, INC.

Seven gables and two dormers create a striking symmetry.

MANORWOOD HOMES

The porch and garage hide the fact that this is a straight ranch.

RITZ-CRAFT CORPORATION OF PA, INC.

A front porch, stone-faced entry, and gable-facing garage give a straight ranch unique appeal.

UNIBILT INDUSTRIES, INC.

©2000-2004 AVIS AMERICA

The recessed entry, cantilevered front, and brick knee wall make this a classic American raised ranch.

The strong vertical lines of the front porch make this straight ranch proud.

©2000-2004 AVIS AMERICA

©2000-2004 AVIS AMERICA

The same raised ranch as shown above (right), in its traditional simplicity.

The two front gables and porch make this gable-entry design exceptional.

two-story homes

The modular industry can build you a colonial, Victorian, contemporary, or most any other style of two-story. You can select a plan that features a Palladian window or one that gives prominence to a stately turret. The manufacturer can build your design with a steep roof and decorative gables or a hip roof and gable dormers. And your dealer can complete it by adding a wrap-around porch and dressing it up in a wood, brick, stone, or stucco finish.

GENESIS HOMES, INC.

A classic gable-entry two-story, complete with front porch, is perfect for narrow urban lots.

MANORWOOD HOMES

This modern T-shaped two-story has character.

EPOCH HOMES

A Dutch Colonial with a portico is an attractive home.

THE HOME STORE

The Home Store's Whately 1 model home is shown with breezeway and garage (see page 56).

©2000-2004 AVIS AMERICA

With all the gingerbread, this is a lovely Queen Anne Victorian.

RITZ-CRAFT CORPORATION OF PA, INC.

Notice the brick facade, the arched window in the vaulted foyer, and the symmetry of the three gables.

GENESIS HOMES, INC.

A spacious three-story winter haven has the look of a ski lodge.

SIMPLEX INDUSTRIES, INC.

The wraparound porch imparts farmhouse charm.

WESTCHESTER MODULAR HOMES, INC.

This classy design uses a sloped property to create a drive-under garage.

WESTCHESTER MODULAR HOMES, INC.

The hip roof and dormers create a stately look as well as usable attic space.

Surrounded by porches, this home supports and invites outside living.

With its tower and stucco
finish, this home boasts
kinship to a castle.

The Victorian features
are crafted within the
modules.

mansions

Many of today's modular mansions are being created by architects. The resulting designs borrow from both classic and contemporary styles. As you can see, these homes are often three stories tall with exteriors that exhibit all manner of flourishes. What you can't see is that the interiors are every bit as custom, with open floor plans and stunning finishes completed by local craftsmen. And these homes are no "McMansions" — all size and show with poor quality — since they are built with modern modular technology.

A simple but inviting entry is complemented by the Palladian window above.

Each wing is large enough to be a home on its own.

Windows flow from top to bottom on this oceanside home.

An aristocratic estate includes a cupola on the guesthouse.

mansions

PENN LYON HOMES CORPORATION

The stone facade makes this three-story home courtly and dignified.

EPOCH HOMES

The stately eloquence of old New England has a modern interpretation.

EPOCH HOMES

Intricate lines catch the
eye from all angles.

interiors

Luxurious master bathrooms and dramatic cathedral ceilings are just a few of the features that grace the interior of today's modular homes. If your dream is an expansive kitchen with fancy cabinetry, you can realize it. If it's a stunning vaulted foyer, you can have it. Or if it's a simple touch, such as an arched passageway, you can enjoy that too.

A sunlit passageway is made elegant with a simple arch and columns.

Classic Greek columns create a generous living room.

A sublime Palladian window frames this dining room.

A vestibule and double staircase are worthy of Scarlett and Rhett Butler.

UNIBILT INDUSTRIES, INC.

This gorgeous master bath makes mornings a sumptuous luxury.

PENN LYON HOMES CORPORATION

PENN LYON HOMES CORPORATION

Surrounded by light, this is a formal yet welcoming dining room.

A spacious kitchen is dressed in beautiful wood cabinetry.

ALL AMERICAN HOMES, LLC

The majesty of a vaulted ceiling creates a breath-taking interior.

5

Selecting a
General Contractor

THREE TYPES OF construction tasks must be completed to build a modular home: the site work, the button-up work, and the construction of site-built structures. Someone needs to be in charge of hiring all of the tradespeople to do these tasks, coordinate their schedules, and oversee their work. That is the job of the general contractor, often known simply as the GC. The GC can be a modular dealer, independent professional, relative, friend, or even yourself. He may have years of experience or this may be his first job. He may have worked on many modular homes over the years or never seen one before. He may or may not be licensed. As a general rule, the best choice for a GC meets the following criteria:

- He is experienced and licensed
- He has prior modular-home experience
- He is also the dealer who sold the home or someone who has worked closely with the dealer in the past

This chapter will explain why hiring a GC with these credentials will serve most customers well. It will also look at several other options that a customer's personal situation or inclination may lead her to consider. The latter road is full of potential pitfalls, which can be largely avoided if the risks are recognized up front and taken seriously.

An Experienced, Licensed General Contractor

On any construction project, the general contractor has a full list of responsibilities and duties, as you can see from the list (page 168). A professional, experienced GC is best equipped to handle each of them competently. When problems occur, as they do on every job, he solves them quickly. Better yet, he anticipates problems as a matter of course and heads them off before they become a threat. A licensed GC knows the building codes, which is important not just because following them is the law, but also because they protect the customer's health and safety.

The General Contractor's Job

The GC has many difficult and time-consuming jobs on a modular home project. Here are just some of the more important ones:

- Obtain competitive bids
- Help determine the scope of work
- Help determine the building specifications
- Select each subcontractor and sign a contract detailing the scope of work, building specifications, and price
- Create and manage the schedule for materials and subcontractors, including the sequence of each subcontractor to maximize productivity and reduce conflict
- Ensure that the job is ready for each subcontractor before instructing him to begin
- Revise the schedule weekly to adjust for the inevitable delays, which can be caused by such things as weather, inspectors, and illness
- Ensure that the work is done to accepted industry standards as well as to the customer's satisfaction
- Ensure that warranty problems that occur after the job is done and after each subcontractor is paid are taken care of

Another advantage of using a seasoned GC is that he is able to call upon a number of subcontractors to obtain competitive bids for each construction task. He is able to get them to perform in a timely fashion, which is no small feat in the building trades. He can command a fair price and a timely response because they depend on the work he provides. An experienced GC knows which subcontractors to avoid because of a history of poor workmanship, unreliability, or unethical pricing. A professional GC is likely to have the necessary insurance to protect against something going wrong on the building site, such as an accident causing a serious personal injury or significant property damage to the home. This insurance is very important in limiting the homeowner's potential liability.

Superior warranty protection is yet another reason to use a professional GC. Although most of the actual construction is done by subcontractors, who are obligated to provide their own warranty, a GC blankets their work with an additional warranty shield. This additional protection can prove invaluable when a subcontractor does not honor his warranty obligations, since the GC must then assume them at his own expense.

Having an experienced GC on your side is also helpful when navigating the approval and permit steps that must be completed prior to beginning construction of a new home. Each state has somewhat different requirements, and towns within the same state may enforce different regulations. Many of these steps include fees and other associated costs. For example, an approved septic design, an engineered site plan, and a building permit are typically required before a new home can be built. A list of approval and permit tasks can usually be obtained from the town's building-inspection department. See chapter 7 for a more detailed discussion of the GC's responsibilities.

Many applications for permits and approvals require sign-offs from local officials, such as the board of health and highway department. In many communities, the person who accepts responsibility for constructing a house in compliance with local building codes must sign the primary building permit application. When a building-permit application requires a signature, it is always best to have the GC sign it. Either the GC, who may charge for the service, or the customer can assume responsibility for first getting the sign-offs from local officials.

The person who signs for the building permit must be thoroughly familiar with the state and local building codes, as well as with other applicable codes, by-laws, rules, and regulations, all of which change periodically. Because most modular homes are approved by an independent inspection agency that is itself approved by the state, the manufacturer will have complied with the state codes and regulations. Other codes and regulations apply to the excavation and completion of the home once it is set on the foundation. The person who signs the building-permit application assumes liability for meeting these legal requirements.

Some states have prohibited inexperienced homeowners from building modular homes without the services of a licensed contractor. They have found that too many homeowners underestimate what is required to complete the home. They want to protect homeowners from modular builders who would mislead them into thinking they only need to "add a little water and stir."

In areas without a formal building-permit procedure, the GC will not be compelled to sign off on the building permit. If you live in such a community, insist that your contract with the GC includes a clause that requires him to build the home in compliance with state and local building codes as well as accepted building practices.

If you are set on acting as your own GC,

do not assume that lending institutions will cooperate. Banks and mortgage companies increasingly will issue a construction loan only to customers who hire a professional GC. They are doing this not to protect the customers so much as to protect themselves. It is hard to blame them. Who would want to lend several hundred thousand dollars to someone who didn't have enough experience to complete the project? Fortunately, you have no more need to act as your own GC than you do to act as your own physician or dentist.

In short, although it is considerably easier to build a modular home than a stick-framed home from scratch, most modular home customers will benefit substantially from the assistance of an experienced, professional GC.

A GC with Modular Home Experience

Any good stick builder can build a garage, porch, or deck. He can also finish the basement in a modular home. A foundation contractor does not need modular experience to make a good modular home foundation as long as he follows instructions. Many of the other modular construction jobs, however, benefit greatly from someone who has solid experience with modular homes. The excavation contractor, for example, must prepare the site for the delivery and set of the modules. Even contractors with years of excavation experience routinely underestimate the logistical difficulties of maneuvering modules that are often long, usually wide, and always heavy.

The button-up work requires the most specific knowledge of modular construction. There are aspects of the plumbing, electrical, and HVAC work that are unique to modular construction, and there are many interior and exterior carpentry tasks that involve modular-specific skills. Skilled GCs who lack experience with modular

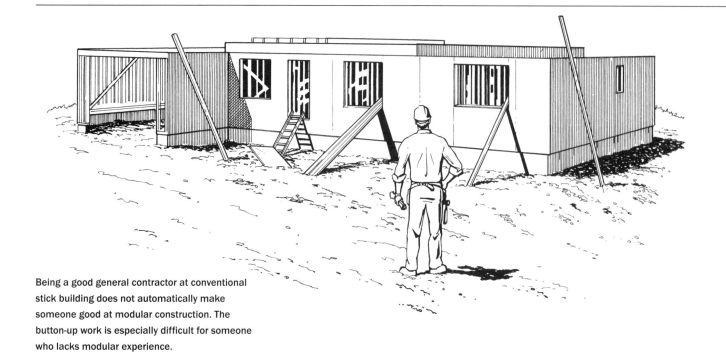

Being a good general contractor at conventional stick building does not automatically make someone good at modular construction. The button-up work is especially difficult for someone who lacks modular experience.

homes often get into trouble when they take on modular jobs.

GCs who have modular experience, however, know which construction tasks must be completed, how to complete them, and in what order. They supplement their trade knowledge by using subcontractors who themselves have modular expertise.

Seasoned modular GCs understand the difference between their button-up responsibilities and true warranty issues. They know that manufacturers expect them to repair and adjust components of the modules that need some additional work because of the delivery and set. Repair of minor drywall cracks, adjustment of miter joints, and realignment of windows and doors are considered normal GC responsibilities. GCs with modular experience know to budget time and money to complete this work. When problems arise, they have solutions at hand. When they need to talk to the modular dealer, they are able to ask the right questions, provide the right information, and execute the plan.

As a general rule, modular dealers do not like to work with inexperienced GCs. They know that if conflict arises between themselves and the GC, the customer will be unsure of whom to trust and support. Many dealers also act as GCs, and an experienced dealer who is just beginning to assume GC responsibilities may also run into trouble, but at least any conflicts that arise will be with himself.

GCs with modular experience can save you money. Because they have done it before, they know how much to charge for specific tasks; they do not have to pad their price estimates to protect themselves from their own inexperience. For example, an inexperienced GC might assume that the manufacturer would supply ductwork for the HVAC system or railings for the stairs. He might be right, but he could be wrong, since manufacturers do not all supply the same things. If the GC guesses wrong and there is no money in the budget for the additional labor and materials, a lot of bitterness and extra expense could result, not to mention legal action.

Many GCs simply overestimate their knowledge of modular homes, which is another way of saying that they underestimate their ignorance of the specialized nature of modulars. There are a few tasks that inexperienced GCs almost always miss, including the need to readjust misaligned marriage walls and interior moldings and repair drywall cracks. A GC who has not built a modular home before is unlikely to include these tasks on his list of button-up jobs.

This is not to imply that an inexperienced GC cannot learn. A well-informed customer, armed with the information presented in this book, could help a GC understand the unique aspects of modular building. If the GC is willing to learn, and accepting of the possibility that he may make some mistakes that he is obliged to fix, the experience can be good for everyone.

The dealer can also help by generating a complete list of button-up tasks required for the home, along with specifications for each of them. With this information in hand, the GC should be able to prepare a thorough estimate.

Think of the manufacturer, dealer, GC, and subcontractors as your team. When you visit your home while it is under construction, you want to feel that everyone is working together. You hope they appreciate your design and selections, but at the least you expect them to approve of your decision to

Speaking from Experience

THERE IS ONE SURE WAY TO MAKE the construction of your new home a miserable time: Hire a GC without modular experience who stubbornly believes that modular homes are an inferior product. You will likely find that he is always looking to say something bad about your home. This was quite common 15 years ago, before stick builders began to have a change of heart about modular homes. My first experience with this was with a family building a two-story Colonial in a typical middle-class neighborhood. Since I did not yet offer GC services, they had to find their own GC. They selected a skilled and reputable local GC who was a friend of the husband's parents. The GC's primary business was building additions and new stick-framed homes. The first time we spoke, he let me know that he was only doing the work as a favor to a friend, and that he thought modular homes were inferior. I tried to explain why they were equal to stick-built homes, but he was not having any of it.

Once my customers' home was set on the foundation, the GC complained about everything, in spite of the fact that my customers loved their home. He

acted as if the home was going to fall apart in the next storm and the proper fix was going to cost a king's ransom. He strongly disagreed with all proposed warranty solutions, asserted that the manufacturer built the modules incorrectly, and insisted that they be taken apart and rebuilt. He then assumed the role of savior, as if he was protecting my customers from some evil modular dealer. Even though I prevailed, and the relatively minor warranty work was completed correctly, he was distrustful of everyone associated with the home. This fueled a series of conflicts each time he discovered something new about the home, even things that made it obviously superior, such as the way the drywall was attached to the framing.

My customers tried to make the best of it, but they were clearly worn out by the end. When I ran into them several years later, they regaled me with a few more stories. Fortunately, they had a sense of humor about it. They even told me that the GC worked on another modular home in a neighboring community a year after finishing their home.

build a modular home. As the story on page 171 indicates, you certainly do not want to work with contractors who disapprove of modular construction and never pass up an opportunity to criticize your home. Unfortunately, people in the construction trades can sometimes be fiercely competitive and self-protective. With modular manufacturers taking away more and more work from local stick builders, it is not surprising that some of them are prone to making unflattering comments about modular homes. You are well advised to steer clear of any GC who is not experienced and comfortable with modular homes.

Dealers Who Are Also GCs

The ideal situation for someone buying a modular home is to work with a dealer who can also function as the GC. From quality, to price, to overall customer service, having one person in charge of the whole project offers the customer a range of benefits.

More Knowledge

As a general rule, a dealer who has been building modular homes for a few years will have acquired the requisite construction knowledge. If he has been with the same manufacturer for a few years, he will know their systems. Since each manufacturer does things a little differently, this will help him do a better job. For example, some manufacturers caulk all of the gaps between the walls and the wood moldings they install around the windows and doors and along the floor; other companies do not. Some manufacturers fill all of the nail holes in the moldings; others do not. And some finish stained moldings with polyurethane; others do not. The dealer's relationship with the manufacturer allows him to know these details, and his general-contracting experience allows him to plan for and complete all of the work needed for a specific home.

Better Quality

A modular dealer usually cares more about the quality of a finished home than anyone except the homeowner. He cares about the quality of the work done by the factory and the quality of the work done by the GC. If a GC does a poor job of completing a home, it will reflect poorly on the dealer regardless of his relationship with the GC. As a consequence, a dealer acting as the GC is motivated to do a first-rate job on every aspect of the home.

Coordinated Contracts

When the dealer acts as the GC, he is responsible for integrating the construction tasks for both the house and the site into one complete contract. If you select a GC who is separate from the dealer, you become responsible for interviewing candidates for both jobs, preparing and signing separate contracts, making sure all required tasks are covered in one or the other contracts, and mediating disputes.

Coordinated Timing

A modular dealer who does the GCs work is responsible for ensuring that everyone's schedules are coordinated. Since the manufacturer is likely to charge the dealer if he is not ready to take delivery of a home when he requested it, he has an incentive to ensure that the foundation and site are ready in time. If the home requires warranty work, the dealer is motivated to honor his warranty obligations promptly, since timely completion of the work helps him maintain his overall construction schedule.

Better Price

Using the dealer as the GC usually results in a better total price for the customer. Because the dealer is also making a profit on the sale of the house, he can often charge less than an independent contractor for the GC work. Of course, these savings only occur with a company that is truly committed to providing GC services. Dealers who lack such a commitment may offer GC services at a premium price or with fewer services, hoping to encourage the customer to purchase the home from them but to obtain GC services elsewhere.

Availability

You may have no choice but to select the dealer as your GC. In many areas, there simply are not many independent GCs who focus on modular construction. And those GCs who acquire a professional interest in modular homes tend to become dealers themselves, because they can earn more money if they sell the homes they assemble.

Warranty Support

Conflict over warranty issues is unavoidable when the dealer and GC are separate companies and the manufacturer and GC both deny responsibility for a particular problem. When this happens, the dealer has two choices. He can agree with the GC and correct the problem at his own expense or he can agree with the manufacturer, which means he is likely to insist that the GC correct the problem. When the evidence is ambiguous, as it often is in these situations, it is tempting for the dealer to side with the manufacturer, which means the customer is often left in the middle holding the repair bill. When the dealer and the GC are the same company, the potential for this conflict is eliminated.

Other General Contracting Options

Many people take pride in acting as general contractors on their own home. For those who do, increasing numbers are finding that modular homes are much easier to manage than stick-built homes, since they reduce responsibilities to a manageable level and afford a greater measure of control. People often enlist the help of relatives and friends in these projects.

Some people do this quite well. They have the right personality, the supervisory and management skills, and the necessary time. Their relatives and friends may have prior construction experience that allows them to bring in the project on budget and on schedule. This earns them the right to celebrate their success by throwing a wonderful housewarming. Often, however, things do not go quite this smoothly, so it is worthwhile to take a look at the potential pitfalls.

Hiring Relatives and Friends

If you are building a modular home, it makes perfect sense to consider hiring a relative or friend who is a licensed, professional GC. It is almost always better to work with someone you know and trust, as long as that person has modular home experience and offers a competitive price.

A more common scenario is for you to have relatives or friends who are subcontractors in the building trades. In this case, you will still need a GC. Once you secure one, you can ask the GC if he is willing to hire your relative or friend. If he is licensed and has experience with modular homes, the GC may be willing to work with him. But he also may not, especially if he has no prior experience with that person and no way of verifying his timeliness or the quality of his work.

Another option is for you to hire your relative or friend as a subcontractor. In addition to circumventing any unwillingness by the GC, this route allows you to avoid the GC's markup on the work of this subcontractor. The GC is not responsible for the work of any subcontractors you hire, but neither does he have any control over their timeliness. This can be a serious problem for both the GC and the homeowner. One

Speaking from Experience

FOR THOSE WHO ARE CONSIDERING hiring relatives or friends as subcontractors, it might be worth noting the experiences of three of my customers.

One of my customers, who acted quite competently as his own GC, hired his brother to complete the electrical work. My customer did a commendable job getting all of the subcontractors to show up on time, and except for one subcontractor would likely have finished the work ahead of schedule. But this one subcontractor (yes, his brother the electrician) did not complete his work until many weeks after the others had completed theirs. The electrician told his brother that since he was doing him a favor by taking the job, he should understand that he needed to finish his other, "real" work first. He figured that his brother would understand the significant delay, since "after all, we're family."

Another of my customers, a single man, hired two friends to complete all of the interior and exterior carpentry button-up on his Cape Cod modular home. Although the friends were professional carpenters, they had no experience with modular construction and did not know all that needed to be done. My customer told his friends what he thought needed to be done, but somehow the friends got it wrong. Maybe my customer, who was also new to modular construction, did not understand the scope of work, or maybe he explained it poorly, or maybe the friends thought they knew what was needed and so did not really listen. Since the three of them were friends, they did not think it necessary to put their agreement in writing. Once the friends discovered that they had more to do than they anticipated, they told my customer that they would not be able to complete all of the work they originally planned to do for the price they quoted. They explained that they had not budgeted the extra time or materials, and they already had given my customer a great price. My customer felt this was unfair. He took his friends' word that the job would be complete and had not included additional money in his budget. The friends, however, felt that their original agreement was based on a misrepresentation by my customer of what was required. In the end, my customer borrowed money from his parents and paid his friends to complete the work. I have no idea how their personal relationships fared after the house was built.

Yet another customer, who used my company to serve as the GC to build her T-ranch home, hired her brother-in-law to complete the plumbing and heating systems. In hopes of saving money, she hired him on an hourly basis ("time plus materials"). Her brother-in-law did a fine job, from everything I could see. I later found out that she paid more than I would have charged her. I suspect her brother-in-law's inexperience with modular homes caused him to be less proficient with the work, which caused him to take longer than a subcontractor with modular experience.

Hiring friends and family to serve as your GC or one of your subcontractors can create problems unless you spell out their responsibilities clearly in advance. Make sure you have a written agreement that explains the scope of work and price in detail. Also, ensure that they will schedule their work in a timely fashion.

subcontractor's tardiness can push the entire project off schedule, costing everyone involved time and money. It is usually best to allow the GC to supply his own subcontractors, since this makes him completely accountable to you. If you choose to do otherwise, work hard to arrange a preconstruction meeting between the subcontractor and the GC to spell out the ground rules and expectations for all parties. You should insist that the subcontractors and GC agree to communicate directly with each other so you are taken out of the middle.

Acting as Your Own GC

Many people are tempted to take on the job of being their own GC by the thought that they will save the 10 to 20 percent markup the GC adds to the cost of the subcontractors' bill as a fee for his services. For most GCs, the fee consists of two-thirds overhead costs and one-third profit.

Some customers perform the role of GC remarkably well. Because they plan ahead and are realistic and cautious in their original estimates, they come in ahead of schedule and under budget. Their home is beautiful, a trophy for all to see and a source of everlasting pride. Their subcontractors compliment them by suggesting that they switch professions and enter the construction trades.

The reality for you, however, might be quite different. Yes, you will save the GC's markup, but you may find yourself paying more to the subcontractors for their services. Subcontractors may not give you the same level of service or low price they give a GC who hires them regularly. They may figure that your inexperience will cause them delays and additional expenses. If you try to save money by buying materials directly, you will avoid the subcontractors' markup, but you may not get as good a price as the subcontractors would, so the money you save on the markup will be lost on the

actual price paid. Also, subcontractors are likely to raise their labor rates when you take away their opportunity to earn a profit on the materials.

Even if you save a little money, there are several risks involved with acting as your own GC. Consider that you will be responsible for any zoning and building-code violations, and you could be liable for any injuries incurred on the job site. Even experienced GCs find their jobs to be difficult, time consuming, and stressful. It is easy for inexperienced GCs to underestimate the toll such a job can take on them. They fail to schedule properly, have trouble recognizing the difference between good and poor work, become overwhelmed with paperwork, and do not understand or acknowledge all of the tasks that must be completely on a modular home.

GC work takes a lot of time. It encroaches on family life and intrudes on work responsibilities. On-site meetings to begin or end the day are typical, as are early morning and late afternoon phone calls. You will need a sympathetic employer and an understanding family. You might think cell phones will help you, and in some ways they will. But they also extend the opportunity for everyone with a problem or question to track you down throughout the day, leaving you little chance to escape the construction demands while at home, with the kids, or in the office. The paperwork cascade is relentless: subcontractor estimates, contracts, change orders, invoices, and payments, as well as supplier estimates, purchase orders, invoices, and payments.

Learning the sequence of tasks from studying construction books is one thing, executing the daily schedule in the face of the routine but unplanned setbacks is another. When the project starts, school is over, as is your chance to practice your moves and be rescued by the undo button when you make a mistake.

Before deciding to serve as your own GC, complete an honest self-evaluation. Consider whether you are a good leader and problem solver. Weigh whether you will come across as competent to skeptical construction professionals. Think about your ability to negotiate prices, specifications, and terms with subcontractors, as well as your ability to arbitrate misunderstandings between subcontractors. Ask yourself if you perform well under pressure and control your anger when events take a turn for the worse. Ask your family, friends, and co-workers if you really are a detail-oriented and organized person.

Better Ways to Save Money

People who think they will save significant money by acting as their own GC are wrong more often than not. Even so, there are steps a budget-conscious customer can take to trim the costs of a new modular home. As discussed in chapter 4, you can reduce costs by altering some building specifications. For example, you can build a smaller deck or porch or use more affordable materials, such as pressure-treated wood instead of cedar. You can reduce the initial investment by delaying items that can be done later, such as a porch, deck, or air conditioning, or holding off on finishing the second floor or basement. These are especially good alternatives if you have relatives or friends who can help you with these projects at a later date. One potential problem with this course of action is that you will not be able to delay completing these tasks if your lender is funding them; lenders usually want the work to be finished before they distribute all of the money. If you choose later on to fund the additional work yourself, you will lose the advantage of having the expenses included in your mortgage. (See chapter 9 for more on financing.)

You can save money by completing some of the construction tasks yourself while the GC is managing the project. Most GCs are willing to let customers do some tasks. For example, you can clear the property of brush and trees, help with the painting and cleaning, and handle your own landscaping. Some GCs may let you install the siding, tile the bathroom, or build the deck. Contributing your own labor, or "sweat equity," will save you both the subcontractor's cost and the GC's fee for the work.

Another cost-cutting option is to hire a general contractor to serve as a construction manager. This arrangement allows you to hire your own subcontractors and purchase your own materials, thus eliminating the GC's markup for these. A construction manager should help you with the following:

- Define the scope of work and specifications
- Select subcontractor candidates for each area of construction
- Review written proposals for each subcontractor candidate
- Select subcontractors
- Schedule and coordinate the work
- Oversee the work with on-site inspections
- Supervise warranty work

The construction manager will charge a flat fee or a percentage of the total cost of the project for his services. While this may save you money, it is riskier than hiring a GC since you must personally sign the contract with each subcontractor. If a subcontractor fails to perform or refuses to honor his warranty obligations, you will have to hire another subcontractor, and you will be responsible for any additional costs. In balance, however, hiring a construction manager will provide you with some expert guidance you might not otherwise have.

Comparing GCs

A competent GC requires a different set of skills from a competent modular dealer, and the screening process needs to take these differences into account. Even if you are using the same person for both jobs, look at his competence in each field separately. It is also important to look closely at your GC's skills even if you have already selected him because he is a friend or family member. After all, you will want to know his strengths and weaknesses. If you are assuming the role of GC, shop for subcontractors using the same advice offered here.

Begin the process of selecting a GC as soon as possible. It can take many weeks or even months to identify candidates, verify credentials, and receive and review estimates. Obtaining a GC's estimate almost always takes longer than obtaining a modular dealer's estimate, since GCs need to solicit bids from each of their subcontractors, which will take time. After obtaining estimates from all of the candidates, negotiate and sign a contract with the one you select. This will allow the GC to schedule commitments from his subcontractors.

If you do not already have a GC picked out, the best way to identify candidates is from recommendations, especially from your modular dealer or people you know and trust. Other sources include your local chapter of the National Association of Home Builders, building inspectors, banks, realtors, and attorneys who serve the real estate and construction industries. Lumberyard employees usually know which GCs have the best reputation, and subcontractors usually have an opinion about which GCs are the most competent. You can also talk with people who are currently building or have built recently. Since the information is a matter of public record, the building inspector may supply some names. The Yellow Pages and the Internet can also produce candidates, but with blind leads such as these, you will need to spend more time checking credentials. Ideally, you should select at least three candidates.

Check out some of each candidate's work. Try to visit homes that the GCs are currently building. If you do not like what you hear and see, move on to the next candidate. Use the checklist on page 178 to guide your interview with each candidate.

The next step is to arrange a sit-down meeting with the candidate. The primary purpose of this meeting is to give the GC the information he needs to generate a written estimate documenting the scope, specifications, and price of the work.

Take note of each GC's personality and see how well it fits with your own. In many ways, your GC will serve as your business partner in the construction of your new home. If you are not comfortable with a candidate, find a gracious way to tell him that you have decided to go with someone else. Do not ask a candidate you have ruled out already to complete an estimate just so you can get a comparison price. Not only is this unfair to him, but it also will not serve your interests. What you need are estimates from candidates whom you would consider selecting.

General Contractor Screening Checklist

When shopping for a GC, complete the following steps:

☐ Select the candidates.

☐ Have the candidates help you design the site-built structures, such as the garage and deck.

☐ Have the candidates help you determine the contracting tasks required to complete your home.

☐ Have the candidates help you select your preferred building specifications.

☐ Obtain a detailed price estimate from each candidate.

☐ Assess each candidate's craftsmanship, price, and customer service, including warranty.

☐ Make the final selection.

Questions for GC Candidates

You will not need to ask each candidate all of the questions below. Tailor the list to suit your circumstances and each candidate's background, including whether he works alone or is part of a larger company.

- ☐ How long have you been in construction?
- ☐ How long have you been working for yourself?
- ☐ How long have you been a GC?
- ☐ What did you do in construction before you were a GC?
- ☐ What is your experience with building new homes?
- ☐ Were you a GC or subcontractor?
- ☐ Tell me about your remodeling experience.
- ☐ How many modular homes have you worked on?
- ☐ What types of modular homes have you worked on?
- ☐ What work were you responsible for?
- ☐ What tasks did you do personally?
- ☐ Which modular manufacturers' homes have you worked on?
- ☐ Have you ever set a modular home?
- ☐ What experience do you have with the type of modular home I am building?
- ☐ Have you built the types of site-built structures I need?
- ☐ What experience do you have supervising excavation work?
- ☐ What experience do you have preparing a site for the delivery and set of a modular home?
- ☐ What experience do you have overseeing foundation work?
- ☐ Are there construction-related tasks you would rather not be responsible for?
- ☐ Do you have other employees? If so, what do they do for you and how will they help me?
- ☐ What subcontracting trades, if any, are on your payroll?
- ☐ Will you personally supervise my project from start to finish? If not, can I meet the person who will?
- ☐ When will you be able to start my project?
- ☐ How long will it take you to complete my project?
- ☐ How many other projects will you be working on at the same time as mine?
- ☐ Will you be directly supervising these projects?
- ☐ Do you already have subcontractors in mind for my job?
- ☐ How often will you contact me once the work begins?
- ☐ How can I contact you?
- ☐ How do you keep track of scheduling?
- ☐ If I have a warranty problem after I move in, what do I need to do to get the problem fixed?
- ☐ How long can I expect it to take?
- ☐ Will you take responsibility for one of your subcontractor's warranty problems if the subcontractor will not?
- ☐ Do you have any partners in your company? If so, what role do they play?
- ☐ When you sign a contract with a customer, do you have a "legalese" section that states the terms and conditions of the contract? If so, can I have a copy to review?
- ☐ How much of a deposit do you require?
- ☐ Under what circumstances is the deposit refundable?

Obtaining Estimates

For a GC candidate to provide a written estimate, he will need to know your house design, the construction tasks required to build your home (the scope of work), and the building specifications you want him to use when completing each task. Most customers will be unable to produce these before meeting with a GC candidate, since it requires that they already know a lot about construction. If you have no prior experience with home building, use your meetings with the GC candidates to obtain an education. Enter meetings with a general sense of what you want, a partial set of drawings, and a list of specifications that are important to you. Present this information as best you can, and encourage each candidate to ask you enough clarifying questions to create his own list of tasks, specifications, and plans.

Instruct each candidate to make his estimate as detailed as possible. The list of specifications should include the type and quality of materials, the construction standards, and itemized pricing, where possible. The detailed specifications and prices will enable you to compare estimates more easily, and, if you are over budget, to omit individual items.

Also ask each candidate to indicate all of the tasks that need to be done that he is not doing. This may seem like an odd suggestion, but there are often several tasks that a GC will not handle. You might need an engineered septic design, for example, or have landscaping needs that the GC is unable to perform. Tasks that are necessary for your house that the GC will not be taking care of should be listed as such in the estimate. This will eliminate a possible misunderstanding about who is assuming the responsibility. For each excluded task, you will need to secure the services on your own, including obtaining competitive bids and arranging for the work to be done.

Each GC candidate should also list those tasks he is not including but that might need to be done depending on what is discovered during the construction process. If, for example, high groundwater were to be discovered when the foundation hole is dug, the GC might need to raise the foundation above the water table. This would require him to bring in additional fill to backfill around the foundation. A good GC will use his professional experience to tell you when this possibility applies to your situation, and he will specify the possibility in writing. You should discuss with the GC whether or not you should set aside funds to cover any of the tasks that he excludes from his estimate.

If the GC is unwilling to include in his estimate a complete list of construction tasks, building specifications, and excluded work, you need to either seek another candidate or create the list yourself. If he is willing, ask him to commit to a deadline for completing the estimate. Then contact him halfway to the deadline to ask if he has any questions and to remind him of the agreed-upon date.

Hiring Subcontractors

If you decide to function as your own GC, you will need to repeat the same steps for each subcontractor that you would otherwise follow for selecting a GC. Since you will likely need to hire several different subcontractors, this will take some effort. Even if you have the time, few subcontractors will care to answer a lot of questions. Since they do not make as much money as a GC completing an entire job, they cannot afford the time. Also, they usually do not need to submit to a customer screening, since they mostly get hired by GCs they have worked for before, by word of mouth, or simply because they are available and return a phone call. You should still attempt to learn something about each subcontractor. Some subcontractors will resist signing a formal contract, but you should press each subcontractor you select to give you one.

Checking References

After you have completed your initial screening and are waiting for written estimates, consider taking some of the following precautionary steps. You can forgo some of these steps if a GC has been in business for a long time and is widely known to have a good reputation.

1. Obtain an insurance binder mailed directly from the GC's insurance company, since this is the only way to ensure that he has a current policy. See if he has sufficient liability and workers' compensation. Your own insurance agent can tell you what the coverage should be. If the GC or a subcontractor is not fully insured and someone gets injured, or if there is significant damage to your property, you could be liable.

2. Check with the state or local building inspector to verify that the GC is licensed and in good standing.

3. Contact the Better Business Bureau and local consumer affairs office to see if there have been complaints lodged against the GC. If there have been complaints, find out if they were resolved satisfactorily for the customer.

Questions for Customer References

Former customers can be a great source of information about a prospective GC. Here are some of the questions that you might like to have answered.

☐ Are you happy with the GC's work?

☐ Did he do what he promised in a timely manner?

☐ Did he hold to his prices? Were there any overcharges?

☐ Were there extra charges because the GC's allowances were too low to get you what you wanted and thought you were getting?

☐ How did the GC respond when you asked to make changes? Were the changes documented and fairly charged?

☐ How good was his choice of building materials?

☐ Did he change specifications from what was agreed to or expected?

☐ How good was his workmanship?

☐ Did he take care of all the little details?

☐ Did he meet his deadlines?

☐ Were there any surprises?

☐ What did you think of the GC's subcontractors?

☐ How was their workmanship?

☐ Were they on time?

☐ Did they follow up on problems?

☐ Did they work in a safe and efficient manner?

☐ Did the GC supervise his subcontractors sufficiently?

☐ Did the GC have any payment problems with suppliers or subcontractors?

☐ How well did the GC clean up and get rid of all of the trash when he was done?

☐ How well has he honored his warranty?

☐ How well did he handle stress?

☐ How well did he handle disagreements?

☐ How available was the GC during the day? during the evening?

4. Ask the state attorney general's office if there are any civil suits that have been filed against the GC.

5. Get references from the GC for his commercial bank, suppliers, and subcontractors, and then contact each one.

6. Get at least four customer references from the GC. Ask for two references who have built in the last year and two who have built two or more years ago. Ideally, all of the references will have built modular homes. Use the list of questions as a guide.

Do not reject a GC based solely on one unfavorable story, unless the reported incident is egregious. Any GC who has been in business for long is likely to have one or two unsatisfied customers, no matter how well he served them.

Assessing the Estimates

If you decide to make a change in your home plans while searching for a GC, which often happens because of suggestions made by a candidate, inform all of the candidates so they can include the changes in their estimates. If one or more of the candidates takes longer than expected to complete their estimate, you will have to decide if they may also have a difficult time keeping on schedule once they begin the project.

Once you have received all of the estimates and compared the prices, make sure that the scope of work and list of specifications are complete in each one. You cannot determine this merely by comparing the estimates. It is possible that all of the candidates unintentionally missed the same task. It is even more likely that all of them have left out the detailed specifications you need to ensure that you are getting what you want for a particular task. These missing tasks and specifications make you vulnerable to significant cost overruns or inferior products.

Missing Construction Tasks

Sometimes a GC will leave out one or two large tasks from his estimate. For example, he may include installing the basement walls and floor but leave out the bulkhead, which gives you access to the basement from outside. More often he will omit several small tasks that need to be done to complete your home. For example, he will list digging the foundation hole but not include perimeter drainage, which helps keep the basement dry. Or he might leave out stump or trash disposal, even though someone needs to take care of these. Some of the small tasks may not cost very much, but if there are several of them, the costs can add up quickly.

You might think you do not need to worry about having modular button-up tasks left out of your estimate if your modular dealer also completes the GC work. It is true that a GC with modular experience is much less likely to omit a button-up task because of ignorance of modular home construction. Even so, GCs can make honest mistakes, and some contractors are more competent at identifying the modular tasks than others. If items are left out of your contract, the GC will not be obligated to complete them without compensation. Since these additional costs are not part of your mortgage, they may require significant out-of-pocket expenditures.

Some GCs deliberately operate this way, providing estimates that they know are too low for the work that needs to be done (thus the term for this practice, *lowballing*). Once you sign an incomplete contract, the GC has you hooked for additional expenses later on. This is a frequent source of cost overruns.

Missing Details

Many GCs do a reasonable job of listing the construction tasks they intend to complete. But unless they also detail the building specifications, you will have no guarantee of what you will get when they complete the construction tasks.

For example, when a GC agrees to "install a forced-air heating system," he should specify the number of zones and whether an air filter and dehumidifier are included. There is nothing necessarily wrong with getting only one zone and no air filter or dehumidifier, but this should be your choice, not the GC's secret.

Many GCs include the following wording in their contracts: "Build an attached 24-foot × 24-foot two-car garage." This assures you that you will get a garage, but it does not indicate if it will include one or two overhead doors, windows, a passage door to the backyard, or a garage-door opener. Without these details being spelled out in writing, you know only that you are getting a garage, not that you are getting the garage you want.

Does your contract cover the garage the way you really want it? Both garages are two-car, but only one has double doors, a side door, and a window.

Some GCs will give a verbal description of the site-built structures they plan to construct without providing specific plans. Signed, scaled, and complete drawings are the only way you can be sure of getting what you expect, and you should insist that the GC supply them.

When a GC knowingly sells his construction services by leaving out the building specifications, he may be engaging in a form of "bait and switch." The GC gets you hooked by offering prices you can afford without mentioning what you will get for your money. Once you have signed his contract, it is OK with him if you discover that you need to pay him more money to switch to the better specifications you want. This is another frequent source of cost overruns.

Low Allowances

Like modular dealers, general contractors do not always finalize the specifications for a particular task. Sometimes it is because their customers have not made a final decision. Other times it is because the costs cannot be known in advance, such as the cost of drilling a well when there is no good way to anticipate the final depth. In either of these circumstances, the GC allocates a sum of money for the task instead of giving you a fixed price. This allowance can be for the cost of the materials, the labor, or both. The final cost is determined either by what you subsequently select or by the actual cost to complete the task.

Although this is a reasonable practice, you should limit the allowances to items that cannot be priced in advance, such as a well. Otherwise, you may find that you are not able get what you want without spending a lot more money than has been budgeted and approved by your lender. The best way to avoid these cost overruns is to take the time to shop for every item you can before signing a contract.

Evaluating Low Estimates

It is not uncommon for one GC to offer an estimate that is significantly lower than the others for what seems to be the same scope of work and specifications. While it might be tempting to jump at the low estimate, it could be a big mistake. GCs often cut their costs by working with the least expensive subcontractors available at the time. Although this can be a good business practice, more often than not it will cause significant problems.

Some subcontractors reduce their prices by using less-expensive and lower-quality materials. Or they work for low rates but keep many jobs going at once to compensate, which can create problems with scheduling and workmanship. If such a subcontractor is offered a better-paying job before he finishes his work on your house, he may rush to finish or simply leave the work undone. When the GC hires subcontractors at the prevailing rates, this scenario is less likely to occur.

A subcontractor who works for a low rate is not going to be enthusiastic about standing behind his work after he has been paid. He may believe that since he gave you more than you paid for, he is justified in cutting corners when it comes to honoring a warranty. A good subcontractor includes a small amount of money in his fee to cover return visits to your home so that he can get the job right. When he is paid a fair wage, he is more willing to come back as many times as needed to honor his contract.

General contractors and subcontractors who avoid charging low fees are more likely to stay in business. They have earned a good reputation and are likely to work hard to maintain it. Their ability to earn a consistent profit means that they can afford to provide customers with the warranty service they deserve.

Another possible explanation for a low estimate from a GC is that he made an

The General Contractor's Contract

A contract between the customer and the general contractor should cover the following topics. Be sure to review each of them with an attorney, as they may provide you with some additional protection.

CONTRACT CONTINGENCIES, CANCELLATION, AND AUTHORIZATION TO BUILD

You might want to commit to a GC even before you know for sure you can build your new home. This would be true if you first have to sell your existing home, obtain financing, purchase a building lot, or obtain a building permit and all requisite approvals (board of health, planning board, zoning board). In this case, your attorney should add contract language that allows you to get started with the preliminary steps without committing you to building the home until you have resolved any contingencies specified in the contract. The clause should state how much money the GC will keep if you are unable to meet all of the contingencies and so are forced to cancel. The contract should also create a mechanism that tells the GC when you would like him to begin. If you have added contingencies to your contract, this mechanism will also indicate that you have either met or waived your contigencies so that you can no longer cancel the contract.

CHANGE ORDERS

The contract should spell out the GC's change-order policy. This should indicate when change orders can and cannot be done, who can authorize the changes, what administrative fees apply, how they are to be paid, and how much of a delay they will cause. The contract should include a statement that all changes, whether or not they increase the cost, must be submitted and approved in writing by both parties.

PRICE ADJUSTMENTS

Since the cost of building materials could increase between the time you sign the contract and the time the GC begins, the contract should state when you are responsible for covering any increase.

PAYMENT TERMS

The contract should spell out how the GC expects to be paid. It should include the deposit requirements and a proposed disbursement schedule. It should also specify how much money you can hold back until the GC completes all of the punch-list items. A fair agreement is that you can withhold 150 percent of the value of any outstanding work.

CONSTRUCTION SCHEDULE

The contract should provide a construction schedule.

INSURANCE

The contract should say what insurance coverage each of you is obligated to provide. The customer is usually responsible for obtaining a builder's risk policy or its equivalent, which provides coverage for theft of building materials and supplies. The GC is responsible for providing general liability insurance and, if he has employees, workers' compensation.

WARRANTY

The contract should declare what is covered and what is excluded for both materials and workmanship. Ideally it should specify the list of standards you and the GC can turn to if there is a disagreement. One popular set of standards is set forth in the *Guidelines for Professional Builders and Remodelers,* published by the National Association of Home Builders.

The warranty should also state the length of coverage. If an extended warranty is included, it should indicate the program, and the GC should give you a copy of the supporting documents. See chapter 10 for more on warranties.

honest mistake. It is more common for a GC to make a substantial mistake than a modular dealer. A dealer may need to add only three prices together — the costs of the manufacturer, crane, and set crew — to determine the cost of his homes. A GC, on the other hand, may have to identify a few hundred costs for his labor and materials. If you receive one very low bid, ask the GC to explain the discrepancy. If it is due to an honest mistake, give him the chance to submit another estimate. You might be tempted to take advantage of the GC and lock him into his low bid, but that approach is likely get both of you in trouble.

Turning the Estimate into an Agreement

When you have completed your research and received all of the written estimates of the cost of your home, you will need to make a decision. Evaluate each candidate's answers to your questions, as well as his reputation, personality, standards of workmanship, and written estimate.

After you select a GC, ask him to turn his estimate into a formal agreement. This will require that he add contract legalese to his detailed list of construction tasks, building specifications, and itemized prices. Review the contract with your attorney before signing it. Make sure that it covers each of the topics listed in the sidebar.

As soon as an agreement has been reached, sign the contract and give the GC a deposit so that he can initiate the first draft of preliminary drawings and line up his subcontractors.

Keeping the GC Informed

When you first present the GC with your modular home plans and specifications, you will be giving him the most current information you have. Some of this information, such as the color of the living room carpet, will have no impact on his responsibilities. Other types of information, such as the type of heat you want, will have a direct effect. You will likely make several changes in both types of information while negotiating with your modular dealer. To avoid costly mistakes, the GC must be kept informed of all changes that relate to his work. This will happen automatically if the GC is also your modular dealer. It is more difficult if he is a separate company. The story on page 186 relays some of the problems that can occur when the GC is left in the dark.

Speaking from Experience

MANY CUSTOMERS FORGET to inform the GC of important changes to their specifications and plans. A few expensive examples from my experience include adding a walk-out bay that required a full foundation, moving the fireplace or a slider to where the bulkhead was going, and reversing the orientation of the house. Serious and costly mistakes are more likely to occur with changes made later in the process, since it is easy to let down your guard while catching your breath.

Once my customers forgot to tell their GC that they reversed the house plan so that left was right and right was left. My customers had a good reason for the reversal, since they wanted to orient the house so that the living area received sunlight rather than the garage. But they had not yet made that decision when they gave their three GC candidates copies of their proposed plans. They then ran into several time-consuming problems with getting a building permit, and did not finally select a GC until six months later.

When the GC was given the revised plans, he asked the customers if the plans had changed much, and the customers said that they had changed the kitchen around, but that was about it. The customers completely forgot that they once had drawn the plan with the opposite orientation.

You would think that the GC would have looked at the plans for himself, but here's what happened. He told his project supervisor to schedule the excavation and foundation subcontractors; they had been awarded the job many months earlier. He then went on vacation the week the work was done. The two subcontractors showed up with the old copies of the drawings they had received when bidding for the work; the GC's supervisor showed up with none. The foundation was poured, incorrectly, with the foundation for the garage on the left rather than the right, and the foundation for the chimney and walk-out bay on the right rather than the left. The lally-column pads were also spaced incorrectly. My customers discovered the mistake when they visited the site while the foundation was being backfilled. The GC blamed the customers for not telling him about the reversal and the customers blamed the GC for not looking at the revised plans. In the end, the GC relented, as he should have, but for the remainder of the project they both suffered through a very testy relationship.

It is critical that you provide your GC with any changes to your modular plans and specifications. Better yet, provide him with a copy of the final draft of your modular plans and specifications, and have a meeting with the GC to review them.

6

Finding and Preparing a Building Lot

WHEN SHOPPING FOR A building lot, you will likely consider its location, purchase price, size, and shape — with good reason. Where you live, the size of your budget, and how much land you have and can effectively use will impact your selection.

The choice of location is likely to be determined by a variety of considerations. The type of neighborhood, tax rate, length of commute to work, and quality of the school system are important. Proximity to family and friends, the distance to shopping centers, and access to recreational activities might also matter. Privacy, traffic noise, and what you can do with the lot (such as put in a swimming pool) may be paramount concerns.

The location of a building lot is seldom in itself of special consequence for a modular home, but the route to the site, as well as its size and shape, can pose delivery difficulties. Also, building lots sometimes come with covenants, deed restrictions, or easements that can restrict what you are allowed to build. For those on a budget, the price of the building lot affects how much money is left over to build a house.

Land costs in many areas have risen appreciably in recent years. In some places, the rise has been astronomical. Buying a piece of property is one thing, but turning it into a usable building lot can be a different matter entirely. There are many factors in developing land that can add thousands of dollars to the cost of building a house. If you are unprepared for these additional costs, they can make a complete mess of your budget.

This chapter helps you understand what goes into making a raw piece of land a viable building lot. Whether or not you already have your land, this chapter will help you assess the total costs of preparing a site, which in turn will make it easier to decide how much you can afford to spend on the house itself. See pages 196–197 for a more detailed discussion of the types of building lots that pose challenges for modular homes and what you can do to address them.

Finding a Building Lot

There are several ways to approach the search for a building lot. You will not be able to take advantage of every lot listed for sale, since many of them are owned by builders who will sell only if they can build your home. If you intend to build a modular home with a dealer and GC of your choice, you need to find a land seller who allows you to do so.

Real estate agents can identify land listed for sale through local agencies. Although this may generate more leads than any other source, most agents prefer to sell homes, not land. They make more money selling a completed home than undeveloped land. Consequently, you may need to talk to several agents.

Another source is a for-sale-by-owner listing in the local newspapers. Selling land is easier than selling a house, so people are more willing to sell on their own. Bypassing a realtor's commission can save you money. Sometimes you can find a lot by driving around the towns you are considering. If you see an inviting property, check with the local assessor's office to find out who owns it. Ask local officials, building inspectors, and surveyors if they know of any land for sale or landowners who might consider selling some land. The modular dealers and GCs you are considering may also have land for sale or they may know where you can find a good lot.

With available building lots scarce in some communities, builders and retail customers alike are turning to older homes.

Instead of renovating them, however, they are tearing them down and building brand-new homes on the site. The financial rationale in these situations is that the homes themselves are worth less than the land they reside on. Modular homes work nicely in these settings as long as the modules and crane can access the site.

When you are shopping for a building lot, it is wise to have the help of an experienced general contractor. He can help determine what you can and cannot do with a particular property and what types of homes would fit best. Most important, he can estimate your site-related construction expenses. Knowing these costs is the only way you can develop a budget for the entire project. Having a budget is a necessity if you are to figure out how much money is available for your modular home.

The best way for the GC to assess a site, if he has any doubts about it, is to conduct some exploratory digging. By digging a few test holes, the GC is able to determine if there is a high water table, rock, ledge, clay, or other potential difficulties. Although the GC cannot determine from a couple of test holes what you will find in other areas of the

When Making an Offer-to-Purchase for a Building Lot

A written offer to purchase a building lot should include language that allows you, the buyers, to build the home you want to build, where you want to build it, and for a price acceptable to you. You should have your attorney include any of the following contingencies that will provide you with the protection you need. If you are not able to meet one of these contingencies, the agreement should allow you to withdraw your offer and receive a full deposit refund.

If the seller rejects a proposed contingency, he might accept a less demanding one. For example, the seller might not agree to a contingency that allows you to first obtain a building permit, since this might take too much time. But he might agree to make the offer contingent on the property passing a percolation test or being approved by a wetlands board. If your research indicates that these are the only potential obstacles to obtaining a permit, your attorney might advise you to submit the offer with these contingencies in place.

The contingencies will only help you, however, if you take the appropriate actions they allow you to do. For example, when shopping for a dealer and a GC, you need to make sure their estimates are for the home you want, built to the correct specifications. This means that the dealer and GC must do their homework, as well. For example, if the property has town water and sewer at the edge of the property, the GC must determine if the hookups can be made inexpensively or require expensive excavation into the street.

Suggested contingencies are:

• The buyers can secure sufficient financing for the home they want to build.

• The buyers can review and approve any easements, deed restrictions, covenants, floodplain designations, or wetland restrictions.

• The buyers can review and approve the applicable zoning regulations.

• The property has a registered survey, the boundary stakes are in place, and the boundaries are as represented by the seller.

• The buyers can obtain a building permit for the home they want to build.

• The buyers can dig some exploratory holes on the property to assess and approve subsoil conditions.

• The buyers can obtain an acceptable written cost estimate from a builder of their choice to build the home they want.

lot, he can at least reduce the chances that you will run into expensive surprises. This job, which requires the use of a backhoe, costs a few hundred dollars, but it is worth it for your peace of mind.

Sellers of land seldom refuse this common request, since the digging usually takes place where the foundation hole is likely to be. When the inspection is completed, the hole is refilled. But most sellers will not allow you to dig on their property unless you also make a purchase offer with a small deposit. Include two clauses in your offer. One allows you to have the property inspected by someone of your choosing. The second grants the inspector permission to complete some exploratory digging.

If the exploration reveals the need for additional expenses, you can use this information either to negotiate a better price or to begin looking for another piece of land. If you still want the land and cannot nego-

tiate a better price, at least you can plan a more realistic budget for the entire project and avoid cost overruns. If you do not thoroughly inspect the land and later discover an expensive problem, such as ledge or a high water table, you might be forced to delay construction until you can raise additional money.

In addition to including an inspection contingency in your purchase offer, you should include other contingencies based upon the information discussed in this chapter. For example, you may want to make your offer contingent on being able to build the type of home you want. This requires you to investigate whether there are any deed restrictions, covenants, easements, wetland restrictions, or zoning regulations that could limit your choices. You may want to hedge your offer until you can determine the costs of the home you hope to build and the general-contracting work

A couple of test holes on a building lot will tell you a lot about what the GC will need to do to build a home on the property.

required to finish it. If you need a loan to purchase the land, you should add that contingency as well.

Fixed and Variable Construction Costs

When comparing one building lot with another, there are a number of construction costs that are relatively fixed and predictable. For example, the costs to complete the button-up work after your home is set on the foundation are seldom affected by the features of the land. A competent modular dealer can give you a good idea of these costs without knowing a great deal about your land.

There are a number of other costs, however, that vary significantly from lot to lot. For example, the expense of excavating the foundation and bringing water and sewage capability to a home can vary widely, depending on the type of lot. These variable costs can add up to some surprising numbers from a contractor. They will only be surprising, however, if you are unaware of what is involved in developing raw land into a building lot. The goal of this chapter is to eliminate that surprise.

Clear Title

Before purchasing any land, make sure that you can have clear title to the property. A lender will demand that you hire an attorney to complete a title search before it lends you the money. This is a smart thing to do even if you are paying cash for the land. If you own the land when you apply for a construction loan, your attorney must verify that you have clear title before the lender will close on the loan.

A title search involves a careful examination of the public land records. Although it is usually a routine formality, your attorney might uncover a title defect that cannot be quickly removed. Some common causes of title defects include the following:

- A mortgage that was never discharged
- A lien for unpaid estate, inheritance, income or gift taxes
- A mechanic's lien by a contractor who was not paid for completing work on the property
- A misinterpretation of a will or an undisclosed heir to the property
- A mistake in recording the documents

Title flaws can appear when the land has not changed hands for many years. They are especially common when the land has been transferred within a family across generations, since these transactions are often completed without an attorney's involvement.

Speaking from Experience

IN OUR FIRST YEAR of selling modular homes, my salesperson sold a ranch to a woman who was going to build a home on a lot her mother was giving her. When her attorney did a title search the day before the closing, he found a title flaw with the property. It turned out that when her mother had inherited the land from her own mother (my customer's grandmother), the grandmother had actually given the land to both of her children, not just her daughter. So the land was really owned by my customer's mother and her brother. No one had paid attention to this because the brother had died many years ago, and he didn't have any children or a will stating what should happen to his share of the land. My customer's mother falsely assumed she automatically inherited her brother's share of the property. My customer had to cancel purchasing her modular home when she was told it could take over two years to resolve this matter in court. A couple of years later I noticed in the local paper that she had taken care of the problem. Since she had already bought an existing home, however, she sold the lot.

Many title flaws are correctable, but it can take a few years and a hefty bill for legal costs, as the story illustrates. It is often a complicated matter when an attorney has to mediate among family members scattered far and wide across the country.

Your attorney should also secure title insurance. This insurance protects the owner of the policy from a faulty title search, which can happen when the public records contain inaccurate information or are missing some of the legally applicable information. For example, a title flaw may be missed when a lien is issued for an unpaid inheritance tax but is not recorded with the deed, or when the deed is executed under an invalid or expired power of attorney. Title insurance in these situations pays for the defense of the title and covers any losses. If you are using a construction loan to pay for your new home, your lender will require you to purchase a loan policy on its behalf. This insurance protects the lender's interets, however, not yours. Whether or not a lender is involved, ask your attorney to obtain an owner's policy, which would reimburse you for losses, including legal fees and lost equity.

Land-Use Restrictions

One factor that is often overlooked by a customer who has never built a new home before is the cost of abiding by any covenants and deed restrictions that apply to the land. Subdivisions almost always include specific restrictions. Complying with them can be expensive, as when there is a requirement that you build a home that is at least 2,500 square feet, cover it with cedar siding, add a two-car garage with a paved driveway, and landscape it with 30 or more shrubs.

Easements and rights-of-way also limit what you can do with your property. For example, a utility company could have the right to access its pipes running across the property, or a farmer could have the right to cross the property with a tractor to get to his field. Some properties carry "view easements," which protect the views of owners and abutters. This can work to your favor, if it protects your view, or to your disadvantage, if it prevents you from locating your home where you want to because it will obstruct your neighbor's view.

Review with your attorney all land-use restrictions that apply to the property before buying it. Do not rely on the word of the seller alone, since the previous owner could have misled her. This work often requires the assistance of a surveyor.

A view easement protects a clear view of the distant mountains for the owners of lot 1 by limiting where the owners of lot 2 can build their home.

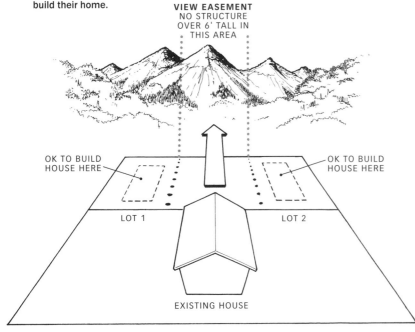

VIEW EASEMENT
NO STRUCTURE OVER 6' TALL IN THIS AREA

OK TO BUILD HOUSE HERE

OK TO BUILD HOUSE HERE

LOT 1   LOT 2

EXISTING HOUSE

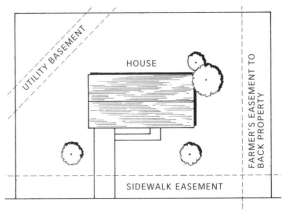

Nothing new can be built in any of the three easement areas.

Zoning Requirements

Most towns have zoning requirements that specify what you can and cannot build. Common zoning restrictions include:

• The minimum lot size necessary for a legal building lot (this is specified in terms of the minimum frontage on the road and the minimum square footage of the lot)

• How close you can build to the surrounding properties (this is specified in terms of the distance that all structures must be from the road and from each abutting property)

• What you can build and what you can use the structure for (for example, whether you can build for residential or commercial use, the number of living units you can build for residential use, and what other structures are allowed)

Zoning requirements affect several things you might want to do, such as where you can locate your home on the lot, how big the house can be, and how tall the roof can be. A zoning regulation, for example, may require you to get a special permit to raise horses, or it may prohibit you from raising horses. If you are not allowed to use your property as you want, you can apply for a variance to the relevant regulation. Obtaining a variance can take a couple of months, or longer, and can cost additional money. Also, many communities are reluctant to grant variances for fear they will undermine the intent of the regulation. The best way to obtain a variance is to demonstrate that the regulation does not truly apply to your situation or that it creates a hardship for you. You may benefit from hiring a local attorney who knows the regulations as well as the local officials, although some local officials might be more responsive to a personal plea from you.

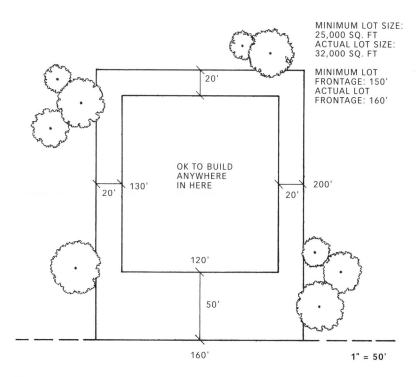

The zoning regulations in this town require a minimum building lot of 25,000 square feet and 150' of frontage. The regulations also require the home to be located at least 20' away from the side and rear property lines and 50' from the road.

Building Codes

State and local building codes can dictate the minimum specifications that must be met when building a home. They can also specify if and when a construction license is required. States differ in the codes they enforce, and some towns enforce stricter codes than neighboring towns in the same state. Ask your local building department if it has any requirements that differ from the state code.

The modular manufacturer is required by law to construct its homes to meet your state's building codes. If your state or local building department enforces any other codes in addition to the state's codes, the manufacturer will comply with the stricter codes as long as you order the upgrades from the dealer.

Environmental Considerations

If you own a site on which pollution is found, you could be responsible for cleaning it up, even if a prior owner caused it. Before you buy land, make an effort to ensure that the site is free of pollutants. Use a title search to see if any businesses have used the property in the past (especially a service station or chemical company). Contact the state agency that monitors known hazardous waste sites. At the very least, you should consider checking the site very closely for illegal dumping, which may require some exploratory digging with a backhoe. If anything suspicious is discovered, you should have the soil tested for toxic substances.

Floodplain

Find out if a lot is in a designated floodplain before purchasing it. Having a lot in a floodplain does not necessarily prohibit you from building on it, but it may influence where you locate the house. It may also induce you to raise the foundation and install the heating system and electrical panel box on the first floor rather than in the basement. If you attempt to purchase flood insurance, which the lender may require you to do, the policy could require these and other precautionary steps.

Wetlands

Before you begin any work on your land, you need to know if it contains wetlands.

Speaking from Experience

I DISCOVERED JUST HOW difficult it can be to judge whether a lot has "wetlands" when customers of mine, a young married couple, bought a piece of open farmland that was covered by what appeared to be grasses. My customers bought the lot in midsummer with plans to build the following year. The property was on a main road and came with municipal water and sewer, so it did not require septic approval and no one thought it necessary to do any exploratory digging. Because my customers had cash, they closed on the property as soon as their attorney completed the title search.

When they applied for a building permit the following spring, the building inspector told them that they needed to get the approval of the town conservation committee. The committee scheduled a meeting at the property in early April. To my customers' surprise, there was standing water several places on the property. Even more surprising was the committee's determination that the property was a protected wetlands and could not be built on. The committee emphasized that even if there was no water on the property, the vegetation made it a protected wetlands. In the end, the only use my customers had for the property was to graze animals. Given that they paid $45,000 for it, this was a devastating mistake.

Unless you will be happy using a piece of land to raise cows rather than build a home, find out if you really can build on a property before purchasing it. Most important, do not take it upon yourself to make this determination. Consult with the proper local authorities.

Failing to do so can be a very expensive oversight, as shown in the story. The determination of what constitutes wetlands is made by the federal, state, and local agencies specifically charged with the responsibility. While lakes, rivers, and marshes are clearly wetlands, other areas that appear to be dry can be designated as wetlands if they contain particular types of vegetation and soil.

The responsible agency can require you to file a special application before it makes an official determination about your site. Whenever public hearings are required and special consultants must be hired, the process can take weeks or months to complete. If the agency determines that your land contains wetlands but still allows you to build, it will specify where you can and cannot build and what precautions you must take to protect the wetlands during construction. Some of these precautions can add appreciably to the time and expense of preparing the site for construction.

Electric Power

For the electric company to bring overhead power to your home from an existing utility pole, the pole must be relatively close to the house, usually within 150 to 200 feet. Otherwise, additional poles or underground cables might be required. Additional poles may also be required if the electric company cannot readily run cables from the existing poles to your house, as is often the case on a steep incline. If more poles and cable are needed, talk with the electric company to determine who covers the costs, which can be substantial. Since this operation can potentially involve thousands of dollars, do not take the word of a seller or realtor, unless she is able to document her assertions in writing.

If you decide to install underground electric service, the distance from the pole or transformer to your home determines the cost. Depending on where you live, the cost could be high. In some regions, a few hundred feet cost a few thousand dollars.

Surveyed Boundaries

Since the zoning regulations limit how close you can build your home to the road, neighboring property, and any easements, it is often critical that you know exactly where the boundaries are. Zoning boards have been known to make a person tear down construction that intrudes only a small distance over the required setback line. Many building inspectors require a survey before they will issue a building permit, and more and more banks insist on a survey before they will finance either the purchase of land or the construction of a new home. You can hire a surveyor yourself or, if you have not yet purchased the land, ask the seller to complete the survey.

Sometimes a property has been surveyed but there are no boundary markers. This makes it difficult to determine where the actual property lines are and, therefore, where you can and cannot build. If you have not yet purchased the land, insist that the seller's surveyor clearly mark the location of each boundary marker. If you already own the land, have a surveyor install the markers before you make a final decision about where to locate your house and garage. If you are building very close to a setback line, ask the surveyor to mark the exact location where you intend to install the foundation before the GC begins his work. You might even ask the surveyor to return to your site after the foundation footings are poured to verify that your GC has not made a mistake. If you discover that the footings are in the wrong location, the cost to redo them in the correct location will be more manageable than if you find the mistake after the foundation is installed.

Choosing the Best Location

When you first start considering where to locate your home on a building lot, you will likely think about how you want it to look from the street, which parts you want to be bathed by sunlight, and what you intend to do with the yard. To meet these personal goals, you need to take into account the topography of the lot, or, in other words, the "lay of the land."

The shape, size, and existing slopes, or "grades," of the lot will influence where best to place a home. Other factors include trees and open spaces, streams and ponds, and existing stone walls and driveways. The lot's location relative to the surrounding properties will also affect your decision.

A property's topography can play a determining role in the type of home you can build. For example, if you want to build a raised ranch with a drive-under garage and a finished family room in the basement, focus your search on appropriately sloped properties. On the other hand, if you intend to build a universal-design home, you may want to avoid a steep lot, although you could also take advantage of a front-to-back slope by creating an accessible walk-out basement in the rear.

You may not be able to locate your home exactly where you want because the topography would make construction too expen-

sive. For example, you might hope to build on a knoll located in the center of the lot to maximize a great view. If the knoll is located on rock or ledge, however, you would face the choice of spending a lot more than expected to place a foundation on the preferred spot or moving the house to a less challenging location. In short, your choice of house location can significantly impact the cost of construction.

Compromises are almost inevitable in locating a house. For example, you might want to both maximize your privacy and provide your young children with a large play area. A rough, wooded area on the property would permit privacy but be a less-than-ideal play area; a flat, open area would be great for kids but not for privacy.

It is not easy to learn about a lot's features when it is heavily wooded or covered in thick brush. In many regions, you can get the most accurate picture in the early spring before the vegetation begins to grow or in the fall and winter after the vegetation dies off. You can also learn about the surface and subsurface water conditions by visiting the property when the water levels are highest.

If you need a septic system, its location can influence where you can place a house as well as the total cost of developing the land. The location of a septic system is primarily determined by a percolation ("perc")

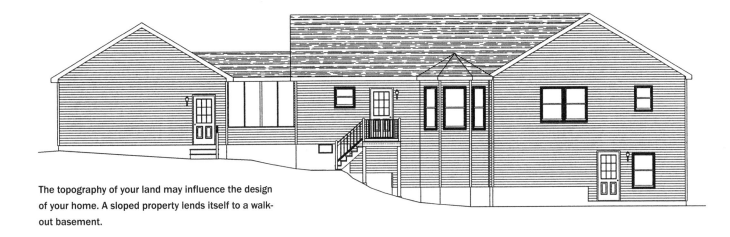

The topography of your land may influence the design of your home. A sloped property lends itself to a walk-out basement.

test, which indicates the composition and permeability of the soil and the depth of the water table. When possible, you should locate the septic system reasonably close to the house, since the farther apart they are, the more expensive the installation. You should also locate the house so that the effluent can use gravity to feed the septic tank. Otherwise, you need to install a pumping system, which adds to the cost. If you do not like the proposed location of the septic system, you can complete perc tests at other locations on the lot with the permission of the seller. You may need to wait to complete the tests, however, since some states and towns allow perc tests only at certain times of the year. In addition, you may need to pay for the cost of the additional tests, and you could find that they fail at the other locations.

If a well is needed along with a septic system, there are additional limitations on where the house can be placed. The well must be located a minimum distance from your septic system as well as the septic systems on your neighbors' properties. The septic and well constraints on a property can significantly limit where a house can be located.

The cost of installing municipal sewer and water lines depends on the length of the lines and whether the sewer line is up- or downslope from the house. In many situations, the municipality alone determines where you can tap into the public utilities. Although you can save money by locating the house closer to this point, choosing such a location may require you to compromise other needs.

The length of the driveway impacts construction costs and is in turn affected by the lay of the land and the location of the house. If your home is set back far from the road, and if the path between the road and your home makes it difficult to build a usable driveway, the cost rises accordingly.

Clearing the Land

Some lots are densely wooded with trees and brush, while others are completely cleared. The cost of cutting down trees, chipping brush and limbs, removing logs, and digging up and burying or disposing of stumps can be substantial. While lumber companies pay for the chance to cut some kinds of trees, most building lots do not have enough of the right type and size to eliminate tree-clearing expenses.

The cost of clearing the land increases if you need to remove trees bordering the road and the land is on a state highway or designated "scenic route." In such cases, you must seek approval from the state, in the form of a permit, before you can cut down the trees. You might even have to compensate the community for any trees you cut down.

Sometimes building a modular home means having to remove more trees than would be necessary with another type of construction. This is particularly true when a tree or two obstructs the delivery or set of your home. In most circumstances, however, modular construction does not require you to remove the extra trees where your house will be located.

Excavation, Trenching, and Grading

After clearing the land, you need to excavate a hole for the foundation. The cost of excavation varies considerably from lot to lot. The difference depends largely on whether the land has ledge, rock, big boulders, hardpan, clay, or other difficult soil and subsurface conditions. Extensive deposits of ledge and rock require blasting, which is expensive, and large boulders and hardpan may require more powerful and expensive excavation equipment. Clay is very difficult to work with, particularly when wet.

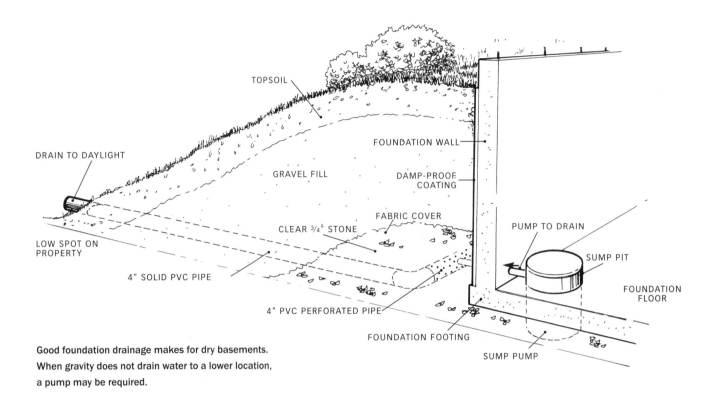

DRAIN TO DAYLIGHT

LOW SPOT ON PROPERTY

4" SOLID PVC PIPE

4" PVC PERFORATED PIPE

CLEAR 3/4" STONE

FABRIC COVER

GRAVEL FILL

TOPSOIL

FOUNDATION WALL

DAMP-PROOF COATING

PUMP TO DRAIN

SUMP PIT

FOUNDATION FLOOR

FOUNDATION FOOTING

SUMP PUMP

Good foundation drainage makes for dry basements.
When gravity does not drain water to a lower location,
a pump may be required.

Large boulders, hardpan, clay, and the debris left over from blasting ledge are usually unfit for backfilling around the foundation or for bringing the grades around the house to the appropriate levels. When these conditions are present, additional fill is almost always required. Buying, delivering, and spreading additional fill can add significantly to excavation costs.

The best way to ensure a dry basement is to install a drainage system under and around the foundation. The type of system depends on the amount of water that is anticipated to surround the foundation during the wettest months of the year. All systems work by drawing water away from the foundation. This is facilitated by putting gravel and stone under the slab. The best systems let gravity drain the water to a lower point, which could be a town's storm drainage system or a lower elevation on the property, assuming this does not create other problems. When gravity does not work, a pump may be required.

In extremely wet areas, the best remedy often requires raising the foundation by digging a shallower hole. Additional fill is then brought to the site to raise the grade when backfilling around the foundation, which would otherwise sit too high out of the ground. Needless to say, as the drainage system becomes more extensive and the amount of fill increases, the cost of building the foundation increases.

Some communities require more extensive drainage systems even when the soil conditions would not seem to call for it. Most state building codes give a lot of discretion to the local building inspector to determine minimum excavation and foundation standards. The only way to know if you might incur some additional expenses is to speak with the building inspector.

When installing a large foundation, you may need to remove some excess fill from the property. While a large site can absorb extra soil, a small one may be able to do so only by raising the grade, which is often unacceptable. In this case, you must add fill removal to the budget.

Excavation work is also required to connect your home to underground municipal water and sewer lines or a well and septic system. Trenches are required if you are running underground utility lines for electrical, telephone, or cable services.

Creating the appropriate grades on a lot is critical to minimizing water problems, making the site usable, and keeping the driveway accessible. Although you must direct water away from your home, you need to do so without causing problems for the neighbors. You also should take into account how water from neighboring properties flows onto your property.

Additional grading may be needed to make the site accessible for the modular transporters and the crane on delivery and set day. If the lot has many rough and uneven hills, the task can be more difficult for the excavator and thus more expensive. If additional fill is required to complete the grading, your costs will also increase.

Municipal Water and Sewer

Some lots come with municipal water and sewer lines already available, often by the side of the road. Others may have only one of these available. Many lots, particularly in rural areas, have neither. The availability of municipal water and sewer connections typically reduces the cost of developing the land. This is because without them you must drill a well and install a septic system, both of which tend to cost more than their municipal alternatives. This is not always the case, however, since some towns have an expensive hookup fee. In addition, towns sometimes require a performance bond, which is a sum of money the customer or GC gives to the municipality until the work is completed to its satisfaction.

Municipal water- and sewer-hookup costs are affected by the distance of the home from the road and the type of land that must be crossed. Land that is heavily wooded may require additional tree clearing. Land that is spiked with rock and ledge may require additional blasting. The costs can also be affected by several other factors, as noted in the list below.

Municipal-Utility Hookups

Customers often underestimate what is involved in hooking up to municipal water and sewer. When the hookups are in a new subdivision, the task is usually easy and uninvolved. But if you are connecting to services that have been in place for many years, you could run into any of the following situations:

• The location of the main lines provided by the town is incorrect, and the excavator digs in the wrong place.

• The main lines are adjacent to a more shallow utility, such as underground electric or gas, or to a surface improvement that prohibits the use of an open trench, requiring the trench to be shored up.

• The main lines are in poor working condition and additional work is required to use them.

• The length of the on-site trench and the connecting pipe is substantial.

• The depth of the main lines is greater than 7 feet, which could mean the trench walls have to be reinforced with steel plates to protect against the possibility of collapse while the crew is installing the pipes.

• Ledge and high groundwater are discovered during the digging.

• The connection cannot be made on the property and instead requires cutting the road, which must then be patched and which could require hiring a police officer to direct traffic.

• The sewer line is higher than the height of the waste line from your home, which forces you to pay for a pump.

• The water line has low pressure at your property, compelling you to get a storage tank and pump.

Septic Systems

If the property does not have public sewer, you need to install a private septic system. A septic system is built according to a set of engineered plans, called a septic design, drawn by a licensed sanitarian or engineer. The design is itself based upon the results of a perc test, which measures permeability by counting the average number of minutes per inch it takes water to percolate through the subsoil in a test pit.

When buying land that requires a septic system, ask the seller for a copy of the perc test results, which you should verify have not expired. Should a prospective lot not yet have a valid perc test, do not purchase it until a test is completed, unless you have compelling evidence that it will pass. Otherwise, you may discover you own a lot that you cannot build on. The expense for a perc test is almost always borne by the seller.

Installation costs for septic systems can vary significantly, depending on the perc rates and engineered designs. If a septic design requires many truckloads of additional fill, the costs can increase by thousands of dollars. Fortunately, an experienced subcontractor can usually give you a reasonable estimate of the costs before you buy the land.

When asking for an estimate, it is best to give the installer the approved, engineered septic design for the land, since it dictates the exact system that must be installed. To obtain an approved design, however, you might have to commission a licensed professional to draw one before you decide to buy the land, since most sellers do not have the design done before they sell the lot. That is because the engineer cannot complete the drawings until he knows how many bedrooms the home will contain, which the seller and engineer cannot know in advance. Waiting for a septic design might take a few weeks, since there is a shortage of licensed septic designers in many communities. There might be an even longer delay if you

Sample schematic for a perc test. A percolation test must be done whenever a septic system is needed.

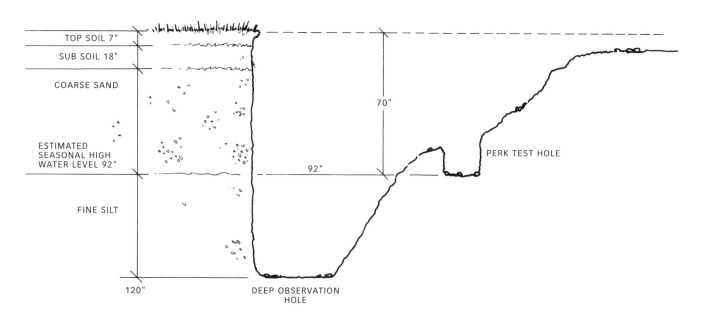

TOP SOIL 7"
SUB SOIL 18"
COARSE SAND
ESTIMATED SEASONAL HIGH WATER LEVEL 92"
FINE SILT
70"
92"
PERK TEST HOLE
120"
DEEP OBSERVATION HOLE

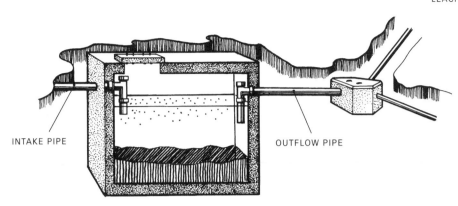

INTAKE PIPE

LEACH FIELD

OUTFLOW PIPE

SEPTIC TANK

Typical layout for a septic system. The septic design is determined by a licensed sanitarian or engineer, based on the results of a perc test.

decide to wait for the community to approve the septic system to be sure it will be allowed. The typical board of health in small towns serves part time and is overworked.

When a septic design is unavailable, an experienced septic installer can often use the data from the perc test to estimate the cost of a septic system on the property. He looks closely at the composition and permeability of the soil and the depth of the water table. Poor permeability on a perc test is detected when the water percolates too slowly through the soil. When you have alternatives, avoid land that has a perc test indicating poor permeability, high concentrations of clay, hardpan, or ledge, or a high water table, even if the test passes the board of health's requirements. If these factors are present elsewhere on the lot, which is likely, expect significantly higher costs for the excavation of the foundation hole and installation of the driveway as well as the septic system. On the other hand, sometimes a seller will discount a lot with a very high perc rate more than enough to justify the additional installation expense.

A good perc rate is a few minutes per inch, meaning that a 5-minute perc rate is better than a 20-minute perc rate. But perc rates can also be too quick. Some states consider a perc rate of less than 1 minute per inch too rapid. They contend that the effluent passes so quickly through the ground that it is impossible for the soil to thoroughly "cleanse" it before it mixes with groundwater. Communities often require a septic design for these conditions that retards the rate of discharge into the groundwater. The resulting septic systems cost more than standard designs.

Drilling a Well

It is almost impossible to estimate how much it will cost to drill a well. Many people mistakenly claim that if you know how your new neighbors fared with their well, you can have a good idea of what you are likely to find with yours. That is often not true. Even an experienced driller cannot predict with certainty how many feet he will have to drill on your property before he reaches sufficient water, nor can he tell how many feet of metal casing he will use, since this will depend on when he hits bedrock. In addition, he cannot foresee how much

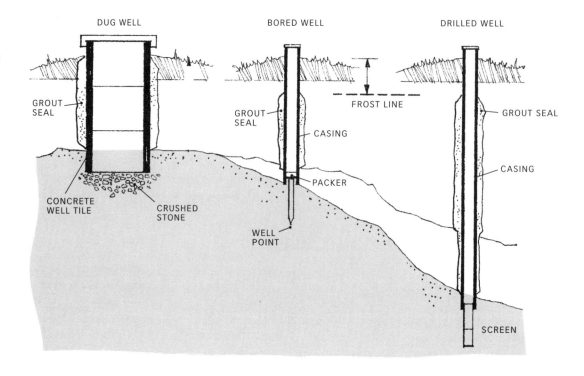

The dug well, though common, is often undependable; a bored well is basically a dug well built with an earth auger rather than a shovel; and a drilled well, the deepest of them all, is least likely to be polluted and most likely to produce a greater yield.

DUG WELL

GROUT SEAL

CONCRETE WELL TILE

CRUSHED STONE

BORED WELL

GROUT SEAL

CASING

PACKER

WELL POINT

FROST LINE

DRILLED WELL

GROUT SEAL

CASING

SCREEN

pressure he will find when he hits water. Without knowing these facts in advance, he cannot know how large a pump will be needed to bring the water to the house. The best protection is to set aside a sizable amount of money as a contingency for excessive well-drilling costs.

Foundation

The cost of putting in a foundation is determined primarily by the size of the home. But there are costs associated with installing a foundation that depend on lot conditions as well. For example, if a foundation is located on a slope, it requires more excavation, a larger area, or both to create a level platform. In cold climates, building codes require the installation of "frost walls" below ground to protect the foundation footings from frost heaves. In much of the Northeast, for example, foundation walls and their footings must be buried at least 4 feet. If the grade is such that one or more of the standard-height foundation

walls will not be covered with 4 feet or more of fill, the bottom of the walls and footings must be extended into the ground to achieve this depth. Building these frost walls, which can make the total foundation wall as much as 12 feet tall when combined with an 8-foot basement, adds appreciably to the cost.

Building a walk-out basement with full-size windows and sliders is usually a better way of handling a slopped property with a full basement, since it makes the basement more usable and access to the yard easier. In cold climates, if the walk-out walls are completely above ground, a frost wall must be built in addition to these 8-foot-high wood-framed walls. The number of windows and doors added to the walk-out walls contributes to the cost, as windows and doors cost more than the foundation they replace.

With a high water table, water can be kept out of the basement with a drainage system and a raised foundation, as discussed earlier. Additional protection can be offered with special waterproofing products that are applied to the foundation. The traditional approach of applying asphalt coating to the walls does not really waterproof the basement. True waterproofing products come with a guarantee. They cost more, but they also deliver better results.

If you are building in an area where soil or seismic conditions create the potential for additional stress to the foundation, you may need to make it stronger. The GC can accomplish this, for example, with thicker walls reinforced by steel bars.

Landscaping

Most aspects of landscaping are decided by personal taste and budget. However, the land does influence which alternatives are most practical. Some land lends itself to inexpensive landscaping while other lots require more extensive work. For example, a flat lot that contains few trees or shrubs might require an extensive lawn. If this same lot has poor soil for growing grass, it might require loam to be trucked in.

REAR ELEVATION

When a property slopes from front to back or side to side, kneewalls (shown on the right side of the house) and full-height walkout walls can be built into the basement.

7

The General Contractor's Responsibilities

A S NOTED IN CHAPTER 5, there are three types of construction tasks that need to be completed to build a modular home: the site work, the "button-up" work, and the construction of site-built structures.

The site work comprises everything done to the land so you can build a home on it. Some of this work is completed before the home is delivered and some of it is done after.

Site-work tasks include:
- Clearing the land of trees and shrubs
- Digging a foundation hole
- Putting in a driveway
- Installing a foundation
- Grading for proper drainage

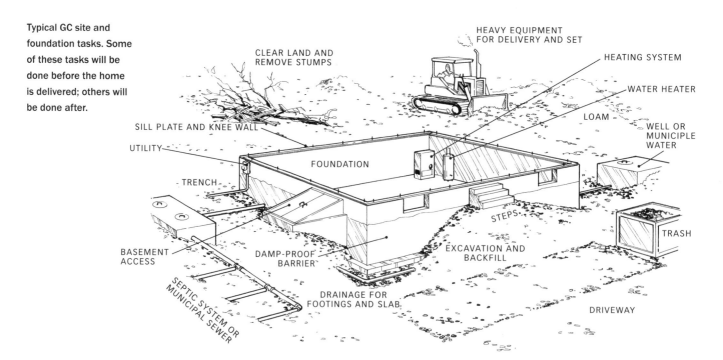

Typical GC site and foundation tasks. Some of these tasks will be done before the home is delivered; others will be done after.

CLEAR LAND AND REMOVE STUMPS

HEAVY EQUIPMENT FOR DELIVERY AND SET

HEATING SYSTEM

WATER HEATER

LOAM

WELL OR MUNICIPLE WATER

SILL PLATE AND KNEE WALL

UTILITY

FOUNDATION

TRENCH

STEPS

TRASH

BASEMENT ACCESS

DAMP-PROOF BARRIER

EXCAVATION AND BACKFILL

SEPTIC SYSTEM OR MUNICIPAL SEWER

DRAINAGE FOR FOOTINGS AND SLAB

DRIVEWAY

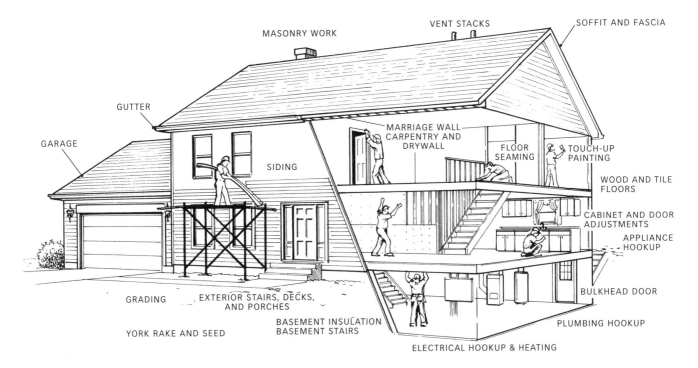

MASONRY WORK

VENT STACKS

SOFFIT AND FASCIA

GUTTER

GARAGE

SIDING

MARRIAGE WALL
CARPENTRY AND
DRYWALL

FLOOR
SEAMING

TOUCH-UP
PAINTING

WOOD AND TILE
FLOORS

CABINET AND DOOR
ADJUSTMENTS

APPLIANCE
HOOKUP

GRADING

EXTERIOR STAIRS, DECKS,
AND PORCHES

BULKHEAD DOOR

YORK RAKE AND SEED

BASEMENT INSULATION
BASEMENT STAIRS

PLUMBING HOOKUP

ELECTRICAL HOOKUP & HEATING

- Installing a septic system or hookup to municipal sewer
- Drilling a well or hookup to municipal water

Button-up work consists of tasks needed to finish a modular home after it is set on the foundation. This always includes:

- Connecting the plumbing and electrical systems
- Installing the HVAC system
- Completing the exterior carpentry
- Completing the interior carpentry

Site-built structures are additions to a modular home, such as:

- Garage
- Porch
- Deck
- Masonry fireplace
- Mudroom
- Finished basement

Each set of tasks requires a different base of knowledge and a different set of construction skills. Together, these tasks require the services of experienced contractors in several different construction trades. For example, you will need an excavator, foun-

dation contractor, plumber, electrician, HVAC (heating, ventilation, and air conditioning) contractor, siding contractor, carpenter, floor installer, drywall finisher, and painter. If you also build a garage, you will need a roofer. Sometimes the same contractor will have more than one skill, so the person who frames your garage might also shingle it. As general rule, however, you will need several different people to complete the work, and most of these will be independent contractors, working for themselves or a small company. Consequently, you will need someone to coordinate the work of these contractors and to oversee and manage the project from start to finish. That individual is the general contractor (GC).

In chapter 5 we discussed how to choose a GC. This chapter covers the work the GC must do to make a modular home livable. It is not, however, meant to be a construction manual on becoming a GC, nor does it intend to train a novice in how to become an excavator, carpenter, plumber, electrician, HVAC contractor, or other subcontractor. A professional GC with modular home experience will be the best-equipped person to handle all of these jobs. Whether you

Typical GC button-up tasks.

decide to act as your own GC or hire a professional, the knowledge you acquire in this chapter will help you have a more successful experience building your modular home.

Before Set Day

Before taking any other steps, the GC should begin with a site inspection of the property. He should complete drawings of the site plan, the basement and attic layouts, and all site-built structures. In addition to assisting with the building permit, he should meet with you to discuss how the project will proceed, including who has which responsibilities.

Site Inspection

Before any steps can be taken, the GC must inspect the site to assess what can be done on the property and determine the scope of work.

Property Plans

Many building departments require a boundary survey and a proposed site plan before issuing a building permit. They need the survey to be completed by a licensed surveyor but will often allow the GC to hand-sketch the critical information on the surveyed plan. Have the GC complete a detailed site plan that refers to the recorded survey, even if one is not required by the building inspector. In this way, you and the GC will have an agreed-upon record of all areas impacted by the construction and how the construction will comply with the applicable zoning regulations. The site plan should indicate the location of any existing and proposed structures, including your home, garage, and deck. It should note the location of the boundaries, setbacks to the lot lines, driveway, septic system, well, utilities, streams, wetlands, and grades. If you are unsure how the plan will actually look

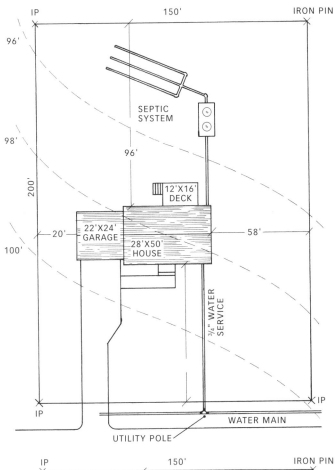

A site plan indicates where the GC intends to locate the house and additional structures.

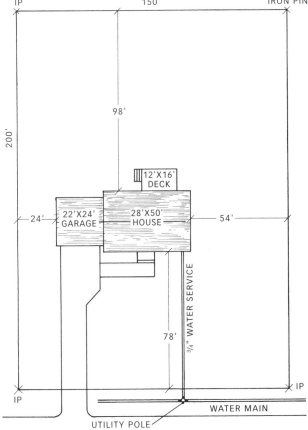

An as-built site plan shows exactly where the GC built the house and additional structures.

on your property, have the GC stake out the important areas before finalizing it.

After the foundation is installed, the building inspector may require proof that its location complies with zoning and building codes. Usually this requires an "as-built" plan drawn by a licensed surveyor, documenting the exact foundation location. Your bank may also require an as-built plan before it will release funds to you and the GC.

Basement Layout Plan

If you will have a full or partial basement that contains some of the mechanical systems, the GC needs to determine the location of such items as the heating system, HVAC pipes and ducts, plumbing pipes, and basement windows. Since this seemingly minor step in the planning process can significantly impact the future use of your home, ensure that the GC discusses the layout with you.

When a GC prepares an estimated price before you have complete plans or a detailed set of specifications, he will often assume the most direct, economical routing of all plumbing pipes and heating and air-conditioning ducts in the basement and attic. These assumptions may not be best for you. This is why it is such an advantage to complete all of your plans and specifications before you ask for GC bids. This usually does not happen, however, unless you work with an architect or home designer.

Assuming you have not worked out every detail in advance, the GC will need to complete the basement plans before you authorize the dealer or GC to begin his work. It helps to understand how various items impact the basement layout and headroom. The checklist at right provides a typical list of such items. Think about how the location of each item might affect your plans for the basement. A water heater, for example, can be located just about any-

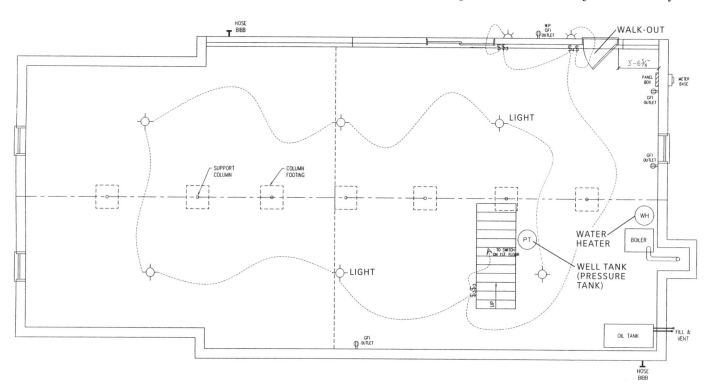

Sample basement layout plan. The stair location was determined by the floor plan, and the walk-out wall in the rear was dictated by the lay of the land. The boiler, water heater, and well pressure tank were set in the front right corner to protect the space in front of the walk-out wall, which will maximize the future enjoyment of the basement. The fuel oil tank was placed in the front corner to allow easy access by the oil delivery truck.

where, but you will not want it located near an area you plan to use for play.

Typically, the hot- and cold-water supply pipes run across the basement ceiling and along the walls. The plumbing drain and waste pipes slope down along the ceiling and walls to allow for gravity discharge to the septic or sewer system. With standard dimensional floor joists, all of these plumbing pipes, especially the waste pipes, reduce the headroom in some places. If your home is built with floor trusses, the supply lines can be run through the webs of the trusses. The location of these pipes affects where you can situate basement windows and doors. If you want windows or doors in a particular place, ask the GC to locate the plumbing pipes elsewhere. Expect to pay more for this new location, should it require more materials or labor than were included in the original estimate.

When a home is built with standard floor joists, the heating and air-conditioning ducts have a major impact on basement headroom. The first-floor ducts are mounted to the basement ceiling directly below the floor joists, reducing headroom by a foot or more. They include a large main trunk down the middle and several smaller trunks branching off to the side. They also affect the location of basement windows and doors. If you must locate the basement windows or doors in a particular place, the GC will need to move the ducts away from this area. If this relocation requires more materials or labor than allowed for in the GC's estimate, it will add to the cost. If the modular manufacturer can build your home with floor trusses, you may be able to avoid at least some of this problem.

Similar issues apply to the venting of the heating system, whether it is done with a power vent, direct vent, or flue. The location of the venting system is determined in part by the location of the heating system, since they must be put relatively close to each other. The venting system, in turn, significantly impacts the location of the basement windows and doors, house windows and doors, decks, porches, and bulkheads, since the building code prevents these from being too close to the exhaust. Choosing to place

Basement-Layout Checklist

Basement plans include many more details than homeowners typically anticipate. Here is a list of the most common items.

PLUMBING
☐ Water-supply line wall penetration
☐ Septic- or sewer-line wall penetration
☐ Exterior frost-free faucets
☐ Under-slab plumbing
☐ Well tank
☐ Water meter
☐ Gas meter
☐ Hot-water tank
☐ Washer, dryer, and dryer vent
☐ Future basement bathroom

ELECTRICAL
☐ Panel box (typically a gable wall)
☐ Electric meter (outside and adjacent to breaker panel)
☐ Central vacuum cleaner

HEATING AND AIR CONDITIONING
☐ Furnace or boiler (including power vent, direct vent, or flue location)
☐ Oil tank (including fill- and vent-pipe locations)
☐ Under-slab oil line
☐ Air-conditioning condenser

FOUNDATION
☐ Basement windows
☐ Bulkhead
☐ Sump pit
☐ Window and door locations for a walk-out basement
☐ Knee wall windows and doors
☐ Drive-under garage overhead doors
☐ Drive-under garage/basement passage door

the heating system and flue in a particular location will affect many other decisions inside and outside the home. Conversely, choosing to locate the basement and house windows and doors, the deck, and the porch in particular places affects where you can put the heating system and flue. This is one reason why you should not finalize your modular home plans until you have also finalized your GC plans.

With your GC's help, you will need to factor in basement stairs, chimneys, site conditions, garages, mudrooms, and other site-built structures, since each of them can affect the basement layout. For example, if the basement stairs come down on the right side of the home, then the two-car drive-under garage will probably need to come on the left. This means that the masonry chimney should go on the right, except the fireplace is in the family room, which is on the left. These kinds of conflicts happen all of the time, which is another reason why an experienced GC can be of enormous help.

Developing a basement-layout plan is even more critical and difficult when you plan to finish the basement, or at least to use it in some fashion. You do not want a waste line hanging from the ceiling in your basement family room or a water heater where you want to place a big-screen TV. If you plan on finishing the basement, tell the GC, and then verify that his basement plans are compatible with yours.

Attic Layout Plan

With most modular manufacturers, the HVAC ducts for the second floor of a two-story are placed on the attic floor joists. Since the ducts include a large main trunk down the middle and several smaller trunks branching off to the side, they significantly affect how the attic space can be used. Therefore, when installing ducts in a two-story with a usable attic, plan the attic layout to route the ducts away from the usable areas.

Site-Built Structure Plans

If you are building a garage, porch, deck, finished basement, or some other structure on your site, the GC will need to complete the required plans. This will include plan views of the structure and foundation, a cross-sectional view, an elevation, an electrical layout, and an HVAC layout. Review these plans closely. For example, the location of windows and doors in a walk-out basement and garage can affect how you use the spaces. The number of lights and electrical outlets, as well as the location of the electrical switches, can significantly impact the enjoyment of your home. Do not wait until after you move in to find out that the basement storage closet is half the size you need, or until half the second floor is finished to discover that the children's bedrooms are no bigger than walk-in closets.

The GC needs to give these plans to his subcontractor candidates, since the plans ensure an apples-for-apples price comparison. The subcontractors will use the plans to ensure that the structures are completed to your specifications.

Site-Determined Specifications

The manufacturer must be supplied with the following construction specifications before it builds your home. These items are discussed in more detail in chapter 4:

- Building-code requirements that differ from your state's codes
- The location of the electrical meter and panel box
- The size and location of any rough openings for work that will be completed on-site, such as a masonry fireplace
- The thickness of site-installed flooring
- The proposed floor plans for unfinished spaces, such as the second floor of a Cape Cod

Got Permits?

Work with the GC to obtain a building permit and determine if you need any of the following supporting permits.

☐ State and local environmental-commission's approval: This is particularly important if the property contains wetlands or protected dunes in coastal areas

☐ Tree-removal permit: If the land is on a state road, federal highway, or designated scenic route, you might need permission from a regulatory agency

☐ Utility permit: Required if you must temporarily take down high-voltage power lines

☐ Demolition permit: May be required to demolish any existing structures; another permit may be required to dispose of the materials

☐ Hazardous-waste permit: Required to dispose of hazardous materials

☐ Well permit: If a septic system is needed, its design might need to be approved before a well permit is issued

☐ Potable-water test: The well water might need to pass a safety test before a building permit is issued

☐ Municipal-water entrance permit: This can be as costly as drilling a well, and the fee does not include the cost for digging the trench or installing the underground pipe

☐ Septic-design approval: Having the design completed by a licensed sanitarian or engineer and approved by the local board of health can sometimes take several weeks

☐ Municipal-sewer entrance permit: This can be as costly as a septic system, and the fee does not include the cost for digging the trench or installing the underground pipe

☐ Driveway curb-cut permit: In most areas this is routine, as long as safety is not an issue, but if the land is on a state road or a designated scenic route, you might need to apply to another agency

☐ Street-excavation permit: If the road must be excavated to bring public utilities to the site, permission must be obtained from the town or state, depending on who is responsible for maintaining the road

☐ Traffic-safety permit: If the road must be excavated or if traffic will be obstructed, the town or state may require special signage or a police officer to maintain traffic control

☐ Fire marshal approval: Required for smoke detectors, oil heating systems, fireplaces, and woodstoves

☐ Building permit for additional structures: Required for all structures built on-site, such as a garage, porch, deck, mudroom, or finished attic

☐ Structural engineering design and plan: Required when a structural beam is being used in the basement to support the home instead of lally columns

☐ Oil permit: May be required when installing oil heat, along with a final inspection to obtain a certificate of occupancy

☐ Gas permit: Required when natural or propane gas will be used for the heating system, an appliance, or a fireplace

☐ Fuel-storage permit: Required when the heating system requires oil or gas storage

Who Pays?

Be sure you know who is paying for each of the permits and fees required to build your home. The cost for permits, utility fees, engineering and survey work, and related items varies widely from town to town, utility company to utility company, and engineer to engineer. For example, building permits can range in price from a hundred to a few thousand dollars, and utility hook-ups can range from no cost to several thousand dollars. Since you are ultimately responsible for the costs, whether you pay for them directly or through your GC, determine the amounts for each item as soon as possible, budget accordingly, and agree in writing with your GC which of you will be making the payments.

Keep Current with Changes

Update the GC in writing about any changes to your plans and specifications. Expect the GC, in turn, to regularly ask for written updates.

Electrical Power

The GC should ask his electrician to contact the electric company to request service to a home at least 30 days before its scheduled delivery. If the utility company needs additional electrical poles or a new transformer, the request should be made much earlier. Scheduling the public hearing that is sometimes required can take several weeks, and the installation can take a few more.

A significant delay in receiving power can prevent the contracting work from getting started. Subcontractors need electricity to operate their power tools. Although they can use a portable generator temporarily, this is an inefficient and expensive way to operate. If the GC discovers that power will be delayed by more than a couple of weeks, he can have the electrician set up a temporary electrical service before the house is delivered.

Preconstruction Meeting

Before the GC begins his work, he should meet with you to discuss the project. He should use the meeting to remind you of the tasks he will be completing along with the schedule he proposes to follow. He should also review your responsibilities for completing these tasks. His objective should be to both educate and prepare you so that the tasks are completed in a timely, professional manner with a minimum of misunderstanding and stress. Use the meeting to ask all of those questions you have wanted to ask about what will happen at each point of the project. Be candid about your concerns and expectations.

Completing the Site Work

The GC must complete all of the site work. Most of these responsibilities were discussed in chapter 6 in order to help you understand the costs of making a piece of land into a viable building lot. This chapter will mention a few additional responsibilities that may apply to your lot. It is difficult to provide a comprehensive list of site-work tasks, however, because there are many responsibilities that a good GC should do automatically when conditions warrant. For example, he should strip and stockpile the topsoil, compact beneath the basement slab, and provide proper grading.

Protected Areas

Before digging, the GC should contact the appropriate authorities to determine if your property contains underground utilities (gas, electric, water, sewer, phone, or cable). If it does, he needs to have their locations marked so that they will not be disturbed during construction. And if you received instructions from a local, state, or federal board that governs wetland protection, the GC must follow the instructions exactly as they are written.

Winter Build

If you are building during the winter months in a region that requires protection against frost and snow, the GC should take steps to reduce the impact of his work. For example, he should blanket the areas he will be excavating before the work begins to minimize frost penetration. This might be done by covering the ground with hay. The excavated areas should stay blanketed until they are backfilled.

Spring Build

If you are building during the spring, find out whether your community allows heavy equipment to travel on the roads during the spring thaw.

Demolishing Existing Structures

If you need to demolish and dispose of any existing structures on your property, the GC must specify how he will dispose of the materials. Most communities will not allow him to bury the materials, and the cost of disposing of the debris at a dump can be significant. He should budget for bringing in enough fill to fill the hole.

Propane Tank

If you are using propane gas to heat your home, consider burying the tank. The GC will need to work with a propane supplier to complete this task.

Radon Protection

In certain areas, it is a good idea to install radon protection under the basement floor. The GC and the building inspector can advise you about this.

Landscaping

Some GCs help with landscaping; others leave it to the customer. Begin by purchasing additional topsoil, if needed. When the finish grading is done, select someone to rake the topsoil and plant grass, shrubs, and flowers.

Preparing the Site for Delivery and Set

The GC is responsible for preparing the site for the delivery and set of the modules. More specifically, he is responsible for preparing both the access leading to the lot and foundation and the area where the crane and modules must be located during

Change Orders

Should you decide that you want to make some changes to your GC agreement after authorizing the GC to begin work, and it is not too late, expect to be charged for the changes. Some items, such as specially ordered doors or materials that require a long lead time, cannot be easily or affordably changed. The best way to avoid these costs is to work closely with your GC to determine exactly what you want before you sign a contract.

the set. If the area is flat, with good soil conditions and relatively wide, straight roads, this responsibility may be without challenges. On the other hand, problems can arise on what appears to be an easy site. Bad weather, poor soil, loose fill, a utility pole in an inconvenient place, a septic system located where the crane needs to go, the customer's refusal to cut down a favored tree or remove a cherished old stone wall: These are the kinds of things that can turn a site into a logistical challenge — even a logistical nightmare.

When the manufacturer's delivery time from its factory to the site is more than a couple of hours, it will deliver the modules at least one day before the scheduled set. In such a situation, the GC should try to create sufficient space on the site to store all of the carriers overnight. Since the most efficient way to set a home on the foundation is to place the crane in front of the house with one carrier on each side of the crane, the preferred storage plan is to create a space wide enough for the carriers to be delivered directly to these positions. Unfortunately, the combination of lot size and configuration, topography, soil conditions, foundation size, and the number and size of the modules can make it impossible to place all of the modules in the right spot and still preserve room for the crane.

When the modules cannot be properly placed, one or more of the carriers have to be delivered to a temporary storage location, which may be at another location on your site or at a nearby parking lot or open field. The carriers will then be moved to their proper positions next to the foundation on set day. Although the GC is responsible for preparing the site to facilitate these efforts, he is not responsible for things beyond his control, such as a heavy rainstorm that washes out the driveway. Nor should you expect him to pay for the required repair.

If the GC does not prepare the site in a satisfactory manner, and this causes the delivery and set operations to be delayed, you and the GC will be liable for the additional costs incurred. The drivers and escorts that deliver the carriers will budget enough time to drive directly to the site or wherever you designate. They will expect to maneuver the carriers into place with reasonable effort and then leave. They will not expect to spend hours waiting for trees to be cut, fill to be delivered, or a bulldozer to arrive. If any of these are required, the delivery company will charge the modular dealer, who will in turn bill you, for the additional time. The crane company will charge the dealer by the hour, with a minimum fee. The longer the set takes, the longer the meter runs. The crew that completes the various set activities will also charge the dealer for any lost time caused by the delays, and the dealer will pass this expense on to you. If the delays are caused by the GC's poor site preparation, it is fair for you to submit the invoices from the delivery, crane, and set companies to the GC for reimbursement.

Many sets take a full day, and some take two or more days. One of the most important responsibilities of the set crew is to protect the home from weather damage as quickly as possible. If the site is not prepared and the set is subsequently delayed by several hours, the set crew may not be able to complete enough of the set to give your home the protection it will need should it rain overnight. If the delay happens before the first module is set, the set crew and modular dealer can cancel the set and reschedule it for the next available day. But if some of the modules are already set on the foundation, with the protective coverings removed, the set cannot easily be stopped and the crew may not be able to take the required steps to protect the home until the cause of the delay is removed. If

Plan and Prepare

General contractors, excavators, and customers consistently underestimate the difficulties involved in preparing a site. Their misjudgments come at great expense and aggravation to the dealer and, ultimately, the customer. The best way to avoid a disagreement between your dealer and your GC is to have them meet on the site to jointly develop a plan before any work begins. The dealer should put the plan in writing so that both parties have written documentation. The plan should indicate what the excavator needs to do to properly prepare the site as well as where the dealer intends to place the carriers and crane. If a problem develops on the delivery or set because the plan was poorly conceived although correctly executed by the GC, the dealer will be responsible for any additional costs. When the excavator is about halfway done with his pre-set work, the dealer should again meet with the GC to review and, if necessary, update the plan; the best-thought-out excavation plans do not always work as well on land as they do on paper. If the dealer has any lingering concerns, he should return for a third visit just before the excavator is ready to leave the site. Either way, the dealer should visit the site for a final pre-set inspection after the GC reports that it is prepared.

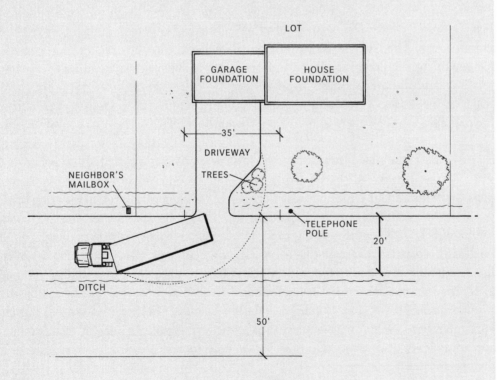

In this example, the delivery crew needs to move a long module onto the property. To do so requires the transporter to cross into the ditch and the neighbor's property on the opposite side of the street. It also requires the group of trees next to the end of the driveway to be taken down.

this delay takes a few hours, the home will be exposed to the elements longer than it needed to be.

Overhead Obstructions

The GC should arrange to have all overhead obstructions removed along the roadway leading to the job site so that the clearance is at least 14 feet. He should contact the appropriate authorities (local, state, or public utility) to handle this work. He should also remove all overhead obstructions along the driveway leading to the foundation and in the area where the crane and modules will be located during the set. He should pay particular attention to tree branches and electric, telephone, and cable lines. Note that freezing rain and snow can cause tree branches and utility wires to sag lower than normal, creating a potential obstruction with little or no warning. If any obstructions strike the modules, they may tear the protective covering installed by the manufacture. The GC will be responsible for repairing the covering. If he fails to repair it and weather damage results, he will be liable for all costs required to correct the damage.

Driveway Entrance and Turns

The GC must make all turns along the driveway wide and long enough to be negotiated by the tractor-trailers that will deliver the modules. With a 60-foot-long module on board, each tractor-trailer can be as much as 90 feet long and 16 feet wide. Such a vehicle requires a substantial swing radius when rounding turns. The GC should take particular note of the driveway entrance and any poles or trees across the street from the driveway.

If neighbors tend to park their vehicles near the driveway entrance, ask them to park at least 50 feet from the entrance. If the modular transporters and carriers can enter your property only by driving across a part of a neighbor's property, ask his permission and offer to repair any damage the activity causes.

Driveway Width

The driveway leading to the site and foundation should be at least 18 feet wide, if at all possible. Alternatively, the driveway can be 12 feet wide, as long as there is 3 feet of clear space on each side to accommodate the overhanging modules.

Driveway Gravel Base

To enable the delivery vehicles and crane to maneuver, the GC should create a compacted and level gravel base along the driveway leading to the foundation. The modular dealer should be consulted if the grade is more than 1 foot in 20 feet.

Ground-Level Obstructions

The GC should eliminate all high curbs, ditches, and steep grades that could cause the modules to drag or the load to stall along the roadway leading to the job site. He will need to do the same for the driveway leading to the foundation and in the area where the crane and modules must be located during the set. If the edge of the property has an asphalt or concrete curb, the GC should protect it from damage by covering it with large planks or gravel.

The GC should remove all obstacles, such as culverts, stumps, large boulders, piles of topsoil or fill, and piles of brush. When necessary, he should temporarily remove any street signs and mailboxes that will hinder delivery of the modules. The GC is responsible for correcting any damage to the modules caused by excessively harsh road or access conditions.

Underground Pipes and Tanks

The GC should wait until the home is erected to install underground pipes or tanks that could be damaged by the heavy loads that will pass over them. If underground pipes and tanks already exist on the property, the GC will need to prepare the site so that the modules and crane can be properly placed without disturbing them. Inform your modular dealer where the pipes or tanks are located so that he can direct the crane and modular carriers away from them.

Garages and Other Structures

If a garage or any other structure is to be built on the side of the house where the crane and modules will sit during the set, the GC should delay the excavation and foundation work for these structures until the set is completed.

Snow and Ice

The access must be kept free of snow and ice. Even though bulldozers can move mountains, their tracks will just spin if they attempt to pull any substantial weight on ice. All snow and icy areas, even small light patches, must be plowed and sanded just prior to delivery. If it snows after the foundation is poured but before the home is set, the GC should remove as much of the snow as possible from the basement before the set. This is often easier said than done, however, and sometimes it is impractical, especially when the slab has not already been poured.

You and your GC will need an agreement about who is responsible for snowplowing, shoveling, and sanding while your home is under construction. Since neither of you can know what the winter will bring in any given year, an allowance, with a reasonable rate per snowfall or per hour, is a fair way to handle it.

Room for the Crane

The area where the crane will be located during the set should be at least 30 feet by 30 feet with a level, compacted-gravel base.

The set crew will need another 20 feet of clearance on both sides of the crane to place each carrier. The minimum necessary length depends on the size of the largest module. The area needs to be large because the crane's outriggers take up a good deal more room than the main body. The outriggers cannot be set too close to the foundation, since the weight can damage the foundation wall.

The GC needs to create this flat spot even if the final grade in the area will be sloped after the home is complete. The GC may need to add gravel for the set and then remove it. If there is a threat of water accumulating in the vicinity, which could compromise the crane's stability, the GC should cut swales to facilitate surface runoff.

Staging Area

If the GC cannot provide sufficient room on-site to store the modules overnight, you or the GC will need to locate an alternative staging area. Select an area where the modules can be stored for several days, in case of a delay in the set. If you decide to store the modules on someone else's property, secure written permission from the owner at least a couple of weeks in advance of the delivery. Your modular dealer or his manufacturer should carry insurance to cover the modules during their stop in the staging area. On the morning of the delivery, visit the area to verify that nothing has changed and that the modules are not blocked.

Bulldozer and Operator on Delivery Day

Some crews will deliver only to the edge of your site. Most will use their transporters to back the carriers down the driveway, if it is well prepared and compacted. A few others will venture onto your site to assist with positioning the carriers, but they will do so only when the distance is reasonable and the path is flat, unobstructed, and relatively compact. The reason more delivery crews

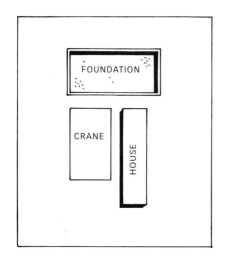

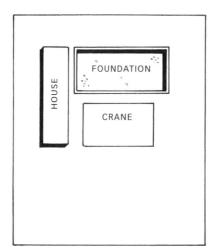

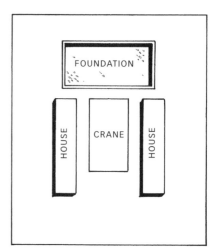

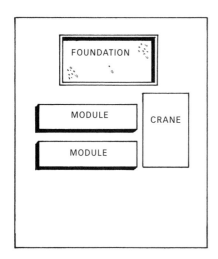

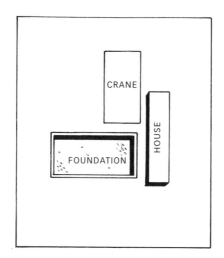

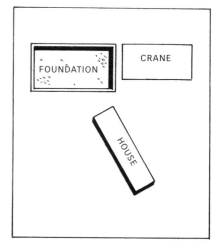

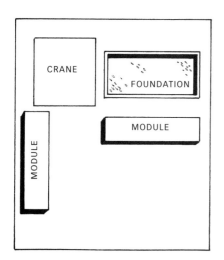

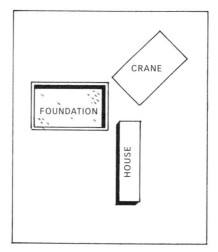

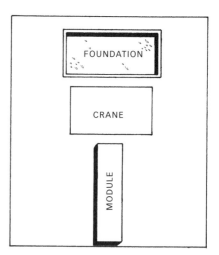

The crane and modules can be positioned in several ways. The top three positions are preferred when the topography of the land and size of the modules allow them. The modular dealer will advise the GC about which situation will work best for your site.

will not help you maneuver the carriers is that their transporters are not designed for off-road activities, and they damage easily when driven onto rough terrain. They are also not designed for the tight turns often required to maneuver across a building site. Nor are they suited for the soft-soil conditions that are usually found once the carriers leave the road and driveway. Consequently, if the carriers need to be placed more than a few feet from the road, either because they must be positioned some distance from the foundation or because the foundation is set back from the property, the GC must provide the necessary equipment to move the carriers the rest of the way. Most of the time, the equipment of choice is a large, 7- to 10-ton bulldozer. A larger piece of equipment, such as a loader, may be required when the lay of the land, soil conditions, or ice and snow prevent a bulldozer from completing the move. Since the GC will be hiring the equipment operators, he should be certain that they are insured and licensed, if required.

Sometimes it might appear that a delivery can be managed without a bulldozer, as long as weather conditions cooperate. Modular deliveries are scheduled a couple of weeks in advance, however, and it is impossible in most areas to predict the weather. The GC should have the proper equipment on call, even if he does not expect to need it.

Transporter or Tow Truck on Set Day

If an off-site staging area has been used to store the modules, your dealer or GC must arrange to have the modules brought from the staging area to the job site on set day. The dealer will determine who is responsible. It would be best if your dealer arranges for a member of the delivery crew to carry out the move, since he will already have the required special permits, license, and insurance. Since the delivery crew almost always returns home with its transporters after delivering the modules, the crew will provide assistance only if the dealer schedules it in advance. If he cannot secure the help, the dealer or GC will need to schedule an insured, licensed operator and vehicle, such as a tow truck with the correct trailer hitch, a pindle hook. The GC might also need to supply escort vehicles to accompany the move.

Bulldozer and Operator on Set Day

If the carriers cannot be placed within reach of the crane on delivery day, the GC will need to provide the necessary equipment to reposition the carriers on set day. He may also need to help the crane maneuver to its proper position. The same equipment options mentioned for delivery day apply to set day. To be safe, the GC should have equipment on call even if he is relatively certain he will not need it. This works best if the equipment operator arranges his schedule so that he can be at the site within one-half hour of being called.

Having the necessary equipment on call will allow the GC to put a set-day problem behind him quickly. Without a backup plan in place, however, he may take a few hours to find the right equipment and operator. The resulting delay will create cost overruns for you or the GC and increased exposure to inclement weather. (See story on page 220.)

Carrier Storage

After the crane sets the modules, it stacks the empty carriers so the delivery crew can return them to the manufacturer. The modular dealer tells the set crew how many carriers can go in each stack. The stacked carriers must be located so they remain secure and safe from damage until the delivery crew can pick them up, usually within a week or two. Repositioning them may require the assistance of the GC's bulldozer. The GC will usually want to dictate placement so he can ensure that the carriers are not in his way. The carriers must be turned so that they are accessible to the delivery crew's transporters. This means the carriers' tongues point toward the road, enabling the transporters to hook up to the carriers without assistance. If the transporters cannot reach the carriers, the GC will have to move them. He will be responsible for the delivery crew's additional costs if the crew must make an extra trip to retrieve the carriers.

Speaking from Experience

WHEN I FIRST STARTED selling modular homes, I had a difficult time convincing customers that they needed to have a bulldozer on set day unless the need was obvious. That changed after one nearly disastrous incident.

My customers were building a two-story house made up of four modules shipped on four carriers. On delivery day we were able to position two of the modules next to the foundation and crane, but the other two modules had to be stored overnight in a staging area. I asked my customers to have their excavator at the site to assist on delivery day, and they complied. It turned out that the bulldozer was not needed that day because the ground was dry and firm. I asked them to keep the bulldozer on-site for the next day's set, but they said they did not think it was necessary. I pointed out that if it rained that night, we almost certainly would have a problem. My customers responded by saying that it would cost them $500 for the second day, and they thought it was a waste of money. When it began raining that night, I called them at home to again ask them to supply a bulldozer. They refused, and I stopped pushing my request.

The set started off well. We got the first module onto the foundation quickly. While we were setting the second, we delivered the third to the site. But the transporters could not get the third module close enough to the foundation no matter what we tried. My customers called their excavator, who arrived two and a half hours later. While we were waiting, a thunderstorm hit hard. My set crew climbed on the roof, in spite of the lightning, and tried to cover the two modules with tarps. They did OK, but while trying to position the tarp, one of the crew slipped and pushed his foot and part of the tarp through the kitchen ceiling. All the water that had pooled on the tarp while it was being installed poured onto a row of cabinets. Fortunately, none of my crew was hurt and the damage was repaired. But the experience taught me that I had to explain to my customers all the things that can go wrong if they do not provide the proper equipment. It also taught me to delay the start of a set if the equipment is not on-site.

When your dealer tells you to provide equipment on delivery and set day, remember that he isn't just protecting his interests; he is also protecting your house.

Installing the Foundation

In many ways, installing the foundation for a modular home is the same as for any other home. There is no need to elaborate on the requirements here, as the GC and his foundation subcontractors should know what they need to do.

Type of Foundation

Modular homes require either a crawl space or a full basement. Pouring a slab on grade, which is common in some areas, does not work for modular homes. For one thing, the slab cannot be used for the first floor, which is the intent of these foundations. For another, current modular construction requires the GC to have access to the mechanicals below the first floor. This limitation may change, however, when manufacturers begin using floor trusses that allow them to incorporate the mechanicals in the floor system.

If you do not build a full-height basement, make sure the height of the crawl space will accommodate the work you intend to do there. If you need only to hook up the plumbing, you can often get by with 3 feet of headroom. But plumbers may charge you a premium because of the difficulty and physical discomfort of working in such a confined space. If a heating system will be installed in the crawl space, it should be tall enough for the heating equipment plus whatever additional space is required by the building code.

Scheduling

The modules will be built on schedule whether or not the foundation is ready. The manufacturer won't be terribly concerned if the GC is two weeks late with the foundation, as long as the dealer pays for the modules in full as soon as they are built. But if you are paying the dealer for the modules through a construction loan and your lender won't pay until the house is set, your payment will be two weeks late. Most manufacturers will penalize the dealer, who will in turn penalize you if your independent GC is at fault. Failing to start the excavation work and foundation installation on schedule can have expensive consequences.

Sometimes the crane and modules need to be located on the side of the house where you are building a garage or other site-built structure. In that case, delay that foundation work until the house is set.

Each module weighs several tons, and your foundation needs to be strong enough to take the full weight as soon as the modules are set. Poured-concrete walls need time to cure properly, and the concrete company should advise your GC on the minimum curing time. The building inspector may override the concrete company and insist on a longer curing period. He may also add a minimum time before the foundation can be backfilled.

If your home will be delivered during the winter in a cold climate, try to have the foundation installed in the fall. The GC should then protect the foundation from any frost.

Dimensional Requirements

Modular homes are manufactured to close tolerances. Each module's length and width will be very close to the plan specifications. If a foundation is 2 inches too long or narrow, a stick builder can frame the house to fit the actual foundation. If this were to happen with a foundation for a modular home and the modules were already built, this adjustment could not be made to the modules. Likewise, a stick-built home can be built to fit a foundation that is neither square nor level. Modules, however, are square and level, and cannot easily be made to fit on a less-than-perfect foundation. With modular homes, therefore, the GC must take special care with his foundation measurements.

When installing a poured-concrete wall, it is difficult to ensure that the foundation's length and width match the foundation plan exactly. If the GC is not able to be 100 percent accurate, he should err on the side of making the foundation slightly shorter and slightly narrower, since this is easy to hide without consequences. If you undersize the foundation too much, however, you will not be able to fasten the modules to the sill plate.

Foundation Plans

The foundation subcontractor needs a foundation plan that includes the dimensions of the house, with all of its bump-outs and jogs, along with any basement windows, doors, and bulkhead. The plan must show the height of each wall and whether any of the walls are made of a material different from the remainder of the foundation. For example, the foundation might have 8-foot-tall concrete-block walls in the front and sides. It may also have a 4-foot-tall concrete block frost wall in the rear with an 8-foot-tall wood-framed walk-out wall on top.

Most manufacturers provide foundation plans specific to each house they build. This ensures that the plans include all of the options selected by the customer. Some manufacturers, however, provide only generic foundation plans, which do not include the options selected by the customer. If your manufacturer uses generic plans, the GC will need to add your options to his foundation plan.

With most manufacturers, the GC will need to add the foundation plan for any site-built component, such as a masonry fireplace, garage, or family room. Manufac-turers are often unwilling to draw these plans because they are not building them. Independent inspection companies are increasingly unwilling to allow the manufacturers to show the site-built structures on their "permit plans." Understandably, they are concerned that showing the structures might falsely imply to a customer and local building inspector that the manufacturer is building them and the inspection company has approved them.

The GC's foundation plan must show the support columns in the exact locations specified in the manufacturer's foundation plan. This means the concrete pads that support the columns must also be located exactly where the columns are shown. The spacing of support columns often varies across the basement because some parts of the marriage wall carry more weight than others. In addition, support columns are often placed directly below sections of the carrying beam that have been spliced together. Consequently, moving the support columns even a foot or two can weaken the structure of a home. This can lead to excessive stress cracks, uneven floors, and the misalignment of doors, windows, archways, and stairwell openings. If you remove any columns without providing another form of support, you could create serious structural problems. For that reason, if the GC changes the support column locations specified in the manufacturer's foundation plan, even with the permission of the building inspector, he relieves the dealer and manufacturer of responsibility for any resulting problems. Should the GC discover that he accidentally misplaced a support column, he will need to add pads in the correct locations. If the mistake is not discovered until after the house is set, the GC can provide temporary supports until he is able to correct the situation.

If you need to remove support columns, a licensed structural engineer can design a beam that will carry the added load. The engineer needs a copy of the manufacturer's foundation and floor plans and must be informed by the manufacturer of the loads where the beam will be placed. The GC can then purchase a beam that meets the engineer's specifications.

An alternative method of support is to install a structural plate, known as a flitch plate, between the modules. Again, a structural engineer must calculate the size of the plate. The dealer must inform his manufacturer that a plate will be used so that it can design the marriage wall perimeter to accept the plate. The set crew will be responsible for installing either the structural beam or the flitch plate.

If you are content with the support column locations drawn by the manufacturer but want the option of removing one or more columns in the future, ask your GC to pour a continuous support column pad along the marriage wall instead of individual pads. If you decide later to remove some columns, the foundation pad will be in place to carry the loads created by the necessary support beam.

When planning the basement layout, you and the GC will want to avoid a common mistake made by those without modular experience. The mistake is to assume that the support column pads are always located halfway between two mated modules. In a plan with side-by-side modules of unequal widths, the columns go where the marriage walls meet, rather than halfway between the total widths of the two modules. This situation would occur, for example, if you were building a garrison Colonial measuring 26 feet on the first story and 27 feet 6 inches on the second story. The first story might be made up of modules with widths of 13 feet 9 inches in the rear and 12 feet 3 inches on the front. This keeps the second-story marriage wall directly over the first-story marriage wall, which is

needed to carry the weight onto the basement support columns, where it belongs. But this means the support columns are located 12 feet 3 inches from the front and 13 feet 9 inches from the rear. A similar caution applies if you decide to create an overhang on the front of a raised ranch with equal-width modules. The GC will accomplish this by shrinking the width of the foundation. The support columns will fall where the modules mate, rather than halfway within the narrowed foundation. As many of today's most innovative modular designs venture away from rectangular plans, modules of unequal widths will be more common.

If your GC is not your modular dealer, you will need to be absolutely sure he gets the final and correct foundation plan from the dealer before he installs the foundation. Ask the dealer to indicate in writing which foundation plans are correct. This reduces the chance of a misunderstanding. Instruct the GC to delay installing the foundation until he has these approved plans. If the dealer does not give you the plans in a timely fashion, he will be responsible for any related delay in the installation of the foundation.

Basement Access

Whether you create a full basement or a crawl space, you must build an entry. If possible, make both an interior and an exterior entry. The interior entry lets you access the basement without going outside and an exterior entry lets you bring things into the basement without entering the home. The exterior entry gives the set crew easier access to install the support columns and bolt the modules together. In addition, it will make it easier for the HVAC contractor and the plumber to do their work.

The most common exterior entrance is a bulkhead. This is a steel door on top of a jog in the foundation that forms a stairwell. A

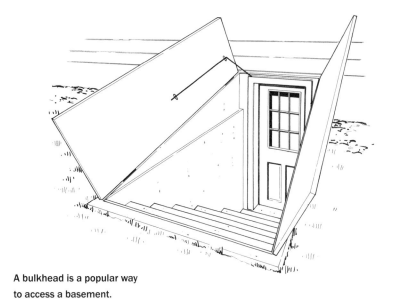

A bulkhead is a popular way to access a basement.

precast bulkhead has the advantage of integral poured-concrete steps. The GC should install an insulated, exterior swing door with a deadbolt at the bottom of the stairs. A bulkhead should not be installed until the house is set on the foundation. If it sits higher than the sill plate, it will obstruct the setting of the modules. In addition, it could easily get damaged.

Another way to create an exterior access to the basement is to build an entry from inside an attached garage. This will require a set of stairs with an insulated exterior door at the bottom. It will also require a half wall or railing to create a safety barrier at the stairwell.

Foundation Specifications

Foundation specifications are determined by building codes, soil conditions, depth of the frost line, and height of the water table. The choice of a poured-concrete or block foundation wall is as much a matter of local practice as the building code. Although they are not common, piers, piers and curtain walls, and piling foundations can also be used. In addition to following the building code, the GC should meet the specifications listed on the manufacturer's foundation plans. If the building codes and manufac-

turer's specifications differ, the GC should meet or exceed the more stringent of the two.

Precast foundation systems are another option. They are made in a factory, delivered to a site, and installed by a crane and crew. Because they are very quick to install and their engineering and design are state-of-the-art, they match up well with modular construction. For example, the Superior Walls system is made of high-strength concrete reinforced with steel. The system makes it easy to finish a warm, dry basement by bonding a layer of insulation directly to the concrete and by making the walls resistant to water infiltration; no additional waterproofing is required. The built-in footers rest on a crushed-stone subfooter, which also helps keep the basement dry. The treated-wood nailers make for easy drywall installation, and the pre-engineered access points make for easy wiring and plumbing.

The minimum basement ceiling height will not be dictated by the building code unless you are using it as living space. Consider building the walls at least 8 feet high to provide sufficient headroom in case you later decide to finish the basement. You might even want to have 9-foot ceilings if your GC is installing ductwork for a forced-air system across the basement ceiling.

Here are a few more foundation details you should attend to:

• Consider making poured-concrete walls at least 10 inches thick when they are longer than 44 feet, even when they are not required by code.

• Since basements tend to be dark, make sure you are happy with the number and size of the windows.

• Add a vapor barrier under the slab and control joints within the slab.

• Add R-10 foam insulation at the edge of the slab, especially if finishing the basement.

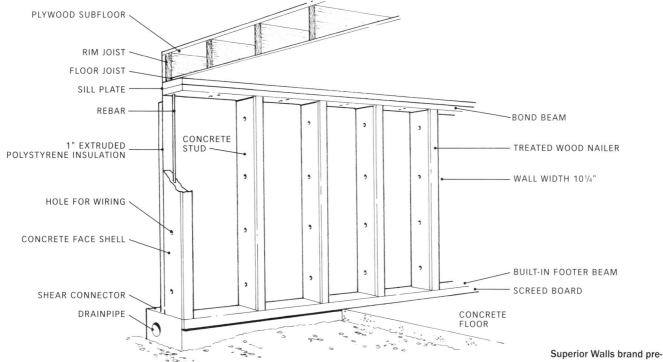

PLYWOOD SUBFLOOR
RIM JOIST
FLOOR JOIST
SILL PLATE
REBAR
1" EXTRUDED POLYSTYRENE INSULATION
CONCRETE STUD
HOLE FOR WIRING
CONCRETE FACE SHELL
SHEAR CONNECTOR
DRAINPIPE
GRAVEL/CRUSHED STONE 4–6"
BOND BEAM
TREATED WOOD NAILER
WALL WIDTH 10¼"
BUILT-IN FOOTER BEAM
SCREED BOARD
CONCRETE FLOOR

Superior Walls brand pre-cast foundation system. The system is made of high-strength concrete reinforced with steel. The built-in footers rest on a crushed-stone subfooter.

• Create a clean look with sleeves in the walls for each of the required penetrations, such as water and sewer pipes.

• If you are putting a sump pit in the floor, have the slab pitched in that direction.

Foundation Floor

The foundation floor can be poured either before or after the house is set. An advantage to pouring it after the set is that the support columns can be cemented in place. The advantages of doing it before the set, however, are substantial. Most important, the GC will be able to start his work in the basement sooner. For example, he can install the stairs to the basement and the heating system immediately after the set. Also, the moisture from the curing concrete will have time to escape into the air rather than into your house. If the GC pours the foundation floor before the house is set, he must bolt the support columns to the floor.

When building in the winter in a cold climate, the GC should try to pour the floor before the temperature drops below freezing, as long as he can also protect it from the frost. If this cannot be accomplished and the floor will be poured after the house is set on the foundation, the area must be covered with hay or in some other way protected from freezing immediately after the hole is dug. Any frost that remains after the house is set must be removed by heating the basement before the foundation floor is poured. Otherwise, the floor may crack substantially. If extensive frost or snow must be removed by heating the basement after the house is set, extra caution should be taken to prevent the house from absorbing any of the moisture.

Backfill before the Set

Whenever possible, the GC should backfill around the foundation before the set. This makes it easier to place the modules close to the foundation on delivery. It also makes it easier for the crane to set its outriggers close to the foundation. A backfilled foundation allows the crew to complete the set quickly and safely. If the foundation is not backfilled, the dealer may need a larger crane, possibly at your expense.

It is usually not possible, however, to backfill a block or precast foundation before the set. Even some poured foundations cannot be backfilled safely when the walls are very long. Sometimes a GC can help support the longer walls by bracing them along the inside.

Additional Concrete Work

If you are installing air conditioning, the GC will set up a condenser outside your home. He will probably want to set the condenser on a concrete pad, although other materials, such as fiberglass, can also be used. He must also create a solid base for any deck or porch he is building.

Sill Plates

The GC is responsible for supplying and installing the sill plate on schedule. You must tell the dealer whether the GC is using a single sill plate or a double sill plate. Otherwise, the siding installed by the manufacturer will not correctly cover the sill plate, and it will be the GC's responsibility to fix the problem.

A single sill plate is all that is required when anchor straps are used. When anchor bolts are used, a double sill plate should be installed. With a double plate, the bolts should pass through the bottom plate into the top plate. Each nut must be countersunk and recessed into the top plate. If necessary, the top of the bolts should be cut off flush with the top of the sill plate so that the modules can sit flat on the plates. The top plate must be nailed to the bottom plate. (If the GC countersinks and recesses the nuts into a single plate, he will weaken the connection between the plate and the foundation.) Wherever there is an opening in the foundation, such as at a bulkhead or a drive-under garage bay, the sill plate should be bolted firmly into place on each side of the opening.

The sill plates for a modular home must be made as level as possible, ideally to within ⅛ inch, plus or minus, on all sections, and no more than ¼ inch. The GC may need to shim the sill plate every few feet to achieve this result. He will need to verify his measurements with a calibrated transit.

This responsibility must be taken seriously. A foundation that is not level can cause the modules to rest improperly on the foundation, resulting in gaps between the perimeter beams and the supporting sill plate and foundation. The gaps will exacerbate the normal "settling" of the house onto the foundation. Since the modules have to flex over a greater distance with unlevel plates, you will see more symptoms of settling, such as drywall cracks. More significant, because the modules are built to minimize flex in their structure, they will take longer to conform to an unlevel foundation. Consequently, the symptoms will continue to appear many months after they are first fixed as the modules slowly succumb to gravity. If you use an independent GC and he does not level the sill plates, the dealer and manufacturer will not be responsible for problems that show up in the house later due to excessive settling.

Knee Walls and Walk-Out Walls

For raised-ranch and split-level models, at least part of the foundation is typically made up of knee walls. These are usually wood-framed walls that are 4 feet or more above ground. The extra height above ground makes it possible to install larger windows than are typically installed in a basement. The larger windows help you create brighter rooms in the basement, should you wish to build them now or in the future. Sometimes the lay of the land makes these above-grade walls 8 feet tall, or even more, which enables the use of full-sized exterior doors in a "walk-out" wall.

Knee walls and walk-out walls must be securely fastened to the foundation. Their tops should be leveled to within ⅛ inch of each other and any interconnecting sill plates installed on full foundation walls. If the knee walls and walk-out walls are wood framed, they should have a double top plate, with the second plate shimmed level to the sill plate on the remainder of the foundation. The knee walls and walk-out walls should also be plumb with the foundation so that the modules sit flush with the walls. This facilitates a smooth installation of the siding. Diagonal measurements of each wall must be within ¼ inch of each other or the house will not sit squarely on the walls. The GC should not expect the set crew to make any adjustments to the walls in order to get them to conform to the modules. If the walls are not square after the modules are set on the foundation, the GC will be responsible for making any of the adjustments required to line up the walls with the modules.

Knee walls and walk-out walls are not as strong as concrete walls and so must be fully sheathed on the outside with the corners securely nailed before the set. Moreover, the inside of the walls needs to be carefully braced. Without proper sheathing, nailing, and bracing, knee walls and walk-out walls are prone to lateral movement, or "racking," which is a dangerous situation when setting a modular home.

Knee walls and walk-out walls have very poor insulation qualities. Even if you insulate the basement ceiling rather than the walls, ask the GC to insulate the knee walls and walk-out walls. Also have him install electrical outlets and drywall on the knee walls and walk-out walls.

Support Columns and Beams

The GC must have the correct number of support columns at the site by the day of the set so the set crew can cut and install them. Concrete-filled 3½-inch lally columns are recommended by most manufacturers, although equivalent support columns can also be used. If there is a long span between columns, larger-sized columns may be required. Expandable telescopic jack posts are not acceptable. As a general rule, it is best to use columns that are taller than the foundation and then cut them to size. For example, use 9-foot columns for 8-foot foundation walls. Although concrete walls are rarely full height (an 8-foot wall is usually less than 8 feet tall), the installed columns are almost always taller than the walls. This is because the support column footings sit below the bottom of the foundation walls if the concrete floor is not already poured and the underside of the modules sit on top of the sill plate, which itself sits on top of the wall.

The main carrying beam in the basement will be made of four to six 2×8 or 2×10 perimeter bands bolted together by the set crew. Because the main carrying beam is several inches wide and made up of separate members, many manufacturers specify oversized plates for the top of the lally columns. Instead of the usual 4×4×¼ plates most contractors are familiar with, they may be as big as 6×12×½, with a welded steel collar. Because of their size, the plates are able to support all of the perimeter bands. Lumberyards will probably not offer these plates, even though they will provide the lally columns. The modular dealer should supply the plates at a reasonable price.

If a support beam or flitch plate is to be used in the building of your home, the GC must have it at the site before the set begins. Otherwise the set crew cannot properly support the home.

Preparing the Basement or Crawl Space

The basement or crawl space must be as dry as practical before the house is set. The GC should remove as much standing water or snow as he can. If a concrete floor will not be poured, which is more likely with a crawl space, a vapor barrier must be installed on the ground.

It is not always possible to remove all of the moisture before the set. In this case, the GC should do what he can to quickly remove the moisture after the set. He should also keep the basement well ventilated until the moisture is removed. If the GC does not minimize the excess moisture, it may be absorbed into the home, causing the wood to expand. This will create additional problems when the home begins to dry out. As the wood shrinks, cracks will appear in the drywall and interior trim. Mold can also develop. These problems are the GC's responsibility.

Delivery-Day Responsibilities

The GC should remain on-site throughout the delivery. If something unexpected happens, he can help the crew. If there are decisions to be made, such as whether to cut down more trees or bring in more fill so the modules can safely enter the site, he can help you decide what to do.

The GC should assist the dealer with inspecting the manufacturer's protective coverings, which are installed to shield the modules against adverse weather conditions. Sometimes the coverings can be torn by low-hanging branches en route to the site. Do not attempt to cut open any of these coverings. If you do, you will be responsible for resealing the modules so they are weather tight or paying for any damage. Water has an amazing capacity to find its way through the smallest of holes. A module can suffer thousands of dollars in water damage in only one evening. In cold temperatures, the plastic coverings are particularly brittle, and tears can be difficult to tape shut.

Set-Day Responsibilities

The GC should remain present throughout the set. He should arrive before the set begins in case there are difficulties positioning the modules and crane or the sill plate or knee wall needs to be modified. If the unexpected happens, he can help. If there are decisions to be made, such as whether to place the modules so the interior stairwell walls or the exterior walls are aligned, he can give input.

Safety Concerns

A set should be an exciting event. The last thing you want is for someone to be seriously hurt. With a crane that can weigh up to 120 tons, swinging a long boom and cable strapped to a module weighing 8 to 15 tons, there is potential for serious injury. Friends, family, and other spectators should be kept away from the set crew and crane.

The GC must do everything he can to ensure that the house is safe as soon as possible after the set. To protect against accidental falls, he should build a set of temporary steps to one of the doors. The homeowner, contractors, suppliers, and inspectors will use the steps until the GC builds finished stairs.

The GC should barricade or sheathe over the rough opening framed in the floor for the basement stairs. If the home has a balcony overlooking a vaulted or cathedral area, the GC should install a temporary barricade or railing system to prevent someone from tumbling over the edge. A temporary cover should be secured to the top of the basement bulkhead.

A crane sets the module onto the foundation. Make sure the GC remains present throughout the set to deal with any problems.

After the set, the site will have piles of plastic wrap that was removed from the modules, some containing strips of wood with nails or staples. The quantity of material almost always surprises customers and GCs without modular experience. The GC should take all of the material left over from the set to the dump or put it into a dumpster immediately after the set. If the site does not have room for a dumpster on set day, the GC can have one delivered the following day.

The GC should have a zero-tolerance policy regarding alcohol or drugs on the job site. The homeowner can support the GC in this regard by not bringing any alcohol to the site for his employees or subcontractors. If you supply alcohol, you could be liable for any resulting injuries that occur on the job site or when the contractors drive home.

Construction-Site Toilet

Like everyone else, contractors are most productive when they can conveniently use a bathroom. If they have to make a trip to the gas station or a restaurant, it costs time; and if they do not have time to make the trip, they may choose to relieve themselves on your property. Eliminate this problem by renting a portable toilet until the plumbing is hooked up. Unless it will impede the movement of the crane and modules, the GC should have it in place for set day.

Trash Removal

As mentioned above, set day will produce a lot of trash. The button-up phase and construction of site-built structures will generate trash daily. Therefore, it is usually best if the GC uses a large dumpster.

The GC may decide instead to use a truck or van to carry the waste from the site. In that case, he should provide a container or at least designate a location for everyone to place his trash. This can work as long as the GC regularly removes the trash. If he does not, your new neighbors may find trash being blown into their yards. You would probably prefer to get to know them under different circumstances.

As you unpack your belongings, you will undoubtedly generate a lot of trash. Resist the urge to use your GC's dumpster. Even though you are paying for it, the GC needs all the room he can get for construction debris. In addition, if your contribution fills the dumpster prematurely, you may receive a bill for a second dumpster, which will cost a lot more than the extra time it would have taken you to use the town's disposal services.

Completion of Roof Sections

Attic gable end windows cannot be installed until the set crew lifts the roof into place. Often, this is also true of skylights. Since a home cannot be made weather tight until these are installed, the GC should do this immediately after the set.

The GC is also responsible for any roof construction that is not completed by the set crew. The following types of roofs sometimes require the assistance of a GC on-site:

- Any roof pitch taller than 12-in-12, since it will likely be too steep to build and ship efficiently

- Decorative gables and gable dormers, which most often sit on the front of the main roof

- A saddle, with its accompanying valley, connecting two sections of roof that are perpendicular to each other, as in a T-ranch

- The small roof that goes over a walk-out bay

- Skylights that the manufacturer was not able to install because of the folding roof

Some set crews do not complete these roof sections, so the modular dealer needs to tell you if he expects your GC to do the work.

Protection of Exposed Modular Sections

There are many types of house plans that cannot be built completely at the factory. For example, mudrooms that do not correspond to module sizes and turrets are often built on-site. This does not pose a problem when the manufacturer is able to frame the modules so the set crew can readily make them weather tight. If a site-built mudroom connects to an installed exterior door on one of the first-floor modules, for example, the home will remain protected from the elements while the GC builds the room. Sometimes, however, a site-built structure connects to an open section of a module. For example, a Victorian turret that is being built on-site by the GC will likely require a large framed opening in the adjacent walls of one or two modules. Although the manufacturer will make every effort to deliver the modules with the openings weather tight, the protective covering is not designed to withstand heavy winds and rain indefinitely. Once the structure is framed and made weather tight, the risk to the modules is removed. Accordingly, the manufacturer expects the GC to begin building the turret immediately after the set. If the GC is unable to do so, which can happen because of scheduling conflicts beyond his control, he is responsible for taking whatever measures are required to protect the modules until he can build the structure.

Panelized Structures

If the manufacturer has the capability, you and your GC may decide to have it panelize those parts of the design that cannot be built as modules because they are too wide or too long to be shipped. Most manufacturers will build panelized garages and many others will build a panelized room, such as an attached angled mudroom or a large family room. Most set crews will erect as much of the structure as possible if the foundation and sill plate are ready. Verify this with the modular dealer. Otherwise, the GC will have to erect the panels. Either way, the GC will be responsible for completing the panelized structures.

Customers sometimes misunderstand what the dealer is including with his panelized structures. Most manufacturers include only the framing, windows, doors, sheathing, siding, and shingles. They seldom include any insulation, electric, plumbing, heating, drywall, painting, or moldings. They sometimes do not include the roof trusses. It is critical that you know exactly what you are getting and what the GC needs to do. The dealer should provide you with a complete set of plans for each panelized structure.

Material Inventory

As explained on page 147, do a careful inventory of all materials "shipped loose" with the home by the manufacturer. The GC should move all fragile and small items to a safe place, such as a closet. If the GC fails to do this, he will not discover that items are damaged or misplaced until he is ready to install them.

In cold weather, most manufacturers ship the paint, stain, and caulk in one of the transporters rather than one of the modules, where they could be left to freeze. These items should be placed immediately in a heated space, such as the GC's office or home. He can bring them to the site as soon as the customer's house has heat.

Security

The GC is responsible for helping to secure the home from unwanted entry. As much as practical, he should cover all potential openings into the modules and basement, and he should make sure that all doors and windows are locked. You and the GC will need to agree on a system for securing the home when no one is using it and for making it accessible when the GC needs to work. Some GCs will hide a key on the property and inform everyone who needs to know where the key is hidden. If this is done with your home, have the doors rekeyed by a locksmith the day you move in. An easier alternative is to keep all of the keys and have the GC replace the lockset on one door with a temporary one. He can then give keys for the temporary lock to everyone who needs them or hide a key on-site. When the GC is done with his work, he can reinstall the original lockset.

The GC should help to secure all materials shipped with the home, whether installed in the home or shipped loose. The dealer will not be responsible for any materials that are stolen or damaged. The customer should find out in advance if his insurance will cover any theft that occurs at the building site.

Button-Up Responsibilities

The GC can now complete the button-up work on your modular home. He will need to finish the interior and exterior carpentry and hook up the plumbing, heating, and electrical systems.

Protection of Interior Surfaces

Some manufacturers do a good job helping the GC protect the factory-installed finished flooring from damage. Recognizing that many people will walk through a home during its button-up phase, they place cardboard and plastic over the carpet, vinyl, wood, laminate, and tile floors, as well as the stain-grade wood stairs. Not all manufacturers do this equally well, however, and some do it poorly. The GC should immediately cover any unprotected flooring.

Most construction workers wear work boots that tend to hold dirt, mud, snow, and ice. The GC should take steps to minimize this threat to brand-new floors. For example, he can mount a heavy brush outside the exterior doors so everyone can clean his shoes before entering, or he can require his subcontractors to wear protective booties when they enter the home. Unfortunately, the booties are impractical for someone carrying tools and materials in both hands.

Other surfaces will also need to be protected. Kitchen countertops are particularly vulnerable, since subcontractors are prone to place their tools on them. Subcontractors are also known to store some of their fragile materials in tubs and showers. The GC should explain to his subcontractors that these practices are forbidden, and that the subcontractors will be obligated to repair or replace any damaged surfaces if they ignore the warning.

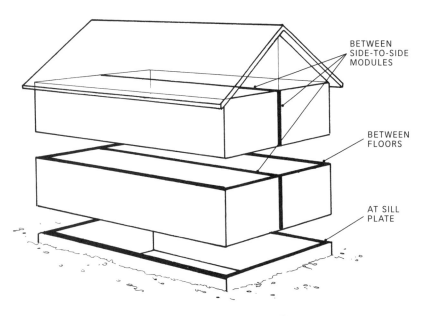

BETWEEN SIDE-TO-SIDE MODULES

BETWEEN FLOORS

AT SILL PLATE

The GC must ensure that the home is sealed against air infiltration where two modules are joined, both side to side and top to bottom, as well as where the modules sit on the foundation.

Air Sealing

One of the most important steps that a GC can take to make a modular home energy efficient is to seal the gaps between modules. The manufacturer can control air infiltration within each of the modules, but when the modules are placed side by side or stacked on top of each other, significant gaps are created. It is just not possible to bring two modules together tightly when a cable is wrapped around each one while being lifted into place. In addition, even when the framing of one module is tightly butted up against the framing of another module, it is not possible to make the joint airtight. Very few set crews completely seal these gaps, so the GC should assume responsibility for the job.

A modular home that is properly sealed is significantly more energy efficient than a typical site-built home. When a modular home is poorly sealed, however, it will leak more than a site-built home. The GC should make every effort to seal the following:

- The basement and attic marriage wall
- The interior marriage wall wherever there is a passageway, door, or clear-span opening
- The exterior marriage wall along the gable ends
- The exterior band between floors on two-story homes
- The exterior sill plate where the first-story modules sit on top of the foundation

On the interior, the best way to seal large gaps is with expandable foam. On the exterior, there is a special technique that should be used before the GC completes the siding. Most manufacturers hold back a small piece of the exterior sheathing where two or more modules join. The GC must insert a piece of sheathing to bridge the space between the adjoining modules. The sheathing connects the modules structurally and creates a flat surface for the siding. By itself, however, the sheathing does not do a good job of reducing air infiltration, even when combined with the sill seal installed between modules by most set crews. But if the GC applies two beads of caulk to the back of the sheathing before nailing it to the modules, this will create what is in effect a gasket seal. The GC can then finish by caulking the outside edges of the installed sheathing to close the remaining gaps.

The GC should carry out a similar procedure where the bottom of the first-story module joins the sill plate and where the top of the modular wall connects to the roof overhang at the eave. He should also foam-seal the gable-end triangle sections under the roof; he might first need to brace the sections from inside the attic. Completing all of these steps will create a significantly more energy-efficient and comfortable home, for a negligible cost.

Additional Siding Preparation

After the GC has sealed the exterior against air infiltration, he needs to pad out the walls on side-by-side modules wherever they do not line up. This will allow the siding to rest evenly. If the house is wrapped in an air-infiltration barrier, the GC must install the remainder of the house wrap before installing the siding.

These steps need to be completed regardless of the type of siding you select or who provides it. If, however, you select wood or fiber-cement siding, whether provided by the manufacturer or the GC, the sheathing will need to be padded out more carefully wherever abutting pieces are misaligned. Vinyl siding is more forgiving of some unevenness.

Install the Siding

The standard vinyl siding offered by modular manufacturers will be partly installed at the factory. The manufacturer will leave off some of the siding for the GC to install so that the finished product will look profes-

sionally installed, with all of the pieces fitting as designed. Manufacturers particularly want to avoid seams in unusual places, which is why the gable end of any home that is at least two modules wide will always require the siding to be installed on-site. Otherwise, you would have an unsightly seam right down the middle rather than the staggered joints you expect to see. Most manufacturers also leave off the siding on the eave sides of second-story modules because they cannot guarantee that the siding will meet perfectly between the top and bottom floors. The manufacturer must leave off the siding wherever it has framed a rough opening in one of the exterior walls so the GC can install an item on-site, such as a door or window. This will make it possible for the GC to hang the siding evenly against the edge of the item after he installs it.

The GC must install the last course of siding along the bottom and top of the eave side of each module. On a one-story, this covers the sill plate and the top of the wall immediately below the eave. On a two-story, installing the section of siding between stories ties them together. While completing the siding, the GC needs to install all of the inside and outside corner posts. The GC also sides any shed dormers, A-dormers, decorative gables, and bump-outs. Finally, if a few pieces of siding are damaged by the delivery or set, the GC needs to replace them. The manufacturer ships extra pieces of siding in case this is necessary, and it compensates the GC if the work involves more than a few pieces.

The GC should stain or paint any site-installed pieces of wood or fiber-cement siding provided by the manufacturer. The nail holes and end cuts need to be touched up even if the siding is prefinished. If the GC installs wood siding, have him seal both sides of the siding to reduce the chance that moisture will be absorbed through the back and prematurely peel the finish.

Install the Soffit and Fascia

In most cases, the manufacturer partially frames and finishes the roof overhangs, but the GC must complete the work. On the eave sides, the overhangs are delivered folded up on top of the roof, except in homes that are only 24 feet wide; these come with the overhangs fully installed. After the set crew folds the eave overhangs back down, it should nail them to the home. The GC verifies that this is done and then caulks where the overhangs attach to the house. For oversized eave overhangs, the GC should do some additional framing to complete the soffits. He may also need to install the eave overhangs on a one-story or Cape Cod home with 9-foot ceilings. The framing for full-sized gable-end overhangs must be completed by the GC, since they cannot be finished until the roof is lifted into place.

The GC usually builds any gable-end returns, since they are not available from most manufacturers. This is also true with a shed dormer having a decorative overhang on the rear or gable ends.

After the overhangs are constructed, the GC can install the fascia, soffit, and drip edge. He should also repair or replace any pieces that were damaged in the set. The dealer should supply the replacement materials and pay the GC a fair price for the additional work, unless the repair is relatively minor.

Additional Exterior Tasks

The GC must complete the installation of the decorative exterior fixtures, such as shutters, door pediments, window mantels, trim boards, and dentil moldings. Some manufacturers install these where they can and others do not install them at all, so you will need to ask your dealer what is required and pass on the information to the GC. For each exterior door, the GC may have to install a J-block, which is needed for mounting an exterior light, as well as a toe kick under the threshold. If you want to add a screen or storm door, have the GC install it.

For any home where the top module is cantilevered over the bottom module, such as a garrison Colonial or raised ranch, the GC must insulate and cover the exposed area under the overhang. Nonperforated-vinyl soffit can be used as the cover.

Bulkhead and Basement Stairs

The GC should immediately build bulkhead stairs for a basement with a bulkhead entrance, unless a precast unit was installed. For a basement with access from inside the home, the GC should build the stairs as soon as he can. Tell the GC if you want the stair risers to be closed and the stairwell to be finished in drywall.

Foundation Ties

A poured-concrete foundation wall will leave behind foundation-form ties inside the basement. Since these spikes are potentially dangerous, the GC should snap them off as soon as possible.

Water and Sewer Pipes

Water and sewer lines, private or public, should be sealed by the GC where they penetrate through the foundation walls.

Heating System

Contractors are less productive when they are cold. Many construction tasks, such as taping and painting drywall, cannot be done until a home is reasonably warm. Consequently, if a home is being built in cold temperatures, the GC needs to get the heating system up and running quickly. He will be delayed, however, if the heating system is going in the basement and the slab cannot be poured because of frost. The installation will also be delayed if the electrical power is not immediately available. In these circumstances, the GC must supply temporary heat, which is more costly and also less effective.

As-Built Mechanical Plans

Once the home is set and ready for the electrician, plumber, and HVAC contractor to begin, the GC needs to make sure that they have a copy of the manufacturer's most current plans. Usually these are the "as-built" or "production" plans, which show what the manufacturer did when it built the home, including any last-minute revisions. The GC and his subcontractors should read these plans closely.

The as-built plans show the locations of ceiling and floor access points created by the manufacturer to enable the mechanical contractors to make their connections. The GC should walk through the home immediately after the set to verify that each access is located where indicated. This is critical because manufacturers occasionally make a mistake and cover an access with drywall, sheathing, or carpet. If you do not find an access, contact the modular dealer, who will in turn contact the manufacturer.

The Electrician's Responsibilities

If the electrical power is coming to the home underground, the GC needs to coordinate the electrician, excavator, and utility company. The excavator digs a trench so the electrician can place conduit in it. Either the electric company or the electrician runs the wire through the conduit. Underground electric service is preferable to overhead, but it almost always costs more.

The homeowner should contact the phone and cable companies to set up accounts. The electrician can run both wires to the home.

Once the home is set on the foundation, the electrician can mount the electrical panel box supplied by the manufacturer in the designated location, mount the meter socket, and begin making the connections in the house. The GC should make sure the electrician has the as-built plans and understands their significance before he begins his work.

Each module contains one or more bundles of wires that the electrician must connect to the panel box. Some modular manufacturers design their electrical system so that all of the wires are run from each module directly to the panel box, while others require some connections between modules, often made with what are called amp connectors. Two-story homes with the panel box in the basement require the set crew or electrician to drop the second-floor wires through a chase or conduit to the basement. The electrician also needs to run the phone and cable lines from each module to their appropriate locations. The telephone company may need the electrician to create a network interface.

After completing all of the connections, the electrician should label each of the circuit breakers in the panel box. He should then conduct a continuity check of all circuits. If he discovers any loose wiring, he should make the needed repair. If one of the circuits does not work, he should contact the modular dealer.

The electrician must connect all interior and exterior fixtures, including lights and ceiling fans, whether supplied by the manufacturer or by the customer. Discuss with the GC whether he expects you or his electrician to supply the lightbulbs or assemble the ceiling fans. The electrician must also connect the smoke detectors, which the manufacturer supplies. The electrician should complete any appliance hookups not done by the manufacturer. This may include the dishwasher, washer, dryer, cooktop, wall oven, garbage disposal, central vacuum, and fireplace blower.

If you are installing a central vacuum cleaner, you need someone to connect the pipes. The electrician can complete this task if you buy the vacuum from the modular manufacturer. Connecting the pipes does not require a plumber's license, and the electrician will wire the vacuum. The electrician should be reminded to make the inter-module connections on a two-story. If you purchase the vacuum locally, ask the supplier to complete the hookup in the basement. The electrician will still need to wire the vacuum, however, and make any inter-module connections.

A basement needs to be wired per code with smoke detectors, lights, and receptacles. If you forget to order a smoke detector for the basement, the electrician must use a brand that is compatible with the system installed by the manu-facturer. If the attic is accessible and has storage trusses, he should install a smoke detector and receptacles there, too. The electrician should also install a light in the attic.

The electrician must wire the heating system per instructions. Wiring a heating system is a lot more complicated than it used to be, and many HVAC contractors do not understand the new systems. This is where a GC can make all the difference, since he can make the two contractors work together. The electrician also needs to connect the thermostats, which should be supplied by the HVAC contractor. The electrician will have an easier time with electric heat, since he has only to tie in the circuits for each room to the panel box.

If you purchase a separate domestic hot-water tank, install central air conditioning, or drill a well, the electrician must wire the equipment. He should also complete the electrical work for any panelized structures shipped by the manufacturer as well as for any structures the GC builds, including the second floor of an unfinished Cape Cod.

The Plumber's Responsibilities

If the height of the septic system or municipal sewer line allows, the GC may want the plumber to place the waste line under the basement slab. This enables you to put a bathroom in the basement without having to install a pump.

Before the plumber begins his work, the GC must make sure the plumber under-stands the as-built plans. When working on a two-story home, the plumber must connect the supply, waste, and vent lines between the upper and lower modules. He may also need to make some connections between the front and rear modules on the second story. The as-built plans will show the access panels for the second-story plumbing in one of three places: the floor of the second floor, the ceiling of the first floor, or the walls of either floor. If the plumber cannot find the access panels, he should immediately contact the modular dealer, who will contact the manufacturer.

Some modular manufacturers install only a minimum amount of plumbing on the second floor of two-story homes. This forces the plumber to do most of the under-floor plumbing for the second-floor bathrooms. To provide access to the under-floor plumbing, these manufacturers leave off all the drywall on the first-floor ceiling below the bathrooms. This results in additional drywall expenses.

All of the first- and second-floor plumbing terminates in the basement above the bottom of the floor joists. The manufacturer cannot bring the pipes any lower, since they would be crushed on the delivery carrier. Consequently, any traps for first-floor tubs and showers that fall below the floor joists need to be installed by the plumber.

After making the inter-module connections, the plumber must connect all of the supply lines that are stubbed below the first floor to the cold- and hot-water supply lines. This includes those from the second floor of a two-story or Cape Cod. The cold-water supply line is in turn connected to the well holding tank or municipal water line and the hot-water supply line is connected to the water heater. The plumber must also connect the waste lines to the septic system or municipal sewer line.

Since the roof on a modular home is folded down for delivery, the manufacturer can only stub the vent stacks into the attic

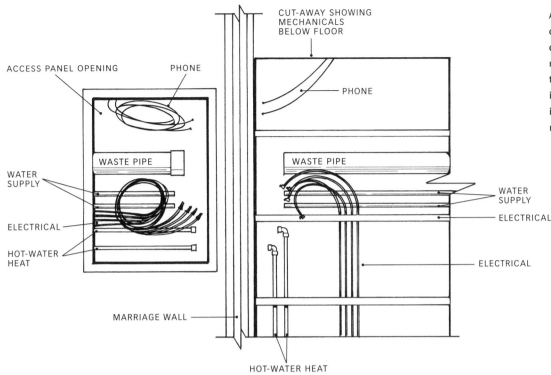

ACCESS PANEL OPENING   PHONE

CUT-AWAY SHOWING MECHANICALS BELOW FLOOR

PHONE

WASTE PIPE

WASTE PIPE

WATER SUPPLY

ELECTRICAL

HOT-WATER HEAT

WATER SUPPLY

ELECTRICAL

ELECTRICAL

MARRIAGE WALL

HOT-WATER HEAT

Access panels, which can be located in a floor, ceiling, or wall, allow the mechanical subcontractors to connect the plumbing, electrical, and heating systems between modules.

floor. The plumber must extend the stacks through the roof. Some manufacturers facilitate this by providing "boots" in the roof. If the boots are not in place, the carpenter may need to assist the plumber by cutting the holes and installing them.

The plumber needs to complete any appliance hookups not done by the modular manufacturer. This can include a dishwasher, garbage disposal, washer, and icemaker. The plumber should install at least two exterior freeze-proof hose bibs and, if desired, a basement sink. To make it possible to fix minor plumbing problems without having to shut off the entire system, the plumber should connect the fixtures and appliances with isolation valves.

Some plumbing inspectors require the plumber to pressure-test the system before they approve it. Typically, this is an air test, but it could be a water test. If possible, opt for the air test, since a water test can damage the home if a pipe is not properly connected. An air test should be performed even if the inspector does not insist on it. It should be completed as soon as possible and

certainly before the access panels are closed. Before conducting the test, the plumber should inspect and tighten all traps, valves, and other connections, including fittings installed by the manufacturer. The plumber should ask the inspector if he would like to be present for the test so he can approve the results. Some inspectors will take the plumber's word on the test results. Since the test requires the plumber to remove the toilets, he needs to reinstall them after the test.

The plumber is responsible for correcting any minor leaks, even ones he did not cause, but the manufacturer is responsible for leaks behind the drywall and any defects in workmanship. The manufacturer will likely ask for the plumber's assistance if he discovers any problems.

The plumber needs to complete any work required for site-built structures, such as a finished basement, a mudroom that contains a washer and dryer, or the second story of a Cape Cod. The plumber also finishes the plumbing for any gas fixtures or appliances. HVAC systems and water heaters are discussed below.

Appliance Hookups

The electrician and plumber are responsible for helping to hook up the appliances, but they cannot assume responsibility for all of the details outlined in the owner's appliance manuals, such as for a downdraft cooktop unit. If there are any questions about how to set up an appliance, ask the appliance dealer.

HVAC Contractor's Responsibilities

For electric baseboard heat, the modular manufacturer wires all of the baseboard units and thermostats. The electrician only needs to tie the baseboards into the electrical panel box.

Some manufacturers provide the boilers for hot-water baseboard heating systems and the furnaces for forced-air systems. This can be helpful to the GC, especially if the equipment is already installed in a utility room. Local HVAC contractors, however, sometimes resist hooking up equipment supplied by anyone but themselves. They may feel deprived of some profit and not want to install a system they have never worked with before. In addition, they do not want to be responsible for warranting the system if it fails, since the failure may be due to an equipment malfunction rather than their installation. They know from experience that these situations often lead to finger pointing about who is at fault, and they do not want to be caught holding the bag. Some modular manufacturers solve these problems by contracting with a national or regional HVAC company that is able to install and service the equipment anywhere. When done well, this arrangement can be a major aid to the GC.

If you purchased a heating system from the dealer, the GC will almost certainly need to wire the system as well as hook up the gas or install an oil tank. Do not assume that the GC and his HVAC contractor will accept this arrangement without discussing it with them first.

When the GC's HVAC contractor installs a hot-water baseboard system, the modular manufacturer and the contractor share responsibility for its design. The manufacturer determines the correct number of baseboard units for each room and installs them. The contractor determines the size of the boiler and installs it. The homeowner should decide how many heating zones to include and which rooms should be included in each zone. The modular manufacturer runs the pipes to accommodate the different zones and provides the thermostat wires as directed by the homeowner.

The GC should make sure that the contractor has the as-built plans and understands their significance. The contractor can then install the boiler and hook it up to each zone. With one-story homes and the first story of two-story homes, he connects the pipes stubbed into the basement for every baseboard unit to the boiler. For two-story homes, the contractor uses the access panels to complete the loops between the second-story modules and the first- and second-story floors. Then he connects the second-story supply and return lines that are stubbed into the basement to the boiler.

Before filling the system with water, the contractor should air-test the system for leaks. After he completes his work, he asks the electrician to connect the thermostats and wire the system. The contactor returns to fire up the system and makes sure it functions as designed.

If you live in a climate with cold winters, you may want the contractor to add antifreeze to the heat loop. You should do this if you are away from home in the winter or if you have a drive-under garage. Leaving a garage door open on a cold day can cause the pipes to freeze.

Another form of hot-water heat that is becoming very popular, in spite of its high installation cost, is a radiant-floor system. Hot water is run through rows of tubes installed either above or below the sheathing. Most people experience radiant-floor heat as very comfortable. It is particularly effective on the first floor when the basement is unheated. Although only a few modular manufacturers offer radiant-floor heat, it can be installed on-site. On the first floor, it is easiest to put the tubing on the basement ceiling. It is more difficult for the GC to install on the second floor, but it can be done on top of the sheathing.

If you select a forced-air system, the manufacturer might agree to install the ducts. If the price is right, this will be a big help to the GC. Otherwise, the HVAC contractor will need to install the ducts and the furnace. When a contractor installs the ducts on-site, he places them on the basement ceiling for the first story of all homes and on the attic floor for the second story of two-story homes. A chase, usually built into the home by the manufacturer, is used to bring the ducts from the basement to the attic. One way to avoid needing a chase is to add a separate heating unit in the attic to serve the second floor. In fact, homes larger than 2,500 square feet are usually better served by two separate forced-air systems. Dual units are often used when hydro-air systems are selected. These systems warm the air by first heating water. They cost more, but are more energy efficient.

Although some HVAC contractors seal and insulate the ducts, this is not a code requirement in all states. Insist that both steps be completed, since they make the system more energy efficient. You and your GC should agree in writing where the registers will be placed in each room. This ensures that they are not placed where you intend to put a piece of furniture. For duct locations, specify which rooms you want on a separate zone and where you would like the thermostats located. Since forced-air systems blow dry air, you should get a good humidifier and a good HEPA filter system. Both are definitely worth the additional expense.

A forced-air heating system makes it easy to install central air conditioning. If you want AC but cannot afford it right away, make sure the ducts are sized correctly. Ducts must be larger for air conditioning than for heat. When you install the AC, ask the GC to select one with a seasonal energy-efficiency ratio (SEER) of 12 or higher. Make sure to tell him not to oversize the compressor, a common mistake made by HVAC contractors. Appropriately sized equipment will lower operating expenses and provide better dehumidification. In addition, ask the GC to supply thermostats that are designed to handle both heating and cooling.

Another option that is gaining in popularity is hot-water baseboard heat with central air conditioning. This will cost you more than a combined forced-air system, but some people find that it offers the most comfortable heating and cooling in a home.

Unless you select electric heat, the GC must provide a fuel-oil tank, a propane tank, or a connection to natural gas. If you are getting a tank, make sure the GC specifies the size. If it is a large propane tank, you may want to bury it.

For an oil-fueled system, the HVAC contractor must install and connect the oil tank. It is the homeowner's responsibility to select an oil supplier and schedule the first delivery of oil immediately after the tank is installed so that the contractor can test the system. You must decide whether to vent the system with a power vent or a flue pipe. A power vent cools the gases before sending them out the side of the home. It cannot be located near a window or door. A flue pipe is made of clay, when it is part of a masonry chimney, or metal. The concrete block of a masonry chimney can be left unfinished or covered in brick, natural stone, or cultured stone. A metal flue can be left exposed or enclosed in a wood-framed chimney chase that can be covered in siding, cultured stone, or brick veneer. Most people prefer a flue pipe because you can usually smell the exhaust gases from a power vent, and the gases tend to blacken the siding around the vent. Power vents are sometimes used, however, because they are considerably less expensive than flue pipes.

Today's tighter homes are best served when a gas- or oil-heating system is outfitted with a fresh-air supply source to help prevent back drafting of combustion gases into the home. A fresh-air supply source is particularly necessary if the heating unit is located in a small area.

If you are building an Energy Star home, upgrade to an appropriately rated heating and air-conditioning system. The greater energy efficiency will pay back its higher initial cost with lower operating expenses for years to come.

Water Heater

Hot water can be generated in a separate system, or the heating system can be used to directly heat the water. The latter system can work with or without an attached storage tank. Depending on the system you choose, the installation can be done by an electrician, plumber, or HVAC contractor.

Stand-alone water heaters are the only option with a forced-air furnace. They can be heated with electricity, gas, or oil. With a hydro-air system, you can use the boiler to produce hot water. A dedicated oil hot-water system has the same venting options as an oil-fired heating system.

When selecting a stand-alone water heater, pay close attention to how fast it produces hot water. This is known as its recovery rate. Electric water heaters have a long recovery rate, which means they take a long time to heat water. There are some stand-alone water heaters that do not have a storage tank. They produce sufficient supplies of water, unlike conventional on-demand systems.

Using the heating system to generate hot water will save money in the long run, espe-cially if you install an on-demand system that does not require a tank. With these systems, water is heated only when you need it, and does not have to be stored. The problem with most of these tankless systems is that they do not generate a lot of water at any one time. If you turn on the dishwasher, washer, and shower at the same time, the system will probably not produce enough hot water. They also do a poor job of producing enough hot water to fill a large soaking or whirlpool tub. For a small household, a tankless system can save money as long as you can live with its limitations.

A better way to use a boiler to produce domestic hot water is to connect the system to a separate storage tank. If your budget can handle the higher initial cost and you get a large enough tank, you should rarely run out of hot water. These types of systems cost much less to operate since they use the same energy that heats your home to heat your tap water.

If you are building an Energy Star home, upgrade to an appropriately rated water heater, one that is more energy efficient and has a good recovery rate.

Product Literature and Warranty Information

The GC should provide you with the literature and warranty information for your heating system, cooling system, and water heater. This means he must obtain them from his subcontractors.

Chimney Chase for Zero-Clearance Fireplaces

Wood-burning zero-clearance fireplaces require a metal flue, which is usually enclosed in a wood-framed chimney chase. The chase can be finished in siding to match the house or in cultured stone. If you purchase the fireplace without a chase, the GC needs to build the chase to the exact size required by the dealer and insulate it below the fire stop. If you use another metal flue for either a heating system or water heater, they can be placed together in an appropriately sized chase.

Most gas fireplaces do not require a full chimney, since they do not use a flue. If they are installed on an outside wall, however, either the manufacturer or the GC needs to build an insulated wood-framed box to enclose the fireplace.

Masonry Fireplace

When you order a masonry fireplace, ask the GC to specify the size of the firebox. You should also find out if it comes with a raised hearth and whether the exterior will be corbeled, which creates a stepped look partway up the chimney. Instruct the GC to outfit the fireplace with an outside air supply line. This outside air will improve efficiency and reduce smoking.

If the GC is building a masonry fireplace, the modular manufacturer should frame an opening in the floor. For an interior chimney, the manufacturer should frame an opening in the ceiling and roof, and the GC should build a cricket on the roof above the chimney to prevent snow and ice from building up. With an exterior chimney, the GC needs to cut back the overhang in that area of the roof.

Completing the Interior

When two modules are joined, the openings between them must be tied together. Most of these openings are doorways, headered passageways, and clear-span ceilings. They may also include stairwell openings and other vertical spaces between stories, such as a two-story vaulted foyer.

One of the GC's first tasks is to bolt together the laminated beams installed by the manufacturer in two side-by-side modules. The beams, which typically replace a marriage wall, are in the ceiling or floor system. The GC must complete the bolting per the manufacturer's plans and specifications. If the two beams cannot be brought tightly together, which is not unusual, the gaps should be shimmed. This will allow the two beams to function as one piece even after the home settles.

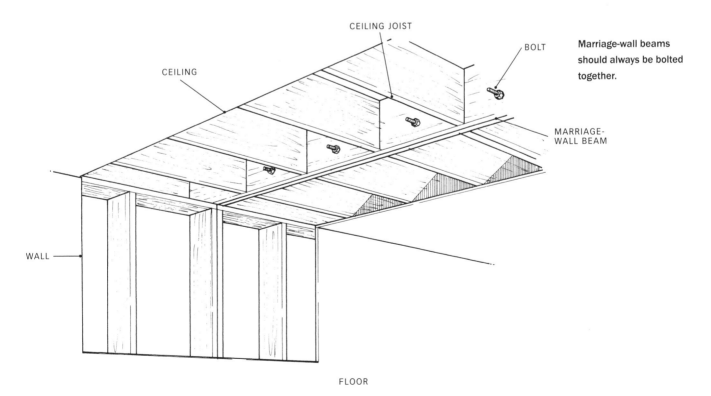

CEILING JOIST

BOLT

CEILING

Marriage-wall beams should always be bolted together.

MARRIAGE-WALL BEAM

WALL

FLOOR

The GC should install the doors, moldings, and drywall in the marriage-wall openings. Before he can do this, he must adjust the openings. Since the two married modules are built separately at the factory, small variations in the size of the openings may occur. The adjustments, which are minor and not structural, will help the materials fit properly.

Adjusting each marriage-wall opening makes its jamb width uniform. The GC pulls the two modules together with clamps until the walls are plumb. Before releasing the clamps, he fastens the walls together with metal strapping spaced 3 feet apart; a Simpson TP37 mending plate works well. He then inserts shims in any gaps behind the straps. The shims will help maintain a uniform distance between the walls regardless of subsequent shrinkage or settling.

A similar procedure is required for a two-story vaulted space and all stairwells between stories. The first-story ceiling system must be married to the second-story floor system. A manufacturer creates a vaulted foyer by excluding both the ceiling of the foyer and the floor of the second-story above. It also leaves off a band of drywall along the foyer walls between floors. The GC installs the rest of the drywall on-site, joining the two stories with one continuous floor-to-ceiling wall. Before the GC can add

Speaking from Experience

ONE OF MY CUSTOMERS, who had hired an independent GC, called me four years after he moved into his home to complain about a bulging drywall seam in his vaulted foyer wall. He told me that he had already fixed the seam three years in a row, but it had reappeared that winter. He wanted to know what was wrong with his home. When we removed the drywall at our inspection, we found a ¼-inch gap between the framing of the first-story ceiling and the second-story floor. The GC had failed to put any shims in the gap, so there was nothing to stop the modules from moving up and down in the winter when the heat was drying out and shrinking the framing or in the summer when the humidity was causing the framing to expand. Since my customer completed his previous repairs to the foyer in the summer when the wood had already expanded, it was not surprising that the problem was occurring in the winter when the materials were drying out. The resulting shrinkage of the framing was causing the second-story floor to press down on the first-story ceiling. This was buckling the band of drywall. Drywall is too brittle to resist that kind of compression. We explained to the customer what the correct procedure was and even helped him complete some of the work. When I spoke to him the following spring, he was very happy to report that the problem had not reappeared.

If you hire an independent GC without prior modular experience, make sure he is advised of the correct procedures for buttoning up a modular home. Your dealer should be able to help with his education.

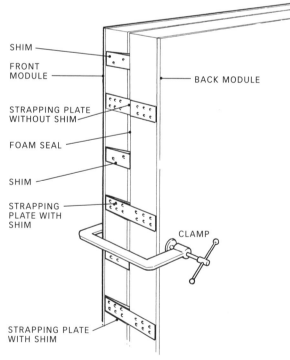

How to join two modules at a marriage-wall opening.

the drywall, however, he needs to ensure that the foyer framing does not move after the drywall is installed.

In all two-story modular homes, the perimeter of the second-story floor system sits on top of the perimeter of the first-story ceiling system, but the two surfaces never match up perfectly. There are always a few small gaps between the two ranging in size from a barely detectable $\frac{1}{32}$ inch to a more pronounced $\frac{1}{2}$ inch. In time, the second-story modules will gradually settle and close the gaps. This does not pose a problem for the exterior of a modular home, but it can buckle the drywall that spans the wall systems of a vaulted two-story. The GC can significantly reduce the chance that this will happen by shimming the gaps and installing the metal strapping along their perimeters in the foyer. Although the drywall might still crack due to changes in humidity, the cracks will be smaller and easier to fix.

After the preparatory work has been done, the GC can install the marriage-wall doors, trim out the passageways, and join the walls and ceilings with drywall. While the floor plan will dictate where the doors are placed, make sure that the GC knows if the headered passageways are to be trimmed out in drywall or encased in wood moldings.

The GC needs to install baseboard along the marriage wall wherever the manufacturer could not finish it at the factory. If the GC is installing some locally purchased flooring in these spaces, he should delay putting in the baseboard until the floors are installed. He may also want to wait if you have unfinished moldings that need to be stained. In this case, the manufacturer should have temporarily tacked the moldings in place, if you followed the advice mentioned earlier. The GC can remove these moldings, stain them along with the marriage-wall moldings, and install them.

Sometimes the most demanding task at the marriage wall is the drywall work. The

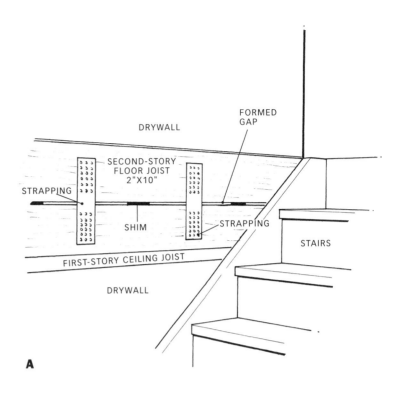

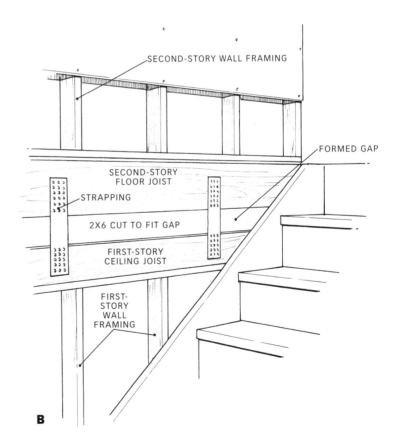

Illustration A shows how to join the two stories of a vaulted foyer when the second-story floor joists sit directly on the first-story ceiling joists. Illustration B shows what to do when the floor joists are spaced several inches above the ceiling joists.

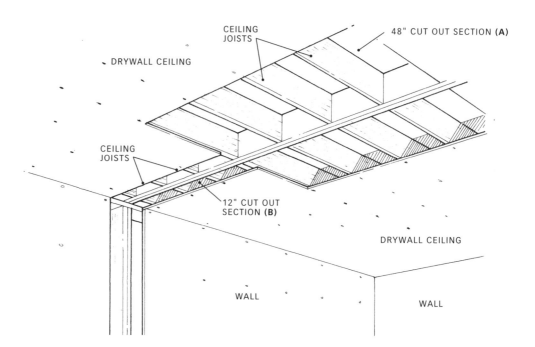

CEILING JOISTS

DRYWALL CEILING

48" CUT OUT SECTION (A)

CEILING JOISTS

12" CUT OUT SECTION (B)

DRYWALL CEILING

WALL

WALL

Installing a 48-inch piece of drywall (A), rather than a 12-inch piece (B), will make it easier for the GC to level the ceiling at clear-span openings along the marriage wall.

GC needs to piece in sections of drywall at all flush ceiling and wall areas in rooms that span two modules. To ensure that the drywall fits properly, the GC must align the married wall and ceiling surfaces as much as possible before adding in the connecting drywall. Usually this is done by padding out the lower surface on one of the modules.

The manufacturer can facilitate this task by holding back the drywall 23 inches on each module wherever the ceilings or walls directly meet on two adjoining modules. This allows the GC to piece in a 4-foot sheet of drywall to cover both modules. The GC can then more easily level those areas, since he has more space to make the appropriate adjustments. Many manufacturers hold back only 6 inches on each module, requiring the GC to piece in a 1-foot strip of drywall, which is often too small to compensate for any unevenness between modules. Sometimes, the result is an unsightly bump. Ask your GC what he would prefer the manufacturer to do. If he agrees with the recommendation that the manufacturer hold back 23 inches on each module, instruct your dealer to do so.

Sometimes this technique will not completely eliminate noticeable unevenness in

a ceiling or wall. In this case, the GC skim coats the area with drywall mud. This is unlikely to make the ceiling or wall perfect, but it is rare for a drywalled ceiling or wall to be perfect with any type of construction. You can usually see slight imperfections if look from certain angles, in certain lighting conditions.

The GC must also adjust the floors. This task should be relatively easy if the set crew did a good job setting the modules. After inserting a piece of sheathing to tie the floors of the two modules together, the GC evens out minor differences in height. When there is a difference of ¼ inch or less, the best way to level the floors is to plane down the sheathing on the high side. This will not jeopardize the structural integrity of the floor system if it is done at the marriage wall, since the floor at the marriage is substantially stronger than required. After the GC has removed most of the unevenness by planing, he can use a floor-leveling compound to eliminate the remaining difference. If the height difference at the marriage wall is more than ¼ inch, the higher floor may not be sitting correctly. The dealer and his set crew should be responsible for correcting this height problem.

The GC should adjust any interior doors, exterior doors, and windows that do not open, close, or seal properly. Some manufacturers do not completely install their exterior doors, since this makes it easier for the GC to adjust them. Before the GC completes the installation of an exterior door, he should make sure the frame is properly shimmed and aligned.

In the kitchen and bathrooms, the GC should adjust the cabinet doors and drawers to open and close properly. He should also caulk where the cabinets and countertops meet the walls, especially the backsplashes.

The GC should eliminate minor floor squeaks. Although it may seem insignificant, he should straighten the electrical receptacle boxes and plates, as needed.

Any miter joints in the moldings that have opened up should be filled with appropriately colored putty if a crack is small and reset if a crack is wide. The GC should offer to fill the nail holes in the moldings and caulk where the moldings attach to the walls, although some customers do these tasks themselves to save money. He should also repair cracks or faults in the drywall and reset any protruding fasteners.

Modular homes sometimes develop long drywall cracks after the delivery and set. The frequency and size of such cracks tend to increase with longer, heavier modules, a difficult road to the site, and a bumpy driveway. The cracks are easily fixed by taping, spackling, and repainting the drywall.

Some independent general contractors are surprised and disappointed that these adjustments and repairs are considered part of their normal responsibilities. They feel that a brand-new home should not need this work, and if it does, the modular dealer should take care of it. Most dealers respond that a modular home is not a finished product when shipped to the site. It is a component of a home, and it is always the responsibility of the GC to assemble and finish the many components.

There is also a practical reason why the GC should do the adjustments and repairs. Such work is most efficiently done by the button-up crew when it completes its other work. For the dealer to send a separate crew to complete only the adjustments and repairs would make the cost of the work unnecessarily expensive. (See story below.)

Speaking from Experience

SOON AFTER I STARTED offering GC services, I had a serious problem with a plumbing leak in a two-story Colonial I built for a young couple. The leak was more like a deluge. The plumber I hired had no previous modular experience, and my supervisor on the job was also quite inexperienced with modular homes.

A month after my customers moved in, the wife invited her parents to stay with them. While my customer bathed her young children in the master bathroom, her father took a shower in the hall bathroom; both bathrooms were on the second story. A few hours later, both of them met in the kitchen and found a puddle of water on the floor. While trying to figure out what had caused the leak, the drywalled ceiling suddenly let go and water poured down. It turned out that my plumber had not made the connections for the hall bathroom waste line in the access area that had been created for this in the second floor. Thus, the water had no place to go but in the bays between floors.

The plumber missed the connection because my manufacturer had carpeted over the access panel, and neither my supervisor nor my plumber had looked at the plans to identify all of the access points. They also hadn't completed a pressure test, which would have told them that a connection had not been made. If you hire your own GC, don't make the mistake I made. Make sure your mechanical contractors look closely at the plans to find every access panel, and make sure your plumber pressure-tests his work.

Mechanical Access Panels

On a two-story home, the manufacturer provides either ceiling or floor access panels so the GC can connect the mechanical systems between stories. Once the connections are made and the inspections have been completed, the GC needs to replace any insulation needed within the access panels and close them up by attaching sheathing to the floor or drywall to the ceiling.

Attic Access

Whether attic access is through a hatch or pull-down stairs, the GC should insulate it. He can build a cover for pull-down stairs using foam insulation.

Uninstalled Materials

Sometimes the GC is required to install materials that the manufacturer normally takes care of. For example, if a kitchen cabinet or a bathroom tub crosses the marriage wall, the GC has to install it on-site.

Vents

The vents for bathroom ventilation fans, clothes dryers, and range hoods are usually installed by the manufacturer. Sometimes the GC needs to cut a hole in a wall or the roof and install a vent extension and boot. The bathroom vents and clothes dryer vents should always open to the outside. Whenever a duct from a ventilation fan or dryer vent is run through an unfinished attic, it should be insulated to prevent moisture from condensing inside the duct. Otherwise, the resulting pool of water will damage the ceiling drywall.

Skylights

The GC has to build the shaft for a skylight. In some cases, the GC must also install the skylight.

Tray Ceilings

If the manufacturer creates a tray ceiling on the first story of a two-story home by removing the center of the ceiling, the GC must drywall the open space. He must also install any crown moldings.

Cathedral Ceilings

The cathedral ceiling systems built by most modular manufacturers must be finished by the GC. The most difficult tasks are to insulate, drywall, and paint the ceiling and the gable wall above the first floor. If R-30 or greater fiberglass insulation is used, the trusses or rafters might first need to be padded out to accommodate the insulation. There will be quite a bit more work if the cathedral area runs to the peak of the roof rather than to a set of collar ties. The GC must install proper vents before adding the insulation. It is common for customers and inexperienced GCs to underestimate the scope of work required to complete a cathedral ceiling. When the cathedral area opens into an unfinished space, as is the case with a typical chalet design, the GC must either finish that space or frame an insulated wall between the two.

Loft Areas

Cathedral and vaulted ceilings often open onto a loft area. The GC will need to either build a half wall or install a railing around the perimeter of the loft.

Interior Trim in Split Entries

The split entry of a raised ranch must be built on-site by the GC. After cutting the temporary rim joist, he must build the entry landing, install the front door, and construct the stairs up to the first floor and down to the basement. The walls framed on each side of the stairs, combined with a door at the bottom, will close off the first floor and stairway from the basement. This step is required by the building code, unless you immediately finish the basement.

Be clear with the GC whether he will finish the split stairwell with a railing or half wall. If you have selected a railing on the

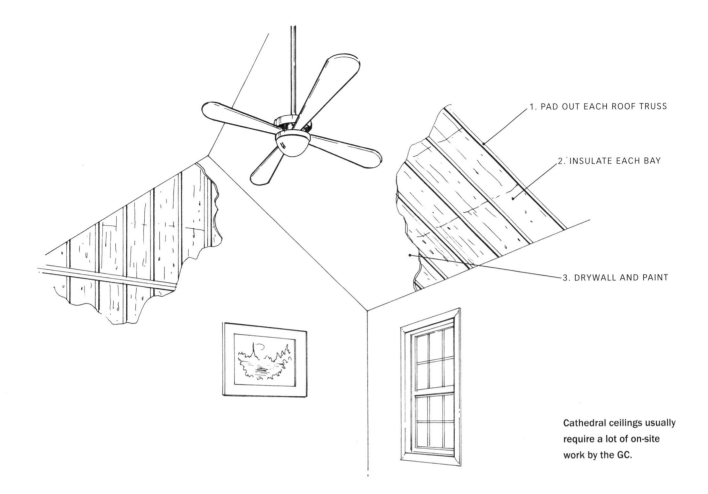

1. PAD OUT EACH ROOF TRUSS

2. INSULATE EACH BAY

3. DRYWALL AND PAINT

Cathedral ceilings usually require a lot of on-site work by the GC.

first floor overlooking the foyer and the manufacturer does not install it, the GC will have to do so.

The electrician must wire the foyer light so it can be turned on from the top of the stairs, the front door, and the bottom of the stairs. He should wire the front-door light to be turned on from the top of the stairs and the front door. The manufacturer should wire the home to facilitate the electrician's work with both lights. The electrician should also add a receptacle at the landing, and the HVAC contractor will need to bring some heat to the foyer.

Stairs to the Second Story or Attic

Modular manufacturers often cannot fully install the stairs between stories. Some-

times the GC must complete the installation after the set, making adjustments as necessary. The GC should follow the instructions sent by the manufacturer. Once the stairs are in place, the GC may also need to install the top treads, nosing, and risers. For stain-grade treads on L- or U-shaped stairs, the GC may need to supply and install the matching materials for the landing. To dress up the stairs, the GC can install scotia or other moldings.

Stair Railings

Once the stairs are in place, the GC needs to complete the railings. Unless the railings provided with the home are preassembled, the GC will have to assemble the individual components.

Steep Roofs and Unfinished Attics

Not all set crews install the attic windows in a home with a 9-in-12 or 12-in-12 folding roof, such as in a Cape Cod. The GC needs to put in any attic windows that the crew does not install.

When the second floor of a Cape Cod is left unfinished and a door has not been installed at the bottom of the stairs, the GC must close off the stairwell. The building code requires you to separate all insulated and uninsulated spaces. This is typically done in a Cape Cod by building a door at the top of the stairs with insulated walls on both sides. Purchase the matching interior door, door handle, and moldings from your modular dealer. You may also purchase the remaining materials, such as framing, insu-lation, and drywall, from the dealer or have the GC get them.

When you are ready to finish the space, the GC must leave all collar ties, knee walls, marriage walls, laminated beams, and other structural members as they are shown in the manufacturer's roof plans. You can only remove or alter these components if you have an engineer redesign the system.

Manufacturers sometimes nail the attic sheathing in place thinking this is one less task for the GC to do. If, however, you need access to the bays between floor joists when you finish the attic, you will have to remove the sheathing. This is harder than it sounds because manufacturers use glue along with the nails. This means you will destroy it while trying to pry it off the floor joists. If

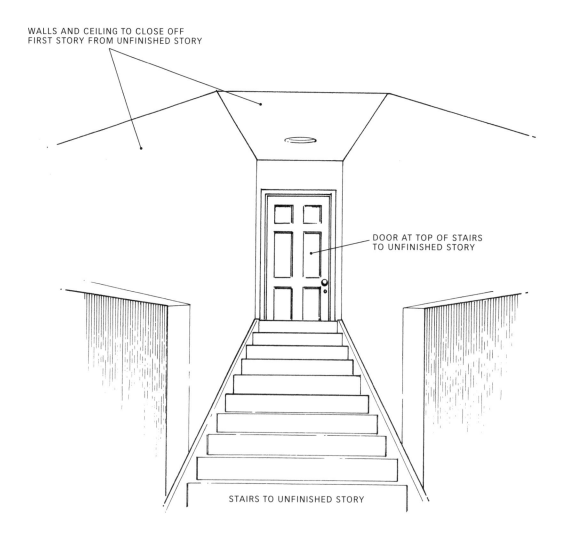

WALLS AND CEILING TO CLOSE OFF
FIRST STORY FROM UNFINISHED STORY

DOOR AT TOP OF STAIRS
TO UNFINISHED STORY

STAIRS TO UNFINISHED STORY

Unfinished Cape Cods require the second story to be closed off from the first at the top of the stairs.

your GC agrees, you are better off asking the manufacturer to screw the sheathing to the joists and eliminate the glue. You can then use nails or screws and glue after the plumbing, heating, and electrical work has been completed in the attic.

Three-Module Capes and Saltboxes

One way to build a Cape Cod design with a shed dormer is to build the dormer as a third module. This reduces the amount of work the GC must do to finish the second-story floor plan. The GC will, however, have to complete the unfinished space under the sloped ceiling in the front of the second story. Since the floor of this unfinished space will be 8 to 12 inches lower than the floor of the second-story module, the GC first needs to build a floor system to match the floor heights. Once the floor is built, he can complete the unfinished area according to your plan.

Basement Insulation

The building code typically requires insulation in the basement. You can insulate the basement ceiling with R-19 or R-30, the interior foundation walls with R-19, or the exterior foundation walls with R-10. If you plan to finish the basement some day, you might want to insulate the basement walls rather than the ceiling. It costs a little more up front, but it reduces the cost of finishing the space later. It also saves money on the heating bill by decreasing the heat loss from heating and plumbing ducts, pipes, and equipment. The GC should also insulate all exposed hot-water pipes in the basement.

Flooring Installation

The GC is responsible for installing any flooring not installed by the modular manufacturer. Some rooms will come with the carpet partially installed, for example,

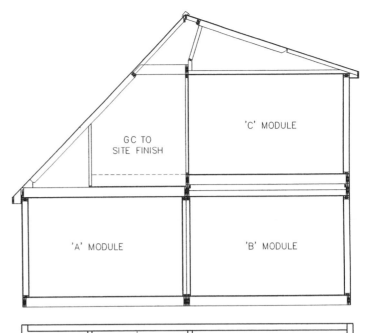

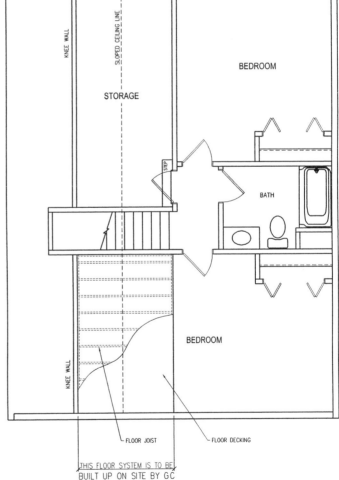

The GC will need to build a floor in the front area of the second story of a three-box Cape Cod so that it lines up with the third module.

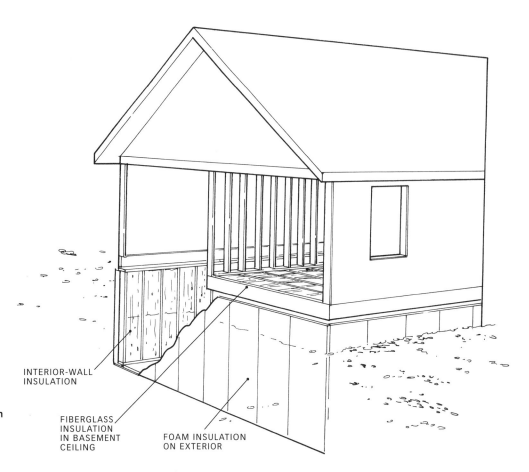

The basement can be insulated by installing fiberglass insulation in the ceiling or on the inside of the foundation walls or by installing foam insulation on the outside of the foundation walls.

INTERIOR-WALL
INSULATION

FIBERGLASS
INSULATION
IN BASEMENT
CEILING

FOAM INSULATION
ON EXTERIOR

where the carpet has been pulled back from mechanical access panels, where the carpet needs to be restretched, and where two pieces of carpet need to be seamed together. For rooms that must be completely carpeted, the GC should first lay out the carpet to be sure it is oriented in the right direction and fits properly. If a piece of carpet does not fit, he should notify the modular dealer before cutting or installing it. If the GC is sure he has the right carpet, he can install the tackless strips, pad, and carpet. After stretching and seaming the carpet within each room, he can seam the carpet between adjacent rooms. Unless you order a threshold, the GC will install a metal strip where the carpet meets a vinyl floor. When it is required, the GC will lay the carpet on the stairs.

Modular manufacturers usually install the vinyl flooring they sell. You may not want them to do so, however, when the same

vinyl pattern runs across two modules. This is because it almost always results in a visible pattern mismatch where the flooring meets. In this situation, your dealer should instruct the manufacturer to install the vinyl in the module that would be hardest for the GC to install and ship the other room's vinyl for the GC to complete.

The GC must install any other flooring that is supplied but not installed by the manufacturer. He must also install the flooring you purchase locally. After the GC finishes the flooring, he then installs the toilets, moldings, or other materials you asked the manufacturer to leave uninstalled or tacked in place.

Paint and Stain

You should expect the manufacturer to paint the walls and ceilings, first with a sprayer and then with a roller. Whether the manufacturer calls the finish a true paint or

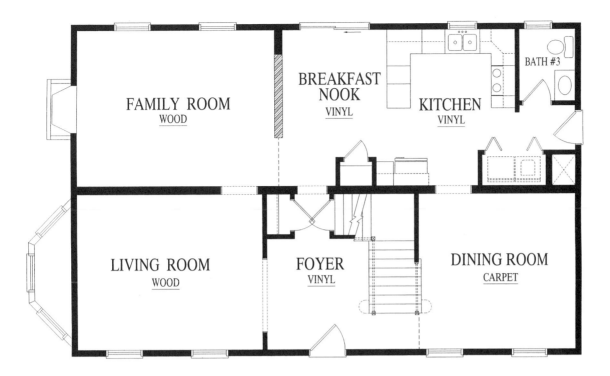

FAMILY ROOM
WOOD

BREAKFAST NOOK
VINYL

KITCHEN
VINYL

BATH #3

LIVING ROOM
WOOD

FOYER
VINYL

DINING ROOM
CARPET

Sometimes the customer will want to install the same type of wood or the same pattern of vinyl or tile flooring in adjoining rooms that cross the marriage wall, such as between the family room and living room or the breakfast nook and foyer of the Whately 1. In these situations, the customer should have the manufacturer install the floor in one room and ship the materials to the GC to the finish the other room on-site.

a primer, treat it like a primer, if for no other reason than only one complete coat has been applied. Some manufacturers point out that they are adding paint to the walls when they "back roll" them after spraying, and they are right. But this is not the same as waiting for a primer coat to dry and then adding a finish coat.

Most manufacturers paint the walls and ceilings in a single shade of white. It is not cost effective to apply a finish coat of paint with different custom colors in each room on an assembly line. Even if it were done, it would have to be redone by the GC after he pieces in drywall along the marriage wall, repairs drywall cracks, hangs doors, and installs moldings. After the GC completes the interior button-up, you should at some point repaint all of the walls and ceilings. You might want to wait a few months, however, to give your home a chance to settle

and the GC the opportunity to make any needed cosmetic repairs. Many customers delay painting for a year or so due to a shortage of money and time.

The manufacturer provides enough extra paint so the GC can touch things up and finish the moldings, doors, and drywall he's installed during the button-up. Do not use this paint for other parts of your home, such as the drywall installed in a site-built family room in the basement or the fire-rated drywall applied to a garage. The GC might need more of the paint than either he or the manufacturer planned, and he will not know this until the home is complete. Save any unused paint at the end of the job for future touch-ups. In both cases, it is best to use paint from the same dye lot as came with your home. You need to be as careful with any stain provided by the manufacturer as you must be with the paint.

Unfinished Interior Surfaces

If you ordered a home with some of the materials unfinished, such as moldings, doors, and a beveled-wood countertop edge, the GC should apply the finishes. In addition, if the manufacturer stains but does not seal any materials, instruct the GC to apply a sealer as soon as possible. If wood is not sealed properly, it can eventually crack, split, stain, or delaminate.

Wood Stairs

The treads and risers of carpeted stairs do not need to be painted or stained. The skirtboards, however, may need to be painted. For stain-grade stairs, the GC stains and seals the treads and then paints or stains and seals the risers and skirtboards.

Completing the Exterior

Exterior steel doors are usually primed by the door manufacturer and require a finish coat of paint on-site. In most cases, the GC must finish the inside and outside of steel doors within 30 days to qualify for the door-manufacturer's warranty. Wood doors especially need to be finished as soon as possible to protect their surfaces. Fiberglass doors should be stained or painted. The GC must also paint or stain any unfinished wood or fiber-cement siding, soffit, fascia, and decorative moldings.

Exterior Entrances

Unless you build a deck, porch, or some other structure that attaches to the exterior doors, you will need stairs or a ramp. This is in part because modular homes are built on a crawl space or full basement rather than a slab. For stairs, pay attention to the size of the landing, which should be wider than the door and sidelights and at least 42 inches deep. Also note whether railings are included on the landing and stairs. Railings on both sides of the landing and stairs

are safer as well as more attractive. Keep in mind that the entry stairs make a significant contribution to the appearance of a home. Finally, make sure the stairs are firmly anchored to the foundation.

You can build a universal-design entrance and eliminate the entry steps by building a ramp. A more attractive alternative is to build a bridge by extending the sidewalk right to the front door, with the last few feet spanning a "moat." This is best done if the entry door is less than 30 inches above the finished grade.

The building code requires a step down from the home to the floor of an attached garage. Many communities will allow you to eliminate the step if you slope the garage floor away from the door to your home. This works best when the overhead garage doors are located directly across from the house entrance. The GC will need to discuss this option with the building inspector. If the building inspector does not allow it and you need to eliminate the step, consider making a concrete ramp when you are pouring the garage floor.

Grills and Screens

The manufacturer is unlikely to install screens or snap-in grills for the windows and doors, since they could be damaged during the delivery or button-up. You or your GC will need to install them.

Gutters

Unless you live in a dry climate, you should have the GC install gutters. Without gutters, water running off the roof can create a ditch around the foundation and find its way to, and through, the foundation. Gutters, installed correctly, can help to keep the basement dry. They also prevent leaks from occurring around exterior doors. Gutters are normally drained to the surface, but a better approach is to connect the downspout to a dry well or storm drain. If the GC does not install gutters, he should be especially careful about grading the site to slope away from the home. In addition, you should landscape around the foundation to protect the soil from erosion.

Storm Doors

When the home is oriented so that one or more of the exterior doors will be subjected to heavy wind-driven rain, the GC should install a storm door on the affected entries. Exterior doors are not designed to withstand the force of heavy rain by themselves; they are water resistant, not waterproof. Any leaks that are not due to a defect in a door or its installation will not be a valid warranty claim.

Building Other Structures

Customers usually ask their GCs to build some on-site structures. Decks, garages, and porches are most commonly requested. Mudrooms, sunrooms, and family rooms are also popular requests. Customers who build an unfinished Cape Cod may ask for the second floor to be finished. Customers who select a raised ranches or split-level design may request that their basement be finished.

If you ask the GC to construct an on-site structure, he needs to specify exactly what he is building and how he is building it. If it is a habitable space, such as a finished basement or an all-season mudroom, he should give you construction specifications and detailed drawings of the floor plan, electrical, heat, and elevations. The GC should provide you with a detailed list of the materials you should purchase from the modular dealer so that the additional structures match your home. Do not downplay the importance of these steps. Many customers are so focused on the modular components that they request little if any documentation for the additional structures.

Energy Star Program

If you build an Energy Star home, the GC needs to take several steps. He should seal basement plumbing and electrical penetrations, the bulkhead, knee walls, and sill plates. In addition, he should put a door sweep on the bottom of the basement door and insulate and seal the attic, including the hatch or pull-down stairs.

If you decide to insulate the basement ceiling rather than the walls, ask the GC to install R-30. The heating system and water heater should be Energy Star–approved. Joints between ducts on a forced-air system should be sealed with mastic and then insulated. Registers should be sealed.

Living spaces built on-site should be constructed in compliance with Energy Star specifications. The GC's plans and specifications for site-built structures should be submitted to the Energy Star program at the same time that the modular home plans and specifications are submitted.

House Cleaning

When your home arrives, it will be fresh from a good scrubbing by the manufacturer. Keeping it as close as possible to this condition during the button-up will require a concerted effort by everyone. The GC should regularly remove the trash from the home and property and sweep the floors.

Just before you move in, the GC should thoroughly clean the interior of the home, paying particular attention to the kitchen, bathrooms, and floors. He should dust all surfaces and wash all windows. If the exterior siding is dirty, you may want to see it power-washed before you move in. Some customers try to save money by doing the cleaning themselves. Many come to regret doing so because of all the other work they need to do preparing to move in. It is usually best to let the GC take care of this.

Certificate of Occupancy

After your home is completed, the GC should obtain a certificate of occupancy. The building inspector can prevent you from moving in until the certificate of occupancy has been issued, although most inspectors will be reasonable if the home is essentially complete and there are no health or safety concerns. Most banks and mortgage companies, however, will not issue their final disbursement until they receive a copy of the certificate of occupancy or its equivalent.

An example of a certificate of occupancy.

Commonwealth of Massachusetts

TOWN OF CHESHIRE

BUILDING INSPECTOR

No. _____

Fee _____

_____ 19 _____

OCCUPANCY PERMIT

"No building nor structure shall be erected, and no land, building or structure shall be used for a new, different, changed, or enlarged use without a Building Permit therefore first having been obtained from the Building Inspector. No building shall be occupied until a certificate of occupancy has been issued by the Building Inspector."

Issued To _____

Address _____

	ISSUED BY	INSPECTION DATE
Wiring Inspector		
Plumbing Inspector		
Fire Department		
Water Department		
Highway Department		
Assessors		
Building Inspector		

THIS PERMIT WILL NOT BE VALID, AND THE BUILDING SHALL NOT BE OCCUPIED UNTIL SIGNED BY THE BUILDING INSPECTOR UPON SATISFACTORY COMPLIANCE WITH TOWN REQUIREMENTS.

_____ 19 _____   _____

Building Inspector

WEPCO · Adams

8 Building a Modular Addition

ODULAR CONSTRUCTION is a great way to build an addition. You get the quality and price advantage that modular homes are known for along with faster build time. Speed is particularly beneficial when building an addition, since the construction temporarily disrupts your family's life, especially if you remain in your home while the work is being done.

Modular additions come in two types. The most popular type is attached to the side of a home to create either a separate living unit, such as an in-law apartment, or additional rooms, such as a new kitchen, dining room, and great room. Some customers build an in-law apartment at the same time that they build a new modular home. The second type of modular addition is set on top of a one-story home to make it into a two-story.

Preliminary Conditions

Before you spend too much time considering an addition, find out whether or not you can build one, and what is required if you can. There are any number of issues that can prevent you from going forward. Not surprisingly, several of the issues that affect your ability to build an addition are the same as those that can restrict what you can do with a particular building lot (see chapter 6).

Covenants, Deed Restrictions, and Easements

When building an addition, you are compelled to abide by any covenants and deed restrictions that apply to your property. Almost all subdivisions have covenants limiting what you can build, and a previous owner of the property might have placed a restriction on what you can do. Although covenants and deed restrictions do not usually address additions, the only way to know for sure is to check. You will also be prevented from building an addition on any part of your property where someone has an easement or right-of-way, unless you negotiate new terms that are recorded on your deed.

Zoning Regulations

Local zoning requirements may affect what you can do with your home. Setback regulations will prevent you from building an

While You're at It, Why Not Remodel?

To make the floor plan of your current home flow nicely into your addition, you may need to do some remodeling. If you add a second-story addition, you need to build a set of stairs to the second floor. But you may also want to rearrange your current floor plan.

You might remodel your home to update some features; kitchen and bathroom renovations are particularly popular choices. Or you might remodel to improve the layout. When building a second-story addition, for example, you might accomplish this by making two of the existing bedrooms into a great room or by carving out a bigger kitchen and formal dining room. When building an attached addition that contains a new master-bedroom suite, kitchen, and formal dining room, you might make the old country kitchen into a breakfast nook and mudroom.

You may want to use the addition as an occasion to dress up the outside of your home. In fact, you might not have a choice, since the siding, shingles, or windows may need to be replaced to match the addition. Replacing or refinishing all of these materials in your existing home can be expensive. On the other hand, it might be the perfect time to do so if they are worn out.

Building an addition is often a good time to add the deck or porch you have always wanted. It is usually more affordable to do these small projects while a general contractor is already completing the other work at your home. Your lender may also be willing to increase your mortgage by the relatively modest amount needed to take on these projects.

Remodeling, however, can be expensive, and the true cost is often hard to pin down in an older home until the work has begun. If you are having a GC do some remodeling, you will want to know what tasks he is agreeing to do and what tasks he is excluding from his contract. You should definitely set aside a good-sized contingency fund.

addition too close to abutting properties or the street. This can force you to build the addition on the side of your home that is less practical and affordable. If your property does not have a current survey, or if the boundary stakes are not in place, you may need to hire a surveyor before you can convince the building department that your addition complies with the setback requirements. It is unlikely the setback requirements will have much bearing on your plans if you are building a second-story addition.

Most communities have specific zoning regulations governing if and when you can add a second living unit to your home, such as an in-law apartment. Some regulations require a special permit or a zoning variance for any two-family unit, regardless of use or size, while others provide an exception for an in-law apartment. Still other communities only allow you to build a two-family home on lots larger than are required for a single-family home. Zoning regulations can dictate how large a home you can build. Some communities restrict how tall the roof can be.

Building Codes

Some building codes mandate that anyone constructing an addition upgrade his existing home to meet current building-code standards. For example, the building inspector might stipulate that you outfit your existing home with approved smoke detectors that connect to those installed in the addition.

Module Access

In order to build a modular addition, the transporters, crane, and set crew must be able to set up in the proper location on your property (see pages 213–220 for more details). Even if your original home is modular and access was not a problem when you built it, you could run into a problem with the addition.

Financing and Appraisal

Before entering into a contract to build an addition, determine how you will pay for it. If you intend to use a lender to finance the construction, you may have a choice of either an equity or a construction loan. To use an equity loan, you must have sufficient equity in your home, since the lender only lets you borrow against that equity. An appraiser hired by your lender establishes the amount of equity in your home. If you owe money on your home but the mortgage is small, the appraisal is less likely to matter, since the lender has sufficient collateral even with a low appraisal.

If you have little equity in your home and need a construction loan, the lender may require a down payment. It will also want the appraisal of your home to include the proposed addition. Before you spend too much time exploring construction costs, speak with a couple of lenders to see what they can do for you.

Hiring a GC

When building a modular home, it is recommended that you hire a general contractor with modular-construction experience. In some respects, this advice is even more important when building a modular addition. There are usually a number of surprises when building an addition, regardless of the type of construction. Most of them derive from the fact that you are connecting a new structure to an existing structure that was not specifically designed to accept it. Surprises are typically more frequent and complex with an older existing home. Construction surprises almost always cost money and time, and they can cause personal stress, especially if you remain in your home throughout the project. The best way to manage the challenges of building an addition is to have a professional GC direct the activities.

Scope of Work

If you are able to build an addition, you need to work with each of your modular-dealer candidates to determine a floor plan, specifications, and price. The steps are essentially the same as for building a single-family modular home, except that you will probably need at least one of the dealers to help you create a custom plan. Although most manufacturers build additions, few offer standard plans that were created specifically for this purpose. Many standard modular house plans, however, can do double duty as additions. For example, small ranches can serve as in-law apartments, and the second story of an appropriately sized two-story home can work as a second-story addition.

Provide prospective dealers with photographs and approximate measurements of the inside and outside of your home. This helps them create a design that meets your needs and fits your existing home. When you sense that a particular dealer can help you, invite him to see your home and take his own measurements.

Take similar steps with your GC candidates. Once you have confidence in a candidate, invite him to visit your home to make sure he can do what you and your dealer are proposing. He should also take his own measurements. He can then finalize the work needed to build the addition and present you with a bid for his services. Add his price to your dealer's price and decide if the project can meet your budget.

The GC tasks are similar to those in building a new modular home, as discussed in more detail in chapter 7. These tasks include completing the site work, foundation, plumbing, electrical, heating, and interior and exterior carpentry. The GC needs to build any site-built structures you want, such as a deck. He is also responsible for completing some construction tasks that are unique to building an addition, which are discussed in the following sections.

Septic System and Sewer Hookup

If your existing home has a septic system, you must obtain approval from the local board of health to use your current system with your addition. The determining factors usually are whether your new home has more bedrooms than your existing home and if the septic system was designed to accommodate them. Without the approval, you will not receive a building permit to go forward. The board might give its approval only if you first enlarge or replace the existing system. Even if the system is adequate as is, connecting to it can create some additional expenses. For example, if your addition has a new bathroom and its waste pipe is below the line connecting to the septic or sewer system, you will need a pump.

Excavation

You should expect that part of your lawn and landscaping will be disturbed by the excavation work. You can preserve some of your shrubbery by relocating it before the construction begins. If you have a paved driveway, it might suffer some damage as well.

Utility Wires and Satellite Dishes

If any utility wires are in the way, the GC should arrange for their temporary relocation before work begins. Since utility companies often require a few weeks notice, this must be scheduled in advance. In addition, other items attached to your home that may affect the construction of the addition, such as a satellite dish, need to be taken down before work is begun and then reinstalled after the addition is complete.

Electrical Work

The electrical service to your home might need to be upgraded, especially if you have an older home with a 60- or 100-amp service. The location of the electrical panel box influences the amount of work the GC must do to connect to the addition. If the box is on the opposite side of the home, it costs more to make the connection. To be safe, the GC should instruct the modular manufacturer to make each electrical run long enough to reach the panel box. Another option is to use a junction box or subpanel. If you are building the addition as a separate apartment, you may want to install a separate electrical service, so that the occupants receive their own electrical bills. The electrician can do this by installing a dual-meter socket.

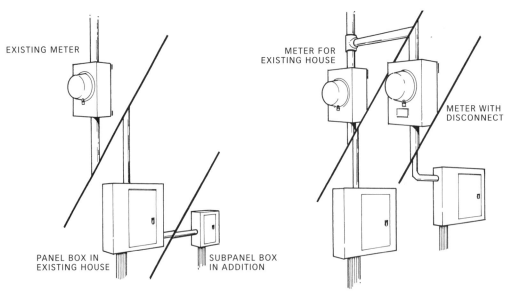

The electrician can use the existing electrical meter or add a second meter when connecting an addition to the electrical service.

EXISTING METER

PANEL BOX IN
EXISTING HOUSE

SUBPANEL BOX
IN ADDITION

SINGLE METER WITH SUBPANEL

METER FOR
EXISTING HOUSE

METER WITH
DISCONNECT

DUAL METER WITH QUICK DISCONNECT

Smoke Detectors

The building inspector or fire marshal might require that you upgrade your home's smoke detectors. You should do this even if it is not mandatory. Have the GC outfit your existing home with hard-wired detectors if you do not already have them. Instruct your modular dealer to coil an extra wire in the basement that connects to the smoke detectors in the addition. The electrician can then pull the wire into the existing home to connect all the smoke detectors. When the two systems are interconnected, a fire in any part of the home or addition triggers the alarms everywhere, which is exactly what you want.

Heating and Air-Conditioning

If the GC intends to tie the addition into your current HVAC system (heating, ventilation, and air-conditioning), your boiler or furnace might need to be upgraded in order to take on the additional demand. You could meet your heating needs affordably if the system is slightly oversized and retains some untapped heating capacity. It is not unusual for contractors to provide more capacity than is actually needed. Otherwise, you will have to consider an entirely new heating unit.

This same point holds for central air-conditioning. You can only tie into the existing compressor if it has sufficient capacity.

You can avoid the expense of replacing the heating system by using electric-baseboard heat in the addition. Putting electric heat in a second-story addition for a home with hot-water or forced-air heat on the first story might even save you money. Since the second story is likely to be made up of bedrooms, you can take advantage of the fact that the temperature in each room can be controlled separately with electric heat. However, if you have central air conditioning on the first floor, you will likely want to extend the ducts into the addition for both heat and air conditioning.

Electric heat can also work well with a small in-law apartment; its small size keeps the heating cost low. If you have a separate electric meter for the apartment, you and the occupants can identify the exact usage.

Another option is to install a separate boiler, furnace, or compressor for the addition. You may want to consider this alternative when your current heating and cooling systems are on the opposite side of the home from the addition, since the distance may cause too much heat and cooling loss between the unit and the addition.

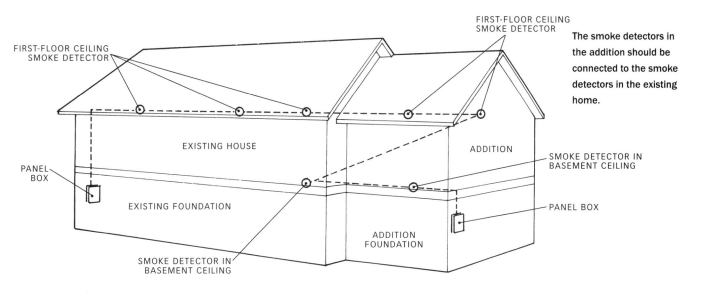

The smoke detectors in the addition should be connected to the smoke detectors in the existing home.

Modulars Not Always the Best Choice

There are times when it does not make good financial sense to build a modular addition. In general, it makes economic sense to build a modular structure only if the modules are reasonably sized and have some value-added features. A one-room addition, such as a great room measuring 20 feet by 27 feet 6 inches, does not meet these criteria. A small multiroom apartment, however, with a kitchen, bathroom, bedroom, and living room, such as a 22-foot by 24-foot in-law addition, does. As these examples show, size is not the only relevant factor. The great room is bigger than the in-law apartment, but it is full of empty space. This kind of structure is better built by a conventional stick builder.

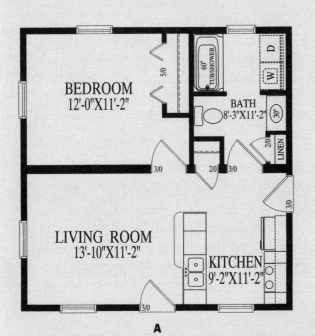

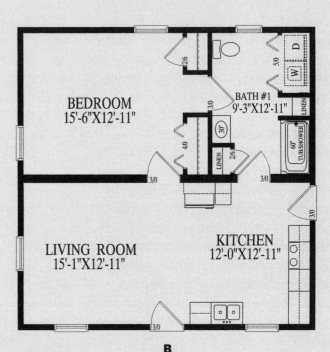

Three in-law addition plans.

Water Heater

Similar issues exist for supplying hot water to the addition as you have with the heat and air conditioning. Unless you have excess capacity in your current system or want to upgrade, you need to add a separate water heater for the addition.

Attached Additions

Attached additions are built to create a separate living unit or to expand living space. Most zoning boards consider any addition with its own kitchen to constitute a separate living unit, which requires that the wall between the two units serve as a "fire stop." The easiest way to accomplish this is to have the modular manufacturer build a fire-rated wall on that side of the addition.

To qualify as a legal addition, your community's zoning regulations will require that it be connected to your home. Detached additions are almost always disallowed. You can connect the two by attaching the addition directly to your home or by joining the addition and your home to another room in between, such as a small site-built mudroom or a large modular great room.

Your property's topography may limit where you can build an attached addition. If one side of your lot is wetlands or con-tains a septic system or municipal sewer pipe, you might not be able to build on that side. You will have the same problem, although to a lesser extent, if one side of your land has a steep slope or an outcropping of rock. The addition is usually located so that its floor plan coordinates with your home's layout. But you may want to consider an alternative if the preferred location would incur substantial extra expense.

In designing an addition, the modular dealer and GC should make sure that the intersecting roofs shed water and snow properly. This is particularly important

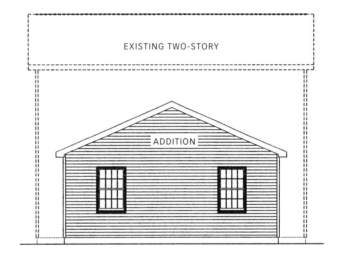

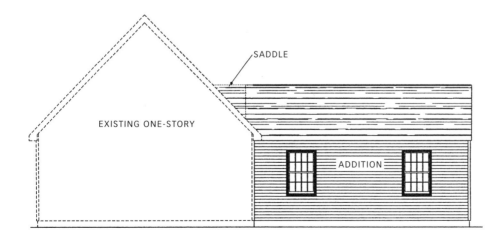

The modular manufacturer will want to take into account whether a proposed addition will impose structural loads on the existing house or the existing house will impose structural loads on the addition.

when the addition is being built in areas with heavy snow. This is because the snow from the higher roof will drift onto the lower roof. Depending on how and where the addition is attached, the manufacturer may ask you to hire a structural engineer to determine what needs to be done to make the two structures work together. The engineer may require the GC to beef up the existing roof so that it will carry the increased load.

Before your addition is built, the GC must measure exactly where the openings into your existing home are located. The modular dealer then uses this information to line up all of the connecting openings in your addition.

Scope of Work for Attached Additions

If you are attaching an addition directly to the existing home, the GC needs to remove the siding on the existing home. Any windows or doors on that wall must be taken out and the resulting holes closed and then finished to match the home. No matter how well the addition is set alongside the existing home, there are bound to be small gaps between the two. The GC should tie the two buildings together on the inside and outside to hide these gaps.

The windows on the left side of the living room and family room in the existing home will have to be removed, and a new opening will need to be added in the living room to receive the door to the addition's kitchen. The two left-side windows on the second floor of the existing home will also need to be moved if the roof pitch on the addition is steep.

EXISTING HOME

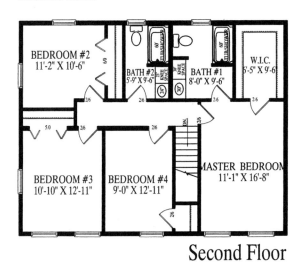

Second Floor

First Floor

Temporary Additions

More and more elderly parents are moving in with their children and their families. Sometimes they move into the existing home and sometimes into a new, attached apartment.

Families are sometimes prevented by local zoning regulations from building in-law apartments in single-family neighborhoods. There are several reasons why communities pass regulations that exclude such apartments. The two most common are to preserve property values and to safeguard neighborhoods from transient residents. In the early 1980s, the American Association of Retired Persons (AARP) proposed a zoning solution that would respect these concerns yet allow families to build an apartment for their elderly parents. The proposal became know as the Elder Cottage Housing Opportunity program (ECHO).

The core idea of the ECHO program was to allow a family to build an apartment on their property as long as they agreed to remove it once their parents moved on. After the apartment was removed, the property would be restored to a single-family residence, bringing it into compliance with the community's zoning restrictions. To make this possible, the apartment would either have to be discarded or relocated. Virtually all advocates of the program recognized that a modular home would make it possible to relocate and reuse the apartment each time the occupants vacated it.

In theory, the ECHO program worked on multiple levels. Many families who want to help care for their elderly parents are not looking to create a permanent apartment. They do not want to maintain it or pay the additional taxes, and they are not interested in becoming landlords after their parents no longer live there. In addition, they would like to reclaim that part of their yard that has been taken over by the apartment. On the other hand, they do not want to dispose of an apartment that is in usable shape. They would prefer to recoup some of their investment by passing the apartment on to another family who needs it.

Unfortunately, the ECHO program has yet to take hold. In many areas, the necessary zoning regulations were not passed. Fortunately, some communities already have zoning regulations that allow for permanent in-law apartments. And they also allow for these structures to be dismantled in the future, as long as you obtain a permit and comply with the regulations. The one caveat, however, is that you must build the addition as if you intend to keep it as a permanent apartment. This means it must comply with all building codes for a new residential home.

An advantage to using a modular home for a temporary addition is that it can be disassembled into a few intact sections that can be easily relocated. This is seldom a viable option for a stick-framed addition. The disassembly of a modular addition requires that the GC carry out the same steps he executed to button up an addition, but in reverse. This should take considerably less time and cost considerably less money than the assembly. The GC would likely do the following:

- Disconnect the plumbing, electrical, and heating systems
- Remove the siding on the gable ends
- Separate the modules at the marriage wall, inside and outside
- Lower the roof on each module
- Lift the home off the foundation with a crane and crew
- Place the modules on rented transporters that have all the necessary permits
- Deliver the modules to their next home
- Restore the property to its original condition, including removing the foundation

The disassembly will render some of the materials unusable for the next owner. Some of the shingles will need to be replaced, for example, as will some plumbing and heating components installed in the basement. If you are building an addition, consider purchasing extra materials so the next owners will have matching replacements when it comes time for them to reassemble the addition.

Second-Story Additions

If you are building a second-story addition, you are most likely doing it to create more living space rather than a separate living unit. The GC turns your one-story into a two-story by removing the roof from your home and setting the new second story with its own built-in roof on top.

The speed of modular construction is a tremendous benefit when building a second-story addition, since the addition can be set within hours after the roof is removed from your existing home. Once the addition is in place, the inside of your home is protected from a sudden storm. A site builder cannot realistically protect your home as quickly. Another advantage is that the second story can be finished faster. This means your family members can use the upstairs more quickly, even if they must wait to enjoy the downstairs until the remodeling is completed.

There are two conditions that must be met before you can build a second-story modular addition. First, the exterior dimensions of the existing home must be compatible with one of the modular manufacturer's production sizes. If your home is too wide, a modular will not easily work. If your home has multiple bump-outs, a modular might work but it may be impractical and expensive. A home can be up to 3 feet narrower than a module, however, and adding a wider second floor creates an attractive, cantilevered garrison Colonial look.

The second condition is having an existing home and foundation that are structurally capable of carrying the additional weight, which is substantial. You should hire a structural engineer to make this determination. He may give you specific instructions on fortifying the structure or the foundation, which might be unacceptable or too expensive. If you decide to carry out his instructions, the GC will complete them as part of his remodeling. Before the engineer writes his final report, he must see exact plans of what you are building and he must receive detailed information from the manufacturer.

Scope of Work for Second-Story Additions

When designing an addition, you must decide where the stairs to the second floor will be located. You should also determine a location for a chase (from the basement to the second floor) to carry the electrical wires, HVAC supply and return ducts or pipes, and plumbing pipes for new bathrooms. If the GC is connecting to a forced-air system in the basement, the chase must be larger, since the ducts take up more space than hot-water lines.

The design of the second-story elevation must be coordinated with the first-story elevation. The window locations should be arranged in a pleasing fashion. This decision should be made early in the design process, since the location of the interior partition walls on the second story must be coordinated with the window locations (you cannot put a wall in the middle of a window). In addition, the window style and sizes should be matched as closely as possible to the existing home.

The exterior elevation of all four sides of the finished home must take into account any first-story bump-outs or structures. For example, the location of an existing bay window, porch, sunroom, portico, recessed entry, or garage can pose special design challenges. The second story must be planned so that it does not affect either the function or aesthetic appeal of these structures. In some cases, it might be necessary to remove a part of the bump-out or attached structure, such as a garage roof, before installing the second story. If the second story is cantilevered, the overhang can pose additional problems with a first-floor bump-out, such as a bay window.

The exterior siding on the second story must fit with the siding on the first story.

Otherwise, the siding on the first story has to be replaced. If you currently have wood siding, you might need to repaint or restain it to create a color match. Similar coordination issues arise for shutters and other exterior trim details.

If you have a chimney on your existing one-story home, you need to make it taller to reach above the roof of the second floor. In addition, all trees overhanging the first story must be removed.

The actual removal of the existing roof as well as any other materials you are replacing in your existing home, such as the siding or windows, will be a task in itself. The cost of disposing of these materials will be appreciable. But when you are done building your second-story addition, it will feel almost like you have just built a brand-new home.

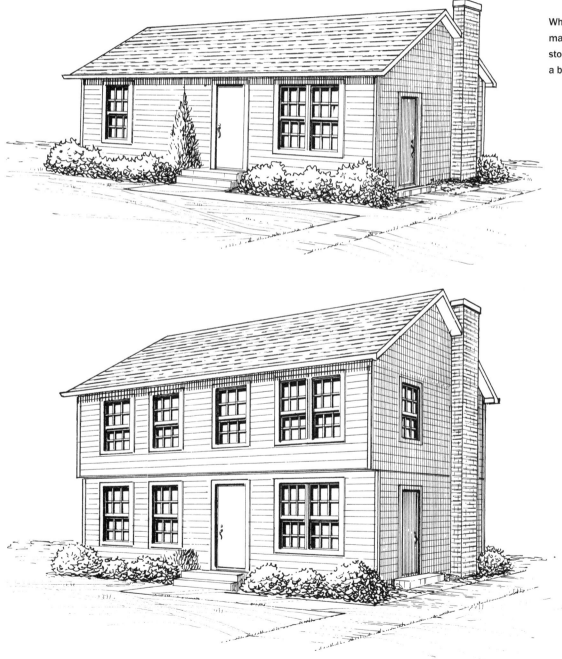

When a ranch is made into a two-story, it can look like a brand-new home.

9

Financing a Modular Home

To BUILD A MODULAR HOME you need to pay the dealer for the modules and the general contractor for his services. If you do not own a building lot, you need to purchase it as well. There are three typical sources of these funds. The first is private funds, such as personal savings, an equity loan on another property, the sale of personal assets, or a family loan. The second is a lending institution, usually a bank, credit union, or mortgage company. The third source is the modular dealer or the GC.

There is one very significant difference between paying for an existing home and paying for a new home. When you buy an existing home, you pay the seller in full before you take possession of the home. If you use a loan to pay for the home, you secure the funds with a mortgage. When you build a home, you make "progress" payments as work is completed. This process protects you and your lender should something prevent the builder, in this case the dealer and the GC, from completing the home. You cannot wait until the home is completely finished to pay the dealer and

GC in full because they need funds to pay for materials and labor as the project moves forward. Receiving compensation as the job progresses also protects the dealer and GC should something prevent you from paying for the finished home.

When you use a lender to build a home, you obtain a "construction loan," which is a short-term loan usually of four- to 12-months' duration. The loan provides for a series of payments as work is completed. Once the local building inspector issues a certificate of occupancy and the lender agrees that the home is essentially complete, the lender pays off the construction loan and issues you a mortgage.

Although you still need to obtain a mortgage, you do not need to secure a construction loan if the dealer or GC finances the construction. They are more likely to do this if the dealer is completing the GC work, but especially if the dealer or GC owns the land. Ownership of the land and responsibility for the construction tasks gives them greater control of the project and reduces their risk should you decide not to purchase the finished home. When you purchase a modular home that is funded in full by the dealer or GC, you are in a sense purchasing an already existing home. In fact, you do not take ownership of it until you pay him when he is done. That is why either will likely require you to provide evidence that you have secured a mortgage or have the personal funds to pay for the finished home. Since paying for the home after it is completed does not require you to make a series of payments, it eliminates the steps involved in securing a construction loan, discussed below. However, you will still need to carry out the steps described in the sections Purchasing a Modular Home (pages 275–278) and Paying for General-Contracting Services (278–281).

Financing with a Lender

With the advent of the Internet, you have many lenders to choose from. You also have more financing products to consider. Ask the potential lenders whether your income, assets, liabilities, and credit history qualify you for a loan. Seek their opinion and perhaps the advice of a financial adviser before deciding whether to select a fixed or variable rate, a term of 30 or 15 years, or a down payment of 20, 10, or 5 percent. Many lenders have a wealth of helpful information available on their Web sites, but it is unlikely that any lender will teach you in much depth about financing the construction of a modular home. This chapter will fill that void.

Not all mortgage lenders offer construction loans, and you need one that does. Customers sometimes run into delays and misunderstandings when they begin looking for existing homes and then decide to build a new home. They assume that the lender who preapproved a loan on an existing home will automatically approve the new home loan. But if the lender does not offer construction loans, the customer will have to repeat the approval process with some-one else. You can avoid this complication by informing each lender you talk to that you are considering building a new home.

Prequalified and Preapproved

When you begin shopping for a construction loan, ask one or two lenders to prequalify you for a loan. This gives you a good idea of the maximum amount you can borrow, which helps you develop a realistic budget. Prequalification, however, is not a promise of a loan. To prequalify, you must answer a few questions about your income, assets, liabilities, and credit history. Assuming the information you give is accurate and no other issues surface when you formally apply for the loan at a later date, you will probably be approved. Should the lender's review of your formal application turn up different information, however, you may be denied loan approval or offered less money than you otherwise would have been.

As soon as you have selected a lender, consider applying for preapproval of your construction loan. Preapproval takes prequalification one step further and results in a written commitment from the lender. Some lenders charge for this service, some do not. Preapproval comes with a list of conditions you must meet for full approval. This usually includes a satisfactory appraisal of your completed home and the requirement that your personal financial situation does not change. If you are preapproved before selecting a modular dealer and GC, you can move ahead more quickly with building your home, since the lender can immediately begin reviewing your construction documents. In order to obtain final approval, the lender must receive an appraisal on the home showing that it is worth what you are paying. It must also obtain copies of signed contracts with the dealer and the GC and, if you are buying a building lot, the purchase and sales agreement with the land seller.

Approval Process

A loan to build a new home takes more steps and more time to process than a loan to buy an existing home. The first step with both types of loan involves approving your credit and income, and the final step involves determining whether the home is worth what you are paying for it. But a construction loan requires an additional step before the appraisal.

From the lender's point of view, building a new home poses greater risks for the buyer and lender than buying an existing home. When your lender loans you the money to pay the seller of an existing home, your lender is relatively confident that you are getting exactly what you are paying for. This confidence is established by the written report of an appraiser who personally inspects the home.

When a lender awards a loan to build a home, it still relies on an appraisal. But the appraisal is based on an analysis of the construction documents showing what you are building, rather than an inspection of an existing home. The lender's first step is to verify that the written plans and specifications contain all of the required information from your dealer and GC and that the prices charged by each are reasonable.

For example, if your drawings show a porch that is not listed in the contract, the lender will likely ask the GC to document that you are getting a porch. If the lender learns that the front elevation was a generic drawing and the porch was only a possible option, it will tell the appraiser to exclude the porch from the assessment of the home's worth. In another example, if the contract shows a cost of $2,500 to install a well and the lender knows that $4,000 is a more typical cost in your area, the lender could ask the GC to justify the price. You might learn that the GC's estimate was only for drilling the well and did not cover the necessary pump and tank. In this case, the lender would insist that the GC revise the contract to include these costs.

In addition to reviewing the specifications, plans, and pricing, the lender examines the legal documents you have signed with the dealer and GC. If you are not adequately protected, the lender is vulnerable as well. If the lender has no previous experience with your dealer or GC, he or she may want to assess whether the two companies are financially viable. Your dealer and GC may resist turning over their financial information. But they may be able to satisfy the lender by providing references that include customers, vendors, and banks or other lenders.

This entire process, which must precede the appraisal of your home, takes time. While lenders can now offer almost instant credit approval, it is not yet possible to process construction documents that quickly. You can help the lender finish its work by submitting a complete and detailed construction package.

Construction-to-Permanent Loan

Most lenders combine their construction loans and mortgages into a single loan with one closing. This saves you money, time, and aggravation. These "construction-to-permanent" loans designate a period of time, almost always less than a year, for the construction phase. Typically, you only pay interest on the money you borrow during that phase. The mortgage kicks in as soon as your home is complete, at which point you begin to pay principle and interest on the total loan amount. A few lenders include the construction phase within the term of the mortgage. For example, if you select a 30-year mortgage and the construction is completed in 6 months, the total length of both loans is 30 years, not 30 years 6 months.

If you choose a fixed-rate construction-to-permanent loan, your lender locks in the same interest rate for both phases of the loan, with the rate determined in part by the duration of the construction. As the name implies, a "rate lock" provides you with protection should interest rates rise after you apply for your loan. If you are concerned that rates will go up substantially during the construction of your home, locking in the mortgage rate is a wise decision. Should rates decrease, you will usually be obligated to stick with the higher rate. Some lenders, however, allow you to switch to a lower rate should one become available during the lock-in phase. A rate lock is usually worthwhile unless you are confident that interest rates will fall.

A rate lock fixes the interest rate as long as you close on the loan within a certain period, usually 45 to 60 days. If you apply for a loan during a time when appraisers are overextended and unable to complete their work before the end of the application period, you could lose your rate lock. To be safe, ask the lender to honor the rate lock even if the appraisal causes you to miss the deadline.

Some lenders allow you to pick the length of the construction phase. Your interest rate is tied in part to this period of time. Since longer construction times usually incur slightly higher rates, you might be tempted to select as short a term as you think practical. If construction takes longer than the allocated time, however, your rate lock could be lost or you might need to pay an additional fee to keep it. To avoid this problem, choose enough time to complete the construction but not so much time that you pay a higher interest rate.

The length of the construction phase is likely to be considerably longer than the actual time the GC takes to complete the home. That is because the modular manufacturer requires you to close on the loan before it puts your home into its production schedule. The construction loan, therefore,

must cover the manufacturer's lead time plus the GC's construction time.

Before deciding the length of the construction period, ask your dealer what the manufacturer's lead time is likely to be after you authorize the modules to be built. Then ask your GC how much time he needs to complete his construction work. Explain to both of them what the consequences will be if they are wrong. This does not make them responsible for delays, however. There are too many factors out of their control for them to provide guarantees, but it should make them cautious. If you are building a complicated custom home with many site-built structures, factor in extra time for you to design it, for the manufacturer to special-order materials, and for the GC to complete his work. Pressuring the dealer or GC to accelerate his timeline beyond his comfort level will put your interest rate at risk if either is unable to meet the schedule.

Other construction-to-permanent loans fix the construction rate at the time of application but do not lock in the mortgage rate until after construction is complete. With this type of program, you do not know what your monthly mortgage payments will be when you close on the loan. You may find that a lender offering one of these loans makes up for that fact by offering more favorable down payment requirements, a higher loan amount, or better service. If you select a financing package that has different interest rates for the construction phase and mortgage, keep in mind that you will benefit more from a low rate on your mortgage than from a low rate on your short-term construction loan.

Appraisal Value

Before the lender approves a loan, it must determine if your home is worth the price you are paying. The lender needs to be confident that it can recoup its money should

Speaking from Experience

SOMETIMES A $15,000 EXPENSE can buy you $30,000 of equity. Several years ago, my customers received an appraisal on their 2,400-square-foot two-story home that was $30,000 less than they were paying us. I was quite surprised by this because most of my customers receive appraisals considerably higher than their cost. I obtained permission from the lender, a local bank that I knew well, to speak directly to the appraiser. I learned that she did not know that the house contained hardwood floors and a masonry fireplace. I couldn't tell whether the bank had neglected to tell her or she missed the information in her review of the construction contracts. But this accounted for only $15,000 of the appraisal shortfall. The main problem, as the appraiser saw it, was that all of the other comparable homes she looked at had a two-car garage, unlike my customers' home. My customers had decided to postpone building the garage for two years. The appraiser explained that the absence of a garage reduced the appraisal value of the house by $30,000. Even when I told her I could build it for $15,000, she held fast to her assessment.

When my customers went back to the bank with this information, the bank qualified them for an additional $15,000 on their mortgage so they could build the garage while their house was being finished. Since the appraisal went up by $45,000, $15,000 for the wood floors and fireplace and $30,000 for the garage, the loan was approved. I learned two things from this experience. First, appraisers don't always know all the upgrades that a customer has selected. When I hear of a low appraisal these days, it's the first thing I ask about. Second, appraisal value is very different from cost. Sometimes, as in this case, spending some additional money can significantly increase appraised value, while at other times, omitting a relatively expensive item can greatly reduce the cost without significantly reducing the home's value.

you default. By limiting the loan amount to some percentage of your home's worth, the lender limits its risk. Most lenders offer loans for 80 percent of the home's value, which means that the buyer needs to make a 20 percent down payment to make up the difference. Some lenders offer loans up to 90 or 95 percent of the home's value. The advantage of a 90 or 95 percent "loan-to-value" loan is that you can invest substantially less of your own money as a down payment. The disadvantage is that you may have to pay for private mortgage insurance (PMI) to cover the lender's increased risk.

An independent appraiser compares your home's location, size, condition, property dimensions, and other features to those of similar homes in the same community. An appraiser could use a table of construction costs to make this determination, but this is not typical. If your home is appraised at a value equal to or greater than the total price you are paying for it, including the land, the lender will conclude that its loan is backed by sufficient collateral. If the appraisal is lower than what you are paying, the lender reduces the amount it is willing to lend you.

For example, if you apply for a $160,000 loan that must cover 80 percent or less of the appraised value, the appraisal should be at least $200,000. If the appraisal is $190,000, the lender will lend you only $152,000, leaving you short $8,000. You can make up this shortfall by increasing your down payment with personal funds, but you might first want to determine why you are paying more for your home than an appraiser thinks it is worth. Ask your lender to verify that the appraiser included the correct information, especially any expensive upgrades. The lender may not have given the appraiser all the information that affects the value of the home, or the dealer or GC may have left out important items from his written plans. Many appraisers are open to readjusting their appraisal if the lender can show them that they did not include the correct specifications.

You can improve an appraisal by making changes to your home plans that add more to its value than they cost. Enlarging a modular home, for example, almost always adds more value than it costs. Following the prior example, if the home is a Cape Cod with an unfinished second story, ask if finishing it immediately will help you qualify for the full loan. If it costs you an additional $20,000 but adds $35,000 to the value, you will be able to borrow 80 percent of the appraised value of the home, which will now be $225,000. Yes, you will have a larger mortgage ($180,000), but you will not have to come up with the $8,000 difference that would be required without the upgrade.

Another option is to remove something from your home plans that costs more than it adds in value. For example, cedar siding costs considerably more than vinyl siding, but in some communities it is no longer seen as a valuable addition. If you choose vinyl instead of cedar, the appraised value might change little, while the cost could drop substantially. Similar results can be achieved by selecting vinyl windows instead of expensive wood windows, or by eliminating cathedral ceilings.

Another response to a low appraisal is to obtain a second opinion. Although there are objective rules that all appraisers must follow, appraisers often come up with different valuations. This most commonly happens when two appraisers use a different group of homes in the same community to establish a comparable value. In an environment of rapidly rising land or lumber prices, you may see significant appraisal differences when one appraiser looks only at recorded closed sales while another includes homes under contract. There is no guarantee a reappraisal will work to your benefit, but it

is worth considering if your lender agrees. If the lender balks, or if the second appraisal is also unsatisfactory, you can apply for a loan with another lender. Some lenders select appraisers who work more aggressively to provide suitable appraisals.

If you are stuck with a low appraisal, you can take a loan for 90 percent of the value rather than 80 percent, since this may enable the lender to make up the difference. In the example, the $190,000 appraisal allows you to borrow $171,000, which is $11,000 more than you need.

Should you fail to get a workable appraisal, ask yourself if the combined price of the dealer and GC is reasonably competitive. If you did any comparative shopping, you should already know the answer. If you knew all along that the companies you selected were more expensive, it may still be worth seeing if they will make any adjustments. Explain your problem to them; show them a letter from the lender explaining your dilemma.

If you conclude that their costs are competitive, then you need to take another look at what you are building. Begin by considering whether your design is too expensive for your community. For example, some contemporary designs are considerably more expensive per square foot to build than more traditional designs. You may have had your heart set on a contemporary design, but then discover that you are unable to build a new home unless you accept something more traditional.

Closing Costs, Points, and Fees

To secure a loan, you need to pay some fees and perhaps some "points." Each point equals 1 percent of the construction loan. Fees vary from lender to lender, so insist on receiving a written list of all costs the lender will pass on to you.

Comparing the lists of different lenders may prove a little confusing because you may find them using different terms to label what in effect is the same charge. For example, a lender who advertises that it does not charge points may charge construction fees, which could cost you the same as another lender's points. A lender with higher costs and a lower interest rate may have the better overall program. Often the best way to evaluate this issue is to ask each of your top three finalists to explain why his program is better than the others.

A House to Sell

Many people building a new home own another one they intend to sell. Most customers in this situation wait to build until they sell their current home so they can be sure they will not have to pay two mortgages at the same time. You may decide to take this course yourself. You may also have no choice about waiting if the down payment is coming from equity locked in your existing home. Even if you have the down payment, you may not qualify for a construction loan until you pay off your current mortgage.

Some customers have enough resources without selling their current home to qualify for a construction loan and make the required down payments to their dealer and GC. They also can handle the monthly interest payments while their home is under construction. Other customers take out a home-equity loan to secure the down pay-

Speaking from Experience

ALTHOUGH THIS STORY MAY NOT seem terribly dramatic, it was quite costly and surprising and taught me a lesson that I will never forget. My customer, a single woman with a young son, was given a small piece of land by her mother. She didn't have a lot of money, so we designed her a small ranch-style home. The total price of the home worked within her budget, but barely. The lot certainly didn't appear, at the outset, to pose any problems. My customer and her mother had spoken to all of the neighbors, whom they both had known for years. They were told that the groundwater was low and there were no large rocks or ledge. The lot was completely flat, with only a couple of trees. There was no concern about the cost of a septic system and well, since municipal water and sewer were available. I remember thinking that this would be an easy job, relatively speaking, which isn't particularly common with New England's rocky terrain.

What a surprise it was, then, when we found solid ledge 18 inches below the surface. The best solution was to blast enough of the ledge to put in a crawl space and connect to the water and sewer, but this would cost several thousand dollars, which my customer didn't have. Fortunately, her bank, which her mother and late father had been using for 30 years, stretched her qualifications and lent her the additional money. Ever since this experience, I never assume that a job will be easy and without surprises. I suggest that you adopt the same attitude as I have and plan for the unexpected by setting aside a contingency fund.

ment and set aside money for construction-loan payments. They, too, must qualify for the construction loan after they tap the equity in their existing home.

If you have a mortgage or equity loan on your existing home and want to build before selling it, some construction lenders help you afford the payments on your two loans by financing the interest-only payments for your construction loan. They do

Planning for Surprises

No matter how vigilant you are, building a new home always produces a few surprises. Unfortunately, these surprises often result in additional expenses. When they appear, you might be tempted to blame someone, but it is not possible for everyone to foresee everything. For example, after the GC begins excavating the foundation, he might discover that your property contains an underground spring exactly where you want to put your home. To prevent a wet basement, you must take some costly and unbudgeted precautions. Unless you had the GC dig a test hole in this area before he finalized his contract with you, neither of you could be expected to have anticipated the problem.

There are other sources of cost overruns. Failing to read your building specifications closely is a common source. This can cause your home to be built to specifications you did not want and to be missing features you expected to receive. Correcting either type of mistake adds to your expenses. Your contract may also fail to budget sufficient money for some preconstruction tasks, such as completing a property survey and paying for the utility hookup or building permit. If you learn about these unfinished tasks after closing on the construction loan, the dealer and GC must stop what they are doing until you come up with the money to complete these jobs.

You may be forced to absorb additional costs if you change your mind and, for example, decide you want hardwood flooring instead of carpet. In other cases, a town official could be responsible for your overruns. For example, the building inspector may require your GC to complete some costly additional work, even though it is not mandated by the building code. You may also incur extra financing costs should the project be delayed by inclement weather, material shortages, or subcontractor scheduling conflicts. Your GC may or may not charge you for these problems. But your lender will almost certainly require you to pay more interest if you are financing your home with a construction loan.

The complexity of building a home presents so many opportunities for mistakes that it is best to plan for them. Although you cannot budget specifically for the unforeseeable, you can create a contingency fund to protect yourself against unbudgeted expenses. For a modular home, 3 percent of your modular and GC expenses will probably suffice. You can keep a reserve of cash for the contingency fund or ask your lender to include a contingency fund as a budgeted line item in the construction costs. Some lenders do this without being asked, even if you would prefer otherwise. Many lenders require this only for stick-built homes, since they are historically more prone to sizable cost overruns.

If your total construction costs already bring you to the maximum amount the lender will give you plus your down payment, the only way you can create the contingency fund is to eliminate other expenses. The best way to handle this situation is to omit something you can readily add later. If you get lucky and avoid drawing from the contingency fund during the project, you can spend the money on the eliminated item. If you are forced to tap into the contingency fund, you will spend money you undoubtedly hoped to save, but you will not be forced to come up with money you do not have.

this by increasing the amount of money you borrow through your construction loan to create a reserve you can apply to your monthly construction-loan payments. Until you sell your current home, you continue to pay its monthly mortgage payment from your monthly income, but you can draw upon the reserve account set up with your construction loan to pay for that monthly payment. Once you have sold your current home, the construction loan ends and the mortgage begins. To qualify for this type of program, you need greater income and a higher appraisal than if you sell your current home first because the mortgage amount increases by the additional money allocated to the reserve account.

If you are going to sell your home before building, you have to make temporary living arrangements. One option is to negotiate a rental agreement with the buyer of your current home. Most buyers are unwilling to allow this, but some are agreeable, especially if you make a major concession to them. The more likely solution, however, is to rent elsewhere or move in with relatives or friends while your home is built. If you can afford to rent, you should do so, since it not only helps you retain the affection of those relatives or friends, but it also helps everyone keep his sanity.

Down Payment, Land Purchase, and Deposits

Building a home requires a lot of money just to get started. Your lender will ask for a down payment of 5 to 20 percent of the cost of the project. If you do not own a building lot, you will need thousands of dollars to complete its purchase. You will also need money to give the dealer and GC a deposit. Most dealers and GCs require at least 10 percent down, although they may let you get started for less. If you have to purchase the land and come up with the necessary deposits, you should probably set aside

more than 20 percent of the cost of the entire project.

If you already own a building lot and it is paid for in full, your lender will apply the value of the land, which is your "equity," to its down-payment requirement. If the value is great enough, your equity could fully cover the entire down-payment requirement. Your lender will also apply any deposits you make to the dealer and GC to its down-payment requirement.

Should you lack sufficient funds to purchase the land, as well as pay the dealer and GC their deposits, you will need help from the lender. Most lenders will assist you if you have sufficient funds to meet their minimum down-payment requirement. Some lenders, however, do not provide sufficient assistance, especially if you are making a 5 or 10 percent down payment. Given the importance of having enough money to purchase the land and cover the required deposits, make sure a lender is willing to give you the necessary help before you select it.

Customers sometimes find that they only have enough money either to give the dealer and GC their deposits or to purchase the land. If you are in this situation, you might assume that you should hold on to your money to purchase the land. Actually, your lender will not care which party gets the money. As long as you have an accepted offer to purchase the land and the seller is willing to wait until you close on your loan, give the dealer and GC their deposits. This will save you weeks, and sometimes months, since the dealer can complete your drawings and get you ready to be put into his manufacturer's production schedule. In addition, your GC can complete his construction drawings and reserve a spot in his subcontractors' schedules. Since the lender will require documentation that you made the deposits, obtain a receipt for each deposit along with a photocopy of your deposit checks. The lender may later require a copy of your bank records showing the canceled checks. Once you provide the proper documentation, the lender will pay for the land using whatever is left of your down payment supplemented with funds from your construction loan.

Lender Requirements to Close and Disburse Funds

Most lenders do not allow you to close on your construction loan until you provide evidence that you own your land and can legally build on it. Other lenders will close on the loan but not disburse funds until you show ownership. A title search establishes that you own the land, while a building permit demonstrates that what you own is a legal building lot. If your property needs a septic system, most lenders also require evidence that your municipality has approved the septic design.

Buying a Modular Home

When customers construct a stick-built home, they usually do not wait until their home is framed, insulated, drywalled, wired, plumbed, and finished with cabinetry, doors, moldings, and flooring before paying their builder. But that is likely what you will do when you build a modular home. Your dealer usually obtains a deposit from you, but does not receive the balance until he has built and delivered your home. As you can imagine, the many thousands of dollars that are required to manufacture a modular home makes the final payment a very significant event for the dealer and his manufacturer.

Most dealers and manufacturers require a 10 percent deposit before they will build your home. Some dealers require a deposit of 25 percent or more for a true custom design, since it could be more difficult to sell than a standard plan should you not

honor your contract. Many dealers also require an additional deposit when you are paying with private funds, as will be explained below.

A few modular dealers give priority scheduling or a small discount if you prepay for the home. But you only want to take advantage of such offers if you are sure the company is financially sound. Normally, you would pay off the balance after the home is delivered to your site or set on the foundation.

When a dealer and manufacturer build a home after having received only a small percentage of the purchase price, they are taking a risk. After all, the manufacturer must pay its vendors, factory production crew, and delivery crew. The dealer must in turn pay the manufacturer, whether or not you pay him, since he has a contract with the manufacturer.

When a customer does not pay for a home, the dealer and manufacturer are compelled to sell it to someone else, usually at a substantial discount. That is why the dealer and manufacturer are very concerned about receiving payment in full as soon as possible after they build a home.

Some lenders insist on making the final payment with a check in the name of the customer only. Others allow the name of the dealer or manufacturer to accompany the customer's name. When the check is in the customer's name, either alone or along with the dealer's or manufacturer's name, the customer must endorse it before the dealer or manufacturer can cash or deposit it. Dealers and manufacturers almost never accept a check in the customer's name alone for an assignment-of-funds payment, and only some dealers and manufacturers accept a joint check. The problem is that when the customer's name is on the check, the customer unilaterally gets to decide if and when the dealer and manufacturer are paid, which defeats the purpose of the assignment. Given that the modules are already on the foundation when the check is issued, the dealer and manufacturer do not want to allow the customer to have this much control. Accordingly, most dealers and manufacturers insist that the assignment-of-funds letter state that the check will be issued in one of their names only.

If your dealer insists on receiving payment only in his or the manufacturer's name, bring this to the attention of each lender before applying for financing. The best way to ensure that a lender and dealer can work with each other's policies is to ask the dealer to give you a sample of an acceptable assignment-of-funds letter before you

Personal Property vs. Real Estate

All manufacturers prefer to be paid cash on delivery (COD), and many insist on it. Most lenders prefer to make the final payment after the home is set on the foundation. The manufacturer wants to be paid COD because once the modules are attached to the foundation, they are legally no longer considered personal property, which is what they are when they are sitting on their carriers. If you do not pay the dealer after the modules are on the foundation, the manufacturer cannot remove them and take them back to the factory, something the laws for personal property allow with a car. They are now real estate, and that difference gives the homeowner a great deal of protection against creditors. The dealer and manufacturer would need to get a court order to remove the modules, and this could take months and many thousands of dollars.

Most lenders take an opposing point of view. They do not want to disburse funds from a construction loan to pay for the modules until they have been set on the foundation. Their view is that they are lending money for real estate, not personal property resting on a carrier. Many lenders, dealers, and manufacturers have reconciled their conflicting demands by relying on what is known as an "assignment-of-funds" procedure, in which an authorized official of the lender writes a letter to the dealer or manufacturer committing to pay one of them an agreed-upon sum after the modules are set on the foundation and inspected by a representative of the lender. This protects the lender and its customer by making payment contingent on an inspection that the home is correct and properly set. The dealer and manufacturer in turn get the security they need by receiving a written commitment from the lender to pay the dealer or the manufacturer once the inspection is complete. In effect, the dealer and manufacturer are relying on the lender's obligation to make good on its assignment rather than the customer's obligation to honor his contract. When done properly, the letter assigns sufficient funds from the customer's construction loan, usually equal to the balance owed by the customer for the modules, to the dealer or manufacturer and promises to make the payment either by wire transfer or with a bank or cashier's check.

select a lender. You can then ask each lender to approve the letter. If a lender asks for some modifications to the dealer's letter or proposes its own letter and the dealer is not agreeable, you will probably need to find a different lender or dealer.

If your modular dealer and his manufacturer require a COD payment and you are unable to find a local lender to assist you, your dealer is likely to know which lenders comply with this requirement. To avoid a misunderstanding, you and your dealer should ask the lender to write you a letter committing to pay for the balance owed on delivery.

You might wonder whether paying the modular dealer and manufacturer in full on delivery or immediately after the set compromises your leverage should you subsequently find something wrong with your home. You certainly do lose leverage. This is exactly why you should shop very carefully for a dealer and not just buy from whoever is the least expensive. Just as you would never buy a car from a dealer who has a reputation for poor warranty service, you should never buy a modular home from a dealer who might not honor his warranty obligations. Regardless of when you pay a dealer, your warranty is only as good as the dealer's integrity and competence.

You also need to ensure that the lender agrees to an acceptable disbursement schedule. This schedule states how much money will be paid at each phase of the construction process. Most of the details are worked out between the customer and his GC. But the customer and dealer are responsible for meeting the full and timely disbursement schedule for the modules.

A lender may agree to an assignment-of-funds procedure but then offer a disbursement schedule that fails to allocate sufficient funds to pay the balance due on the modules. Since the dealer is unlikely to agree to a partial payment, you need to

Speaking from Experience

WHEN A CUSTOMER selects a lender I have not worked with before, I contact it immediately after the home is ordered to ensure that the lender agrees with our assignment-of-funds procedure. It seems that every year one of my customers completes his application, gets approved for his loan, and is ready to close before we all realize that the lender will not follow the procedure. This is in spite of the fact that we send the lender and the customer a copy of the assignment-of-funds letter soon after the customer orders the home. Even a follow-up phone call to the loan officer to review the procedure does not prevent the problem. We've found that some loan officers say yes to our procedure without first running it by their manager. So we now ask the officer to discuss it with whoever is empowered to make the decision.

Sounds simple, right? Well, recently, we followed this procedure only to run into problems when the manager was overridden by the lender's attorney. Not even giving the lender the names of the many other lenders in the area that were comfortable with the procedure was enough to change its mind. My customers had to start over with another lender, which caused them to fall almost two months behind schedule.

Perhaps the lesson to learn from this is to make sure that the lender's attorney is on board with the assignment-of-funds procedure. But another lesson is that no matter how vigilant you are, you still may be hit with a frustrating surprise. My best advice is that if this happens to you, follow the example of my customers, who to their credit were able to hold on to their patience and good humor even though they lost seven weeks applying with another lender.

inform prospective lenders about the dealer's payment requirement. If a lender's schedule does not provide you with sufficient funds at the right time and you call this to the loan officer's attention before you sign the loan agreement, a lender will usually adjust the schedule to accommodate your needs. After you sign the paperwork, however, a lender will usually resist changing the schedule, which will likely force you to find a new lender.

Keep in mind that it takes longer to receive your home if you use a construction loan because the manufacturer waits for the lender to write its assignment-of-funds letter before putting your home into the production schedule. And the lender usually waits to do this until you have closed on the loan, which likely cannot happen until you have a building permit. As you approach the closing on your loan, do everything you can to prepare your lender to write the letter immediately after the closing. A couple of weeks before the delivery of your home, ask the lender to schedule its representative to inspect and approve the modules and disburse the balance due. The inspection and payment are required whether the payment terms are COD or assignment of funds.

Payment with Private Funds

When you use a private source of funds to pay for some part of the balance due on a modular home, the dealer and his manufacturer are likely to require you to pay for the modules when they are delivered. A COD payment must be made with a bank or cashier's check made payable to the dealer or manufacturer, as instructed by the dealer. Needless to say, you are not obligated to pay for the modules if the dealer and manufacturer build you the wrong home. It is very unlikely, however, that this will take place if you select a reputable dealer.

If you are paying COD, your dealer may require an additional deposit for each module before he schedules your home to be built. These funds serve as insurance for the dealer should you fail to pay when the modules are delivered. He uses this additional deposit to defray the expenses incurred if he must return the modules to his manufacturer or sell them to another customer at a discount.

Change Orders

Before your dealer instructs the manufacturer to build your home, he requires you to sign off on your selections. If you make any changes to your order after that and are paying with a construction loan, you need to make special arrangements to pay for the changes. If possible, instruct your lender to include the full cost of the change order in its assignment-of-funds letter before it is written. If the letter is already written, ask your lender to write a revised letter. Of course, asking your lender to cover the cost of the change order does not work if all of your construction funds are already allocated to the project. In that case, you will need to use personal funds. Your dealer will likely require prepayment for the changes before allowing your home to be built, since this is the only way he can be sure you have the funds.

Paying for General-Contracting Services

If you have no construction experience, many lenders will not let you act as your own GC. The GC's responsibilities are substantial, and many lenders are unwilling to take the risk of having an inexperienced customer carry out these duties. Lenders are concerned that your budget will not include all of the required work, which means you will run out of money before completing the home. They are also worried that you will not hire good subcontractors, which means you will end up with poor workmanship. In addition, they are afraid you will not keep the project on schedule, which means your interest payments will mount beyond what you budgeted. The delays could also cause you to lose your rate lock and make it harder to afford the home. Customers sometimes try to fool their lender by claiming that an experienced relative or friend will be the GC when they plan to play the role themselves. This is not a wise strategy. If you really want to be your own GC, find a lender who allows it.

If you hire a professional GC, his responsibilities begin after you sign a contract and give him a deposit. He may ask for a 10 percent deposit, which is what most modular dealers do, or he could request more. The GC has greater financial responsibilities than the dealer. He uses the deposit to secure commitments from his subcontractors. Once work begins, he uses it to purchase materials and pay his employees and subcontractors. These expenses will often substantially exceed your deposit. Should you lack the funds to pay the GC his full deposit, find a lender who will advance you the money at the closing or negotiate with the GC for a reduced deposit.

If you are serving as your own GC, you need to have sufficient funds to line up subcontractors. If you do not have adequate funds to secure a commitment from subcontractors, make arrangements with the lender to advance you the funds. You also need either a charge account with a local building-supply store or cash to purchase materials, since the lender is unlikely to advance money for materials.

You should make sure that the lender will disburse funds to your GC at the required times. The disbursement schedule for a GC is more involved than for a dealer because his work is completed in a series of stages

Speaking from Experience

HERE'S ANOTHER STORY THAT seems to occur at least once a year with my company. Recently a young couple decided to act as their own GC so they could save money. They went to their favorite local lender, which apparently told them that the general-contracting work on their home had to be supervised by a professional GC. They said they understood and proceeded with their application. About a month after they applied, the loan officer asked for a copy of the GC's insurance policy and contractor's license, which is required in the customers' home state. Neither of the customers had a license or contractor's insurance. After asking around for a couple of weeks, my customers found a licensed GC who was willing to assume the official role of general contractor for a fee of a few thousand dollars. He also agreed to allow them to complete most of the work and hire their own subcontractors.

Unfortunately, the lender was not satisfied. It now wanted written quotes for all of the carpentry work. My customers were really surprised by that requirement, since they intended to do much of it themselves with the help of family and friends. When they finally were able to secure written estimates, the lender said that the total cost of the project was now greater than what it could lend them, so they were turned down for the loan.

Fortunately, I was able to recommend a lender that would allow them to act as their own GC and complete some of the work. But, as you would expect, my customers lost a couple of months and suffered a great deal of unnecessary stress. As this chapter recommends several times, you need to learn what a lender's policies are before selecting it.

over several weeks. The schedule must allow for funds to be disbursed in several smaller payments to coincide with his progress. If you are using private resources, this should be relatively easy, assuming you have ready access to your money. If you are using construction financing, ask your lender how many disbursements are included with your construction loan. Most lenders include about five in addition to the disbursement that is done at the closing. Each disbursement costs the lender money, since it must send an inspector to verify that the work has been completed.

Your GC may ask for more disbursements than the lender has included, especially if the scope of work is substantial. A $100,000 GC contract paid over five disbursements should be manageable for most GCs, but a $300,000 contract may warrant additional disbursements. Most lenders will agree to additional disbursements if you agree to pay them an additional fee, which should range from $50 to $100 per disbursement.

The GC and lender must also agree on the amount the lender will hold back until the final disbursement. Most lenders insist on holding back a percentage of the loan, which can range from 2 to 20 percent. Some lenders apply the percentage only to the GC's contract amount, while others apply it to the entire loan amount for the land, dealer, and GC. The differences in these approaches can be significant. Avoid problems down the line by ensuring that the GC and lender understand the details of the holdback. If the GC will not accept the lender's standard holdback, the lender may compromise, but only if you negotiate before closing on the loan.

Serving as your own GC does not eliminate the need to ensure that your lender's disbursement schedule keeps pace with the construction work. Instead of securing the agreement of one GC, however, you need a written agreement with each subcontractor. Setting realistic payment expectations for each subcontractor motivates him or her to stay on schedule. Should you fail to do this and one of your subcontractors becomes concerned about receiving his payments in a timely fashion, he may delay his return to the site. This can create a situation in which the lender will not disburse more funds until the subcontractor completes more work, but the subcontractor will not return until he is paid. If a GC finds himself in this difficult position, he often uses his own funds to pay the subcontractor to keep the job on schedule. Since you may not have sufficient funds to do this, you should ensure that your subcontractors will accept your lender's payment terms.

The first disbursement to the GC is due after he completes the site work and foundation. Since most GCs complete this work a couple of weeks before the modules are set, the payment is typically made at the same time the lender pays for the modular units. However, if the GC completes the work several weeks before the dealer sets the modules, the lender needs to complete a separate inspection and disbursement. The remaining payments are disbursed one to three weeks apart, depending on the required scope of work and the GC's completion schedule.

For each progress payment, the GC should provide you with an invoice listing the work he has completed since his last invoice. He should apportion a percentage of your deposit to each invoice. Just as you cannot afford any delays on the completion of your home, your GC cannot afford any delays in receiving his payments. Review the invoice as quickly as you can, but make sure the GC did not bill you for items he did not do. When you receive the final invoice, make sure that everything has been done.

If you are acting as the GC and using a lender, make sure your subcontractors do not include the cost of uninstalled materials in an invoice, even if they are expensive and delivered. Few lenders disburse funds for materials that are not yet installed.

If you are using a lender, authorize the GC to fax a copy of the invoice to the lender, or fax it yourself. Inform the lender that you have approved the invoice, since your lender might not authorize an inspection without your approval. After the lender receives the inspection report, it may complete a title search to verify that no contractor has filed a lien. This usually adds a day or more to the disbursement timeline. Should the lender not act in a timely fashion, intervene on your GC's behalf. Should the lender disburse funds directly to you, get them to the GC as quickly as possible.

There would not be as much pressure to expedite the invoice if the GC was stick building a home over a period of six months. But one of the many advantages of modular homes is speed of completion. Such speed requires timely payments to keep the construction process running efficiently and on schedule. Although your GC should be willing and able to provide the funds necessary to move forward on each stage of the project, he cannot allow himself to move several stages ahead without being paid for work he has already completed. Otherwise, his other customers, who are making timely payments, as well as his subcontractors, who have financial obligations of their own, will suffer unfairly. You and your lender must do your part so that your GC can do his.

Once the GC has obtained a certificate of occupancy and completed all of the contracted work, you still need to pay him in full before you can move in. If the GC has not completed a few punch-list items, hold

back some money until he has done so. Many GCs will agree to a holdback equal to 150 percent of the value of each item.

Once the GC has finished all of the punch-list items, pay him the balance in full, even if you discover a couple of warranty service items. Your payments to the GC are for completing your contract with him. They are not for correcting warranty issues that appear after he has completed his contracted work. Warranty service work is a normal occurrence with all new homes; it is not grounds for holding back any of the balance owed the GC on your new home.

Lien Waivers

When making the final payment to your GC, ask him to sign a lien waiver indicating that he has been paid in full and accepts full responsibility for paying all his subcontractors and suppliers. This protects you from having the GC, a subcontractor, or a supplier place a lien on your property, which would give your lender reason not to turn your construction loan into a mortgage. However, you should not ask your GC to waive his rights to place a lien on your property unless you have paid him in full.

Change Orders for General-Contracting Services

When you sign a change order that incurs additional costs, most GCs require you to prepay for the work, especially if you are using a construction loan and cannot provide evidence that the lender has approved the extra funds. The prepayment assures the GC that you can pay for all of his work before you run out of money.

10 Warranty Service

WARRANTY-SERVICE WORK is required at some point on virtually every new house. The imperfections of materials and people inevitably create the need. The hundreds of different materials that make up a house do not always respond as they were designed, sometimes causing undesirable results. The human beings who build homes bring different abilities and attitudes to their work, occasionally creating unacceptable quality. Dealers, manufacturers, and general contractors work hard to prevent quality problems and avoid mistakes, but they also have created a mechanism to address them. That is the function of the warranty. What distinguishes good companies is the extent to which they assume full responsibility for correcting warranty-service problems in their customers' homes. Since this policy is costly, most of these companies also continually try to improve their organizations to reduce the number of warranty items. Over time, they build a company that has fewer warranty claims than other companies.

Most builders of stick, panelized, and log homes provide the minimum warranty mandated by law. In most states, this is one year. Most modular manufacturers offer an extended warranty, which usually adds a two-year warranty for mechanical systems (plumbing, electrical, and heating) and a ten-year warranty for structural systems. These extended warranties express a great deal of confidence by the modular industry in its homes. Although it is comforting to know you have this extra insurance, the likelihood is you will never have reason to use it for a serious problem. You will, however, almost certainly need warranty service coverage for other matters.

The first time you might need help is immediately after your home is delivered and set. This is your opportunity to see the home and identify any mistakes, defects, or damages. The button-up work creates a second phase of warranty-service coverage, which requires the dealer, manufacturer, and GC to work together to resolve any issues that develop. The final phase begins after you move into your new home.

Comparison of Conventional and Extended Warranty Coverage

	COVERAGE: WORKMANSHIP AND MATERIALS	COVERAGE: MAJOR MECHANICAL SYSTEMS	COVERAGE: STRUCTURAL SYSTEMS	COVERAGE: TRANSFERABLE	FHA, VA, AND RHS ACCEPTED	BINDING ARBITRATION
Conventional Warranty	1 year	1 year	1 year	1 year	No	No
Extended Warranty	1 year	2 years	10–15 years	10–15 years	Yes	Yes

Immediately after the Set

When you buy a modular home, you expect it to arrive without mistakes, defects, or damaged materials. If you discover any, you expect the manufacturer to repair or replace them. You also expect the manufacturer to do this at no cost to you. Manufacturers usually understand these expectations, but they have a few of their own. They accept responsibility for problems found when your home arrives, but they expect your dealer, as well as you and your GC, to accept responsibility for any damages incurred after that. This seems fair, and in principle it is. When you purchase your home from a dealer who completes the GC work, your expectations are likely to be met. When the dealer and GC are separate companies, however, the situation can trigger contention and distrust.

A modular home is typically built with most of its interior complete. Walls, cabinets, tubs, doors, moldings, and electrical outlets are almost always installed at the factory. All of these products can be damaged accidentally, and this can happen as easily at the factory as at your site. Your home is thoroughly inspected before it leaves the factory. The manufacturer tries to repair or replace any defective or damaged goods before shipping the home. When that is not possible without causing a delay, the manufacturer documents the problem, makes plans to fix it at your site, and informs the dealer so that you are not surprised. Either way, the inspection enables the manufacturer to document any warranty problems with your home.

The inspection, however, does not preclude disagreements among the manufacturer, dealer, and GC. If you discover any damage to your home after it is delivered and set, it could have been caused by the manufacturer, even though it is not listed on the inspection report. It could also have been caused by someone on your site.

The manufacturer could have missed an item, or an employee could have caused the damage and failed to report it. The same damage, however, could have been caused by one of the GC's subcontractors, who may or may not have been aware of it. You or a friend could have unknowingly caused the damage.

Under the circumstances, it is not surprising that there are occasional disagreements over who is responsible for damages. The modular industry has developed a procedure for handling these situations. Modular manufacturers attempt to minimize these misunderstandings by requiring their dealers to identify and report in writing any warranty-service issues right after the set. You can expect your dealer to insist that you complete a warranty-service inspection and sign the resulting written report. If the GC is separate from your dealer, ask him to sign the report along with you. You should receive a copy of the warranty report that is also signed by the dealer.

The exact time allowed for the inspection varies from manufacturer to manufacturer, with some giving the dealer 24 hours after the set and others allowing him a few days for some items and a few weeks for others. Items that are easily damaged on-site, such as installed vinyl floors and carpeting, are less likely to be covered beyond a few days unless there are extenuating circumstances. This procedure allows the manufacturer to limit its responsibility to preexisting conditions. Consequently, if you find a damaged item after the reporting period expires, the manufacturer assumes that the damage was caused by someone on your site, and will not accept responsibility for correcting it.

Since the set-day activities can cause accidental damage to a home, some manufacturers require the dealer to complete the inspection as soon as the modules are delivered. This is common with manufacturers that ask their dealers to select an independent set crew. Since the dealer selects the crew, the manufacturer wants the dealer to assume responsibility for any set-day damages. The manufacturer secures this accountability by having the dealer complete its inspection before the set. While this may seem reasonable, a delivery-day inspection is unfair to the dealer and the customer. It is impractical to complete an accurate inspection on delivery day, given the poor lighting available in each plastic-wrapped module. It is also difficult to inspect a module when it is stuffed with ship-loose materials. Waiting until after the modules are set allows for a more accurate inspection. If at all possible, resist a delivery-day inspection.

Material Inventory

Before the manufacturer ships your home, it places some materials into one or more of the modules. The GC uses these ship-loose materials to complete the button-up of the modules and the construction of site-built structures, such as a garage or porch. These materials include items such as shingles, siding, flooring, interior doors and moldings, and light fixtures. The dealer should take a written inventory of the ship-loose materials immediately after the set. If any materials are missing or damaged, the manufacturer should replace them, but you should take responsibility for replacing any materials that are damaged or stolen on your site. Both you and the GC have an interest in helping the dealer complete the inventory. The dealer should require you to sign the inventory list, and if the GC is separate from the dealer, you should ask him to sign the list as well. You should receive a copy of the inventory that is also signed by the dealer.

Dealer and GC

Unless your GC is part of the set crew, the only thing he might have to do during the set is to assist with maneuvering the crane and modules into place. He should also watch the set and help the crew, if needed. If your dealer is with a different company from your GC, he should use this time to discuss several important issues with your GC, beginning with warranty procedures. He should tell the GC whom to contact with questions or concerns and how he should submit warranty claims. He could use the time to discuss the button-up procedures with the GC. If the GC does not have experience with modular homes, he might appreciate hearing some tricks of the trade, such as how to adjust a marriage-wall opening or how to prepare the exterior of the house for the siding installation. The dealer and GC may not use the set day in this way unless you make the suggestion. If either the dealer or the GC says that his set-day responsibilities or some other conflicts do not permit this, ask them to have the meeting on another day immediately after the set. If they claim it is unnecessary, tell them you want them to do it for your peace of mind, and insist on it.

During the Button-Up

Once the GC begins the button-up of your home, he must inform your dealer if he discovers any missing, defective, damaged, or poorly installed materials that were not identified during the set-day inspection. The dealer then notifies his manufacturer, who determines the corrective action. The GC should not do anything until he hears from the dealer. If the GC completes the unauthorized work and presents you with a bill, the dealer and manufacturer are unlikely to reimburse you.

The GC may prefer to report a warranty-service issue directly to you so you can pass the information to the dealer. This has the advantage of keeping you informed, but it has the disadvantage of keeping you in the middle. Ask your dealer and GC to talk

Speaking from Experience

Subcontractors who finish the siding installation on a modular home should have all of the materials accessible when they begin their work. Sometimes these materials, which are shipped loose with the house, are buried beneath other materials. The subcontractor for one of my customers was only 15 minutes into his work when he called the customer claiming that no J-channel had been shipped with the house. He told my customer, who had hired him directly, that he needed the J-channel delivered within two hours or he would leave and not return for several weeks.

My customer called me in desperation. Unfortunately, I did not have any in stock and the local suppliers did not carry a suitable match. The only alternative was to "borrow" some J-channel from another house that wasn't ready to be sided and have our modular manufacturer send us replacement materials as soon as possible. Before taking this step, I checked the ship-loose inventory, which showed that my set-day supervisor had found the J-channel and my customer had signed for it. I then called my supervisor to ask him what he remembered. He told me that my customer had not been paying attention when he conducted the ship-loose inventory, even after my supervisor repeatedly pressed him to do so. My customer told him, "Don't worry about it. I'll sign for it when we're done," which he did. After reviewing the situation, I suspected the materials were there, but I couldn't be sure. So I had my supervisor track down and deliver the replacement materials. Unfortunately, it took him four hours to make the delivery, and the siding contractor was gone by the time my supervisor arrived.

Before leaving the customer's house, my supervisor searched for the missing materials. He really wanted to know whether he had made a mistake. Sure enough, he found the J-channel buried beneath a couple of rolls of carpet. I learned later that the siding contractor was good to his word and didn't show up for several weeks, so my customer's mistake cost him a lot of time. It is worth mentioning that if the J-channel had been missing, many dealers would have understandably informed the customer that he was responsible for replacing it. After all, he had signed a form saying he had been given the materials. My recommendation is to give the set-day inventory your full attention.

directly with each other. As two construction professionals, they will probably communicate more effectively without you as a translator. Insist, however, that the GC tell you if he reports any warranty-service issues to the dealer. You can then decide for yourself whether to speak to the dealer.

Warranty Service for Modular Materials

An independent GC should wait for the dealer and manufacturer to replace any damaged or missing materials. This ensures that the correct materials are used, and it allows the manufacturer to make a warranty claim against the company that supplied the materials or parts. Even if the manufacturer has no warranty claim to make, it can often replace any parts or materials more affordably than the GC, because of its volume-purchasing power.

Warranty-Service Labor on the Modular Home

Once the GC reports a warranty claim, the dealer and his manufacturer must decide how to address it. Typically, they take care of the work themselves. Modular manufacturers have crews whose primary responsibility is modular-service work. As modular experts, each crew will have mastered the remedial procedures and specialized skills through experience, allowing them to complete the work in less time and for less money. What may look like a huge and expensive problem to an inexperienced GC may be a quick fix for an experienced crew. So it is both smart and fair to let the manufacturer use its service crews.

Giving the manufacturer responsibility for remedial action serves your interests. Should the correction not work or create a different problem, you have only one company to turn to for additional help. For the same reason, you do not want anyone but an automotive dealership authorized by the manufacturer to repair your car while it is still under warranty.

In some situations the manufacturer asks the GC if he wants to do the warranty-service work, especially if he has the interest, time, and skills and offers a suitable estimate. In fact, the manufacturer is more willing to ask the dealer to complete some of the warranty work when he is also the GC, since the manufacturer has confidence in his modular construction skills and trust in his warranty-service charges. In addition, the manufacturer saves money, the dealer saves time, and you can move into your home earlier. This is another advantage of buying a modular home from a dealer who acts as the GC. If the GC does not have prior modular experience, the manufacturer may be reluctant to have him handle warranty-service work. Make sure the GC, if he is not also your dealer, understands that only the dealer and his manufacturer can make this decision.

When the dealer and manufacturer authorize your GC to do warranty work, he should complete it as soon as possible. After the work is finished, he should give you his bill, along with copies of supplier and subcontractor invoices so you can submit them to the dealer. Once the paperwork is received and approved, the dealer will compensate you, and then you can in turn pay the GC.

These procedures do not always work out as they should. Often, a GC or his subcontractors go ahead and fix a warranty problem without first obtaining authorization from the dealer. The impulse is understandable. Subcontractors are accustomed to fixing problems, and they do not like to wait for either authorization or a warranty-service crew to show up. It can take more time to obtain replacement materials from a modular manufacturer than to purchase them locally. Also, it sometimes takes longer for a manufacturer's warranty crew to make the trip to your home and correct the problem than it does for a local subcontractor to do the same.

The biggest risk the warranty-service procedures pose for you is that your independent GC discovers and reports some damage several weeks after the official inspection period. Fortunately, most manufacturers are reasonable about warranty claims. Regardless of their official policy, they tend to judge warranty claims on a commonsense, case-by-case basis, allowing for the fact that some damages caused by their own people are often hidden from view by protective coverings and poor lighting. They also recognize that some defects cannot be discovered until a product is used, which will not happen until it is made functional by the GC. With uncertain claims, they may even assume responsibility for damage they did not cause. This helps manufacturers build loyalty with their dealers, which helps dealers earn loyalty from their customers.

Warranty Service for Modular Emergencies

The only time a customer should authorize warranty-service work without first getting the dealer's approval is in an emergency, when someone might be endangered or the damage worsened if prompt action is not taken. If you authorize work just for the convenience of the subcontractor or GC, you may get stuck with the bill.

Warranty Service for GC Work

If you discover any mistakes, defects, or damages involving your GC's work during the button-up of your home, report this directly to the GC. He will want to correct the items during the course of his work, since he cannot bill you in full for his services until he has done so.

Warranty Disagreements

It is possible you will disagree with your dealer or GC about whether something in your home is defective, damaged, or poorly installed. When a warranty-service prob-

Speaking from Experience

SOMETIMES A GENERAL CONTRACTOR with limited modular experience can turn a "couple of hours" fix into a "couple of days" project. I had a customer whose master bedroom ceiling was not level at the marriage wall. My set-day supervisor should have noticed this and told the customer that we would come back to fix it. But he missed it, and my customer, who was acting as his own GC, decided to fix the problem on his own. He ended up tearing down half of the bedroom ceiling before he contacted us for help. The problem, we discovered, was that the ceiling framing on one of the modules was hung up slightly on the other module, preventing the first module from settling all the way down. All that needed to be done was to free up the first module, which we were able to do in two hours. When we finished, the ceiling framing was even, but my customer now had to replace a lot of drywall.

It is very important that you and your GC understand that he should not attempt to fix any warranty problem without first contacting the dealer for authorization, at least not if you want to take the right corrective action and be compensated for your efforts.

lem involves a building-code violation, the burden is usually on the dealer and GC to correct the problem. Installing the wrong type of smoke detector is something the dealer, through his manufacturer, must correct. Using undersized framing for your garage or deck is the kind of mistake the GC must correct. The dealer is not, however, automatically responsible for meeting specifications that exceed the state building code. For example, if your local building inspector insists that an air-infiltration barrier must be installed under your siding but this is not required by the state building code, your dealer would be accountable only if he had accepted responsibility for verifying whether the local inspector enforces any special codes. If you agreed to assume this responsibility but failed to obtain the correct information, then you would be responsible for the additional material and labor, including, in this case, the cost for removing and reinstalling whatever siding was already installed by the manufacturer.

The GC is not responsible when the scope of work for a task was not included in his original contract. For example, the fact that you need a set of stairs from the door to the backyard does not obligate the GC to provide them if you excluded them so you could build a deck in the future. Nor is the GC responsible for providing clean backfill to place around the foundation if the building inspector declares that the soil removed from the cellar hole cannot be used as backfill. If the GC uses the fill before the building inspector instructs him not to, the GC is responsible for removing it, since he should have known the building code. You are still responsible, however, for the cost of the replacement fill, since the poor soil was not the GC's fault and you needed it to be replaced regardless of the GC's mistake.

Cosmetic issues are often sources of disagreement. A customer should receive the degree of finish he selected and paid for, but this is often different from what he may have seen in a model home. As mentioned on pages 150 and 184, one way to handle disputes of this kind is to have your contract include a set of quality guidelines for materials and workmanship that can be used to help settle differences. Keep in mind, however, that guidelines and standards spell out the minimum acceptable workmanship and product performance. Your personal standards will likely exceed these standards in some areas.

Walk-Through Inspection

Before making the final payment to your GC and moving into your new home, do a walk-through inspection with the GC to be certain it is complete and finished to your satisfaction. Use the walk-through to point out any items that need additional work or correction, including exterior finishes and site work. If you have been keeping an eye on the work as it progressed, you should find few new items. Have the GC document any problems in writing, and try to have the work completed before you move in. Decide on a monetary value for the correction and hold back 150 percent of this amount from the final payment until the work is done. Any nonemergency items discovered after the inspection, but before you move in, should be added to a new list that can be completed along with other items you discover after you move in.

During the walk-through inspection,

What Is "Good Enough"?

Most construction warranty guidelines for materials and workmanship assume that it is neither realistic nor fair to expect a GC to remove blemishes that are not readily visible, and can be seen only in unusual light or from very close range. Finished drywall, especially, almost always shows minor blemishes in the right light and from the right angle. For a customer to insist that such small items be addressed under warranty is to create a potentially antagonistic relationship. The customer wants her dealer and GC to take seriously those things that are most important to her. She does not want to create an atmosphere in which the dealer or GC feels compelled to deny assistance by appealing to some technicality in a set of guidelines that relieves him of responsibility. In other words, if the customer can exercise some flexibility over defining what's "good enough," she should expect the dealer and GC to adopt a similar attitude. You could ask your dealer to replace a pine bifold closet door with a tiny dent on the inside, one that can be seen only when the door is fully open. If you do so, however, do not be surprised if he tries to hide behind a technicality for some other item that's important to you.

have the GC show you how to use any new or unusual features and systems in your new home. Use the sample checklist below and on page 290 as a guide. Ask the GC to explain how to turn off the master switch on the electrical panel box and where to find the emergency shut-off switch for the heating system. Many of the products and systems installed by your GC come with their own component warranties. The GC should provide you with the written warranty documentation for each product. Don't forget to complete and mail the warranty registration cards included with the materials. The GC should remind you to finish any jobs you assumed responsibility for, and he should discuss your maintenance responsibilities as homeowner.

If your dealer did not give you a product demonstration on set day as part of his

Use this sample checklist as a guide during the walk-through inspection with your GC. (See page 2 of this form on the next page.)

Walk-Through Inspection Checklist

During your Homeowners' Walk-Through, our representative will demonstrate and/or discuss the items or services we provided you. However, we make no representation to be experts in the use and maintenance of these items and services. That is why we also provide you with copies of the warranty and instructional information that accompanies them. We strongly advise you to read this information closely to ensure that you use and maintain the items correctly.

HS – Indicates the item or service was provided by The Home Store. The Home Store representative will demonstrate or discuss each of these items.

OWN – Indicates the item or service was provided by the Homeowners. The Home Store recommends that the Homeowners consult with whoever provided them with the item or service.

N/A – Indicates the item was not applicable

Plumbing

HS	✓	Emergency water shut-off
HS	✓	Plumbing leak
HS	✓	Plumbing clog
HS	✓	Well system
HS	✓	Cleaning of faucet aerators (wells)
HS	✓	Water saving showerheads
HS	✓	Frost free garden faucets

Heating System

HS	✓	Emergency shut-off
HS	✓	Thermostats
HS	✓	Zoning
HS	✓	Radiator or vent controls
N/A	✓	A/C
N/A	✓	Oil tank gauge
HS	✓	Filter cleaning and replacement
HS	✓	Maintenance

Domestic hot water tank

HS		Amount of hot water

Electrical

HS	✓	Panel and sub-panel boxes
HS	✓	Circuit breakers
HS	✓	GFI's
HS	✓	Switched outlets
HS	✓	Smoke alarms

Interior Finish

OWN		Painting, staining, or clear coating of unfinished surfaces
HS	✓	Touch-up paint and stain left with house
N/A	✓	Finish & maintenance of beveled edge countertops
HS	✓	Drywall blemishes
HS	✓	Drywall touch-up
HS	✓	Use of correct roller mat (no brushes)
HS	✓	Color differences when touching up

Condensation

HS	✓	Causes
HS	✓	Control

Revised September 24, 2004

Page 1 of 2

warranty-service inspection, he should provide you with one before you move into your house. He should demonstrate and discuss the materials, equipment, and appliances or services he provided, and he should give you copies of the respective warranty and instructional information that accompanied these items. Again, use the sample checklist shown on page 289 and below as a guide.

The GC should review his warranty procedures with you, paying particularly close attention to how you should deal with emergencies and report routine items. He

The Walk-Through Inspection Checklist continued.

"Settling" of Home

HS _____ ✓ Natural contraction & expansion of wood: drying, shrinkage, and gaps
HS _____ ✓ Drywall cracks
HS _____ ✓ "Nail pops"
HS _____ ✓ Gaps between backsplash and wall
HS _____ ✓ When apply finish paint and/or wallpaper
HS _____ ✓ Wood and pellet stoves

Flooring

HS _____ ✓ Vinyl floors
HS _____ ✓ Hardwood floors
N/A _____ Tile floors
HS _____ ✓ Carpet seams

Cleaning of Surfaces

HS _____ ✓ Tubs
HS _____ ✓ Plastic laminate countertops
HS _____ ✓ Cultured marble countertops
HS _____ ✓ Dust after first move in, especially with hot air heating systems

Windows and Doors

OWN _____ Painting exterior doors
HS _____ ✓ Exterior door thresholds
HS _____ ✓ Tilt windows
HS _____ ✓ Garage door openers
HS _____ ✓ Keeping garage doors closed in winter

Fireplaces

HS _____ ✓ Zero clearance use and maintenance
N/A _____ Masonry use and maintenance

Exterior

N/A _____ Painting, staining, or clear coating of unfinished surfaces
HS _____ ✓ Gutters, down spouts, and extenders
HS _____ ✓ Vinyl siding, soffit, fascia, and shutters
N/A _____ Storm doors

Roof

HS _____ ✓ Shingles and wind
HS _____ ✓ Ridge vent systems
HS _____ ✓ Attic access

Appliances

HS _____ ✓ Dishwasher shut off
N/A _____ Garbage disposal reset button
HS _____ ✓ Icemaker shut off

Foundation

HS _____ ✓ Cracking concrete
HS _____ ✓ Opening windows or vents in spring
N/A _____ Crawl space access

Site

HS _____ ✓ Settling of soil around foundation

Revised September 24, 2004

Page 2 of 2

should leave you a form that lists the names and phone numbers to call in case of an emergency. Put this in an accessible place, such as the inside of the kitchen-sink cabinet door.

After Moving In

After moving into your home, you will almost certainly find problems that need fixing under your warranty. Contact your dealer and GC right away if you discover an emergency or other time-critical item. For less critical problems, however, it would be better to compile a list and submit all of the items together about 30 days after you move in. Plan to submit a final list two months before your warranty expires. You should expect the dealer and GC to make a warranty-service visit about two weeks after receiving your list, when they can complete all of the work at once.

Always submit a warranty-service list in writing. This increases the likelihood that records are both complete and accurate. Verbal messages are often more likely to be misunderstood, recorded inaccurately, and misplaced.

The dealer and GC may be able to complete their service work during the first visit or they may require other visits. Either way, you should allow the service crew to perform nonemergency work during its normal business hours. That is the time of day when the crew can reach the dealer or manufacturer for answers and shop for materials. The more efficiently the crew can work on the first visit, the less likely it will need to return.

If you have an extended warranty on your home, it usually provides an additional year of coverage on the mechanical systems installed by the manufacturer and an additional nine years of coverage on the structural systems built by the manufacturer. The coverage does not apply to those components completed by your GC, unless the GC purchases extended coverage on his work from the same insurer who covers your modular home company.

Homeowner's Warranty Obligations

Homeowners have a few warranty obligations that must be taken seriously, especially those relating to normal maintenance and care. A good overview of the homeowner's responsibilities can be found in a booklet by the National Association of Home Builders titled "Your New Home and How to Take Care of It."

One responsibility often ignored by homeowners is the obligation to contact the appropriate party in a timely fashion when a warranty situation is discovered. Even a simple warranty issue can become serious and require an expensive fix when you delay reporting it. For example, if your front door

Wood Contraction and Expansion

The first heating season removes excess moisture from the wood and concrete in a new home. After that, the home undergoes an annual cycle of moderate expansion and contraction, especially if it is located in a climate that experiences a lot of variation in temperature and humidity throughout the year. The framing, doors, trim, and wood floors shrink under conditions of low humidity (most often the winter) and expand under conditions of high humidity (most often the summer). This may cause wood fittings, such as miter joints, to temporarily tighten or loosen. Usually, these conditions return to normal when the humidity returns to normal.

During the first year, use of a coal, pellet, wood, or other such stove at very high temperatures can lead to excessively fast drying. This can cause an unusual number of drywall cracks and nail pops as well as excessive warping, cupping, and shrinking of wood and other materials. Neither your modular dealer nor your GC can be responsible for any damage caused by these excessive heat conditions. Waiting a year before using these products at very high temperatures helps the wood and other materials in your home to dry slowly and normally.

leaks a little water every time it rains because the threshold needs to be adjusted, the finished flooring and framing can quickly become damaged.

Modular homes are strong, but they are not indestructible. Expect your home to show signs of normal wear and tear over time, and accept responsibility for fixing the inevitable results.

You will want to restore your home to as-new-as-possible condition after the first heating season, since most of the settling and drying of wood will have occurred by that point. In a typical home, completing this tune-up usually takes a day or two by someone who has carpentry, drywall, and painting skills. Some of these normal changes will reappear in subsequent years, but they should be less noticeable and easier to repair.

If your modular dealer was also your GC, it is reasonable to expect him to correct these problems before your warranty expires. It is less clear, however, who should make these corrections when the dealer and GC are separate companies. Some of the drywall and moldings will have been installed by the manufacturer and some by the GC. You could insist that each correct what he built, but this assumes that all changes in a particular part of your home are due to the company that completed the work, which is not always the case. If your home has excessive drywall cracks in a few different areas, for example, they could have been caused by the way the manufac-

Condensation

Most customers notice moisture condensation on the windows at the beginning of the first heating season. As much as a ton of moisture can be released by the lumber, concrete foundation, and drywall as they dry out. The condensation can appear as fog on the windows and can even freeze on the glass. Condensation is most likely to appear on windows because glass surfaces have the lowest temperature of any interior surface in a home. When the warm, moist air comes in contact with the cooler glass, the moisture condenses. The same action occurs on the outside of a glass of iced tea in the summer and on the bathroom mirrors and walls after you take a hot shower. If condensation occurs, you need to provide ventilation to dissipate the moisture. Turning on the kitchen and bathroom ventilation fans each day or briefly opening a few windows, especially during the first heating season, should take care of the problem.

Moisture can also build up in a home after the first year because of normal living. If the problem continues, you should remind members of the family to use the bathroom ventilation fan when they are bathing and the range hood fan when they are cooking. Today's tight homes are more prone to retain moisture from cooking, bathing, drying clothes, operating humidifiers, heating with fossil fuels, and breathing. Proper ventilation, however, maintains the right amount of moisture in your home to balance comfort and safety. If an abnormally wet situation exists, use a dehumidifier. Otherwise, problems may result, such as peeling paint, rotting wood, buckling floors, insulation deterioration, mold and mildew, and even moisture spots on walls and ceilings. The homeowner is responsible for any problems caused by improper ventilation.

Excessive moisture and condensation can also be caused by conditions outside of the home itself, such as high winds during a heavy rainfall or snowstorm. Dampness in the basement, caused by poor exterior grading, a high water table, or other site conditions, can also lead to moisture problems in the home. Again, in an abnormally wet situation, use a dehumidifier.

turer built your home or by the way the GC leveled the sill plate. If there is a lot of shrinkage of the wood moldings and floors installed by the manufacturer, it could have been due to the materials used by the manufacturer or to excess moisture that entered the home during the button-up. The best course in this situation is to contract with your GC to complete all of the tune-up, regardless of who built the different parts of your home.

Your GC may balk at taking on this responsibility. Since he did not build the modules, he might fear that he is exposing himself to too big a risk. In addition, if he has no prior modular experience, he may feel unable to predict the amount of time required for the tune-up. A fair way to handle this is to agree to pay him for his actual time and materials. An alternative would be to take on the work yourself, if you have the skills.

When completing the tune-up, the GC should retape any cracks in the drywall or the tape covering the drywall. He might be tempted to cover them with compound or caulk to save time and money, but the cracks will reappear if he does. On the other hand, fine cracks in the mud covering the drywall tape can be filled with a high-quality, paintable caulk. Small, open miter joints or other small gaps between pieces of wood can be filled with wood filler or caulk; larger gaps should be corrected by removing and reinstalling the wood. Popped drywall fasteners should be driven farther into the framing, when possible. Otherwise,

additional fasteners should be used. A small gap between a wall and a kitchen or bath countertop should be filled with caulk.

After these corrections are completed, the reworked areas can be touched up, ideally with paint or stain left over from the original button-up. If the GC has to buy new paint or stain, he may not be able to obtain an exact color match with the previous application.

Although you do not need to, you might want to wait until your home has finished settling and drying out before painting the walls and ceilings with custom colors. If you do not wait, you should save some matching paint to complete the tune-up. However, you may still need to paint an entire wall or ceiling in a room when you do the tune-up to avoid shadows caused by slight variations in color.

You might also want to wait until your home has finished settling and drying out before wallpapering or stenciling. Regardless of when you apply it, you are responsible for repairing any damage to the wallpaper due to settling or drying.

After your warranty expires, you should continue to take care of normal homeowner maintenance tasks. Most people find it is best to develop a list of tasks to do each season. For example, they clean the gutters and downspouts every spring and replace the batteries in the smoke and carbon-monoxide detector every fall. Your modular home will last many generations if you maintain it well.

11

Building on Schedule

BUILDING YOUR HOME ON schedule takes every bit as much planning as building it on budget. In fact, if you put a lot of effort into planning both, you are likely to find that keeping on schedule is considerably harder than staying on budget. That is because you can determine what you will get and how much you will spend before construction begins. Indeed, almost all of the required decisions for building a modular home are reasonably predictable and within your control. This is not true for scheduling. There are many players involved who will not always bend to your will, including the surveyor, engineer, utility company, building inspector, dealer, manufacturer, and general contractor. You may even find that your own personal and family obligations become an obstacle to maintaining your schedule.

As you will see, there are two very separable timelines when building a modular home. The first includes all of the tasks you, the dealer, the manufacturer, and the GC must complete before your home is set. You are primarily responsible for making final decisions about your modular and general-

contracting drawings and specifications. You are also responsible for completing those tasks related to obtaining a building permit and financing. The second timeline involves those tasks the GC must complete after your home is set. This chapter identifies all the players, the tasks they must complete, and the sequence in which the tasks should be done. It also explains your responsibility for each of the tasks.

Getting Started on Time

Most people building a new home are prepared for the construction to take longer than planned. They have heard that subcontractors, inclement weather, utility companies, and inspection officials all contribute to delays. Few people, however, anticipate how long it takes to complete those tasks that must be done before construction can begin. Consequently, they budget too little time for these tasks and then try to compensate by skipping some tasks and rushing through others. When this strategy fails, they miss their desired move-in date and pay for it with stress with their family, conflict with their dealer and

GC, and cost overruns with their budget.

It can take you as little as five weeks or more than a year to complete all of the tasks that must be done before your modular home is delivered. Your responsibilities can take as little as one day. But you will need to order a standard modular plan with no changes, select only standard features, agree all decisions are final, have cash to pay for everything, have a GC lined up and ready to go, and have a building permit in hand or not need one. If this is true for you, you will be an exception.

More likely, you will want to customize your modular and GC drawings and specifications and take some time to consider your decisions. You will also need to wait for the lender to approve your loan and the building department to issue your permit. You may even want to revise your drawings and specifications two or more times. Consequently, you will likely need several weeks before you are done.

Even if you are able to make final decisions about your drawings and specifications in one week, the manufacturer cannot build your home, and you do not want it to do so, until you have obtained a building permit and secured financing. These tasks can take a couple of months. Closing on a construction loan often takes six to eight weeks. Completing the preliminary steps required to apply for a building permit can sometimes take several weeks. And receiving a building permit after submitting the application can take up to 30 days. One of the most important variables affecting whether you will be done on time is how quickly you begin your efforts. If you wait two weeks, you will not be able to make up the lost time by asking your dealer, GC, lender, or building department to work faster.

The start of your building schedule will also be delayed if you have not finished the following tasks before you order your new modular home:

Your Modular Dealer and Financing Tasks

- ☐ Sign a contract with the modular dealer
- ☐ Apply for financing
- ☐ Receive the dealer's first draft of the preliminary plans and specifications
- ☐ Meet with the dealer to revise the plans and specifications
- ☐ Specify where the GC wants to locate the electrical meter on the modular home
- ☐ Design the second-story floor plan for an unfinished Cape or attic
- ☐ Tell the dealer what special building codes are enforced by the building department
- ☐ Tell the dealer what matching materials the GC needs to complete the site-built structures
- ☐ Tell the dealer what rough openings the GC wants framed in the home
- ☐ Tell the dealer what site-installed flooring and baseboard specifications the GC wants for the home
- ☐ Receive the dealer's second draft of the preliminary plans and specifications
- ☐ Meet with the dealer to sign off on the plans and specifications
- ☐ Authorize the dealer to complete the permit plans
- ☐ Pay the dealer the balance of the required deposit
- ☐ Receive the dealer's permit plans
- ☐ Deliver a copy of the bank's commitment letter to the dealer
- ☐ Deliver a copy of the building permit to the dealer
- ☐ Deliver the bank's assignment-of-funds letter to the dealer
- ☐ Authorize the dealer to build the home
- ☐ Send the dealer a certificate of insurance
- ☐ Pay the dealer the balance due in full
- ☐ Remain present during the delivery and set
- ☐ Complete a walk-through inspection for warranty work and material shortages

- Secured a building lot
- Surveyed your building lot
- Resolved any deed and zoning issues with your building lot
- Resolved any wetland issues with your building lot
- Obtained a valid perc test and at least applied for an engineered septic design
- Selected a GC and/or subcontractors

Your Permit and Preconstruction Tasks

- ☐ Sign a contract with the GC or separate contracts with the subcontractors
- ☐ Submit the septic design for approval to the board of health
- ☐ Order the land boundary stakes for the property
- ☐ Inform the GC of any easements or deed restrictions that apply to the property
- ☐ Review with the building inspector any zoning regulations that apply to the property
- ☐ Ask the building inspector if he enforces any special local building codes
- ☐ Determine the cost of all permit fees that apply to the property
- ☐ Give the building permit application to the GC to fill out
- ☐ Ask the GC where he wants to locate the electrical panel box on the home
- ☐ Ask the GC to help design the second-story floor plan for an unfinished Cape or attic
- ☐ Ask the GC to specify the matching materials he needs to complete the site-built structures
- ☐ Ask the GC to specify the rough openings he wants for the home
- ☐ Ask the GC to specify the site-installed flooring details for the home
- ☐ Receive the GC's first draft of the preliminary plans and specifications
- ☐ Get back the completed permit application from the GC
- ☐ Obtain the necessary signatures on the building-permit application
- ☐ Have the surveyor complete the boundary stakes on the property
- ☐ Receive the septic design approval from the board of health
- ☐ Meet with the GC to revise the GC plans and specifications
- ☐ Receive the GC's second draft of the plans and specifications
- ☐ Meet with the GC to sign off on the plans and specifications
- ☐ Ensure that the GC drills the well
- ☐ Authorize the GC to draw the permit plans
- ☐ Pay the GC the balance of the required deposit
- ☐ Receive the dealer's GC permit plans
- ☐ Submit the well-water test results to the board of health for approval
- ☐ Apply for the building permit
- ☐ Receive the building permit
- ☐ Authorize the GC to begin his work
- ☐ Meet with the GC, his excavator, and the dealer on the site
- ☐ Ensure that the GC begins the site work and prepares the site for delivery and set
- ☐ Schedule the surveyor to complete the as-built plan
- ☐ Schedule a bulldozer and tow truck
- ☐ Ensure that the GC installs the foundation
- ☐ Locate a staging area to store modules
- ☐ Obtain the as-built plan and get a copy to the bank and the building department
- ☐ Ensure that the GC completes his site and foundation preparation
- ☐ Ensure that the GC brings the necessary heavy equipment to both the delivery and the set
- ☐ Pay the GC for his excavation and foundation work

Once you complete your responsibilities, the manufacturer needs a minimum of five weeks to complete its tasks. Typically, this involves at least three weeks to finalize your production drawings and order your materials, one week to build your home, and one week to ready it for shipment. It does not matter whether you complete your responsibilities in one day or one year; the typical manufacturer still needs a minimum of five weeks. Furthermore, if you select materials that must be "special-ordered" or are "back-ordered," the manufacturer needs even more time. And if the manufacturer is enjoying a sales boom, which is common today, its backlog of orders can add weeks to its schedule and your delivery date.

Step-by-Step Guide to the Building Process

The following section will help you understand the sequence of events required of a homeowner — from the time a modular home is ordered until it is built and ready for occupancy.

Phase 1

☐ Enter into a contract with a modular dealer
☐ Enter into a contract with a general contractor
☐ Apply for a construction loan
☐ Start your preconstruction tasks

The first steps you should take are to sign contracts with your dealer and GC. If you are acting as your own GC, you should sign contracts with your subcontractors. They will then complete the modular and GC drawings, which you can submit to the lender, to process your loan application, and to the building department, to process your permit application. Many customers delay this step while trying to resolve other issues, such as selling their home or closing on the purchase of their building lot. This can be a mistake, since it forces them to rush through decisions regarding drawings, specifications, and colors. The more careful you want to be, the sooner you should begin with your dealer and GC.

While you are waiting for the plans to be drawn, apply for a construction loan, if you are using one, and begin working on the permit and preconstruction tasks. The manufacturer will not schedule your home for production until you have closed on your loan and obtained a building permit. The following are typical permit and preconstruction tasks:

• Submit the septic design for approval

• Order boundary stakes for your lot

• Provide the GC with a copy of any easements that apply to your lot

• Provide the GC with a copy of any deed restrictions that apply to your lot

• Review a copy of the town's zoning regulations and consult with your building inspector

• Ask the building inspector if he enforces any local building codes that are different from the state's codes

• Determine the cost of the permit fees

• Get a copy of the town's building-permit application

You and the GC need to act quickly with all of your permit-related responsibilities, especially in obtaining signatures from the various town officials and boards. In most states a building inspector is legally allowed to take up to 30 days to issue a permit, and the clock does not start running until the application is complete and correct. Fortunately, most inspectors issue a permit within one to two weeks, and a few do so the same day. Do not assume that small building departments are faster than large ones. Sometimes they are significantly slower because they are understaffed and overworked.

Since the manufacturer cannot complete your permit drawings or build your home without first knowing all of the GC's construction specifications, you need to provide the dealer with the following critical pieces of information:

• The electrical meter and panel box location

• The design of the second-story floor plan for your unfinished Cape or attic

• The specifics about any materials you are having the GC install in your house, such as the rough opening dimensions for a special window and the type of underlayment that you want installed under the tile floor

You can obtain this information from the GC, who in turn can acquire it from his subcontractors. You should also ask the GC to provide you with a list of the matching materials he needs to build your site-built structures. Since the GC needs some time to collect the information, you must request his assistance immediately after you order your home.

Phase 2

☐ Receive and review the first draft of the modular and preliminary drawings
☐ Complete some preconstruction tasks

Both the dealer and GC should give you a copy of the drawings in advance of your next meeting with them. The GC should also tell you the panel box location, the design of the second-story floor plan for your unfinished Cape, and the requirements of any materials he is installing on-site. Review the

drawings closely before your meetings so you can bring your corrections and changes. Come prepared to discuss any changes to your building specifications. By now you should have made progress with your preconstruction tasks, including the boundary stake installation, septic design approval, and building permit application.

Phase 3

☐ Meet to revise the first draft of the modular and GC drawings and specifications

You must meet with the dealer and the GC to review the first draft of your drawings, specifications, and pricing. You should also begin to make your color selections, such as for siding, roofing, flooring, cabinets, and countertops. Most people like to think about their color choices for several days, so the sooner you start reviewing them, the better the chance you will have to make selections without a sense of panic. Be sure to give the dealer the GC's information requested above.

The dealer and GC should complete a second draft of preliminary plans if you make substantial changes to your first draft. If you are able to finalize all of your decisions at this meeting, you can skip the next phase.

Phase 4

☐ Receive and review the second draft of the modular and GC preliminary plans

Both the dealer and the GC should give you a copy of the revised drawings in advance of your next meeting. Review the drawings, specifications, and prices closely before your meeting so that you can bring your corrections and changes. Take extra care with all of your reviews, since ideally you'll be making final corrections and changes.

Phase 5

- [ ] Meet to sign off on modular and GC drawings and specifications
- [ ] Sign the authorization for modular and GC permit drawings
- [ ] Complete more preconstruction tasks

If you have completed all of the previous steps, this will be your final meeting to make changes and corrections to the modular and GC drawings and specifications. To formalize your decision to move ahead, the dealer and the GC should ask you to sign off on all of these documents. Should you not be ready to make final decisions, take all the time you need and let the dealer and GC know when you are ready.

If you and the dealer have signed a contingency-based contract, your commitment to have the permit plans drawn does not commit you to build a modular home. You retain the right to cancel your contract and receive a deposit refund minus the design fee until you have either met or waived the contingencies.

Occasionally people balk at signing off on the details, and some even call their dealer in a panic a day or two after they have done so. Committing to a definite floor plan or a particular color is daunting for some people, no matter how much time they have thought about it. Other people find it next to impossible to set aside enough time to think through their decisions and only the reality of the impending critical deadline forces them to commit. The dealer and the GC should never make you decide before you are ready. You cannot stay on schedule, however, unless you finalize your decisions in a timely fashion. If you need more time, take it. The dealer and the GC will put your order on hold, and you can adjust your move-in date accordingly. If you want to avoid a lot of anxious second guessing of yourself, review all of your modular and GC specifications and drawings every time changes are made to them.

Since you will soon be applying for a building permit, obtain any remaining permit signatures from the pertinent town officials. If you need a well and must demonstrate that you have potable water to obtain a building permit, you must drill the well and submit the water for testing.

Phase 6

- [ ] Receive the modular and GC permit drawings
- [ ] Complete the well test
- [ ] Apply for a building permit

If you receive the modular and GC permit drawings and complete all of the permit tasks, immediately apply for a building permit. Take the modular and GC permit drawings to the building inspector along with the application form containing the required signatures. In some states, the building inspector must also receive the following: a copy of the manufacturer's site-installation manual, a letter of approval from the state to the manufacturer authorizing it to build in the state, and certification that the crew setting your home is approved by the manufacturer. The dealer should provide a copy of these documents. If all is progressing according to plan, you should be submitting your well results to the board of health for approval.

Phase 7

- [ ] Provide the dealer with the lender's commitment letter
- [ ] Instruct the lender to write an assignment-of-funds letter

If you are using a construction loan, provide the dealer with a copy of the lender's letter of commitment stating that your loan has been approved. After you receive the commitment, ask the lender for a realistic closing date. Share this information with the dealer so he can help you stay on schedule. If your construction loan is progressing normally, you should be closing soon. Should the loan be processed more quickly than this schedule suggests, try to put this task behind you as soon as you can. To keep your home on schedule, ask the lender to write the assignment-of-funds letter in advance. That way it can send the letter to the dealer or manufacturer immediately after the closing.

Phase 8

- ☐ Receive your building permit and deliver a copy to the dealer
- ☐ Close on your loan
- ☐ Provide an assignment-of-funds letter
- ☐ Sign the authorizations for modular home construction and GC work
- ☐ Pay the balance of the deposits

If you are ready for the dealer to put your home into the manufacturer's production schedule, you must immediately complete several tasks. Provide the dealer with a copy of the building permit and assignment-of-funds letter along with any balance owed on your deposit. Authorize the GC to begin his work, and make sure that the GC agrees to have the site and foundation ready on time. Once you have completed all of these tasks, you can then authorize the dealer to build your home.

With your authorization in hand, the manufacturer can prepare your home's production drawings, order its materials, and assign it an actual production date. If the manufacturer has no production backlog, it will schedule construction of your home to begin in approximately three weeks. Since it takes one week to build your home and an additional week to get it ready for shipment, your home can be delivered in approximately five weeks from the date of your authorization. If the manufacturer has a longer production backlog, which is typical of most good companies, or if some of your building materials are back-ordered, the production will take longer.

Phase 9

- ☐ Provide your dealer with a certificate of insurance
- ☐ Schedule the as-built plan

The lender requires you to have insurance in place when you close on the loan. If you are financing your home with your own funds, have coverage in place before the GC begins any work. For insurance, secure either a homeowner's policy or a builder's risk policy. The advantage of a typical builder's risk policy is that it automatically provides coverage for theft of building materials and supplies as well as vandalism. You should direct your insurance agent to provide this additional coverage even if you opt for a homeowner's policy. Since your personal circumstances may differ and your agent may offer other alternatives, consult with your agent.

If you are paying for the modules with funds from your lender, which means you are paying by the assignment-of-funds method, instruct your insurance company to mail or fax the dealer a certificate of insurance a few weeks before the scheduled delivery. This proves that you have the necessary coverage. The effective date should be set at least 48 hours before the scheduled delivery date and remain in place for at least a week. The certificate should state, "[Dealer's company name] is loss payee as interest may arise." The certificate protects the dealer and manufacturer should your modules suffer damage after they are set on the foundation but before your lender pays the dealer. This might happen, for example, if lightning were to strike the modules the first night of a two-day set. Should this unlikely event occur, the certificate ensures that your insurance company would compensate the dealer so he can pay the manufacturer. Once the dealer is paid for the house, he no longer has any insurable interest, so your insurance coverage reverts to you and your lender.

The manufacturer's insurance covers the modules while they are being delivered to the site. The dealer's and crane company's insurance covers the modules while they are being lifted onto the foundation. You should ask the dealer for certificates of insurance documenting his coverage, and you should ask your insurance agent if the coverage is appropriate. If your lender is paying for the modules after the set, the dealer's insurance should be responsible while the modules are parked on your property before the set, since you will not yet own them. If the dealer does not provide coverage, you should direct your insurance agent to provide it.

If you are using private funds to pay for the modules upon delivery, your insurance should provide coverage when the modules are parked on your property, since you will already own the units. You should verify this. Your insurance is less likely to provide coverage when the modules are stored away from your property in a staging area. If you cannot obtain coverage for your situation, ask the dealer for help.

If either your building inspector or lender requires an "as-built" plan completed by a licensed surveyor, schedule this to be done as soon as the foundation is installed. There is often very little time to complete this task, since the surveyor cannot take his measurements until the foundation has been installed, which is often only a week or two before your home is delivered.

Phase 10

- ☐ Begin the site work and prepare for the delivery and set
- ☐ Meet with the dealer's delivery coordinator

If all goes according to schedule, the GC will begin your site work and foundation. He should start earlier if there are many trees to clear or if there are any site conditions that might unduly prolong the work. To ensure that the site work is done according to the dealer's requirements, the GC should meet with the dealer when he is ready to begin work.

Phase 11

- ☐ Schedule the bulldozer and tow truck
- ☐ Locate a staging area
- ☐ Pour the foundation
- ☐ Learn the projected delivery and set days

If directed by the dealer, the GC should schedule a bulldozer and tow truck to assist with the delivery and set. If the excavator cannot provide sufficient room on your site to store the modules overnight, you must locate a staging area. The GC should install the foundation several days before the set to allow the concrete to cure.

The dealer should give you projected dates for the delivery and set at least two weeks in advance. Do not think of these dates as carved in stone, however, since many things, almost all of them out of the dealer's control, can cause the dates to change. These include:

- A delayed delivery due to a last-minute factory production problem, a mechanical problem with the transporter, and a closed road due to a car accident

- Inclement weather the day of the set

- Equipment breakdown the day of the set, especially with the crane

- Ill health of the crane operator or a member of the set crew

- A set scheduled the day before your set that is delayed or requires the set crew to stay another day

You are likely to support the dealer if he cancels your set in the face of threatening weather and reschedules your set for the next available day. But you might not be prepared for what can happen when a set that is scheduled the day before yours is canceled. When this happens, your set will be postponed, hopefully to the next day. In fact, your set might even need to be rescheduled to accommodate a set that was canceled three days earlier. This occurs whenever the set crew or crane has scheduled a few back-to-back sets, which can sometimes happen when everyone is trying to make up for the canceled sets, as could result from several days of rain. Needless to say, this juggling is impossible to manage without inconveniencing everyone. You will probably have made special arrangements with your job to be at the set, and you may have invited family and friends. But delays do happen, and it is best to prepare yourself for the possibility.

Phase 12

☐ Complete the as-built plan

Make sure that the "as-built" plan, if required, is completed as soon as the foundation is poured, and give a copy to the lender and the building inspector.

Phase 13

☐ Prepare the site and foundation

The GC must complete the following tasks so your site and foundation are prepared for the delivery and set:

- Grade the site so the crane and modules can be properly placed

- Back-fill the foundation, if possible, so the set crew can safely set the modules on the foundation

- Install the sill plate and build the knee walls according to the dealer's specifications

Phase 14

☐ Arrive at the site before the modules arrive
☐ Pay for your home, if required
☐ Leave protective coverings undisturbed

You must arrive at your site before the modules are delivered and remain until the delivery is complete in case there are any last-minute decisions. These can include cutting down a tree, removing a stone wall, and bringing in additional gravel. If you do not heed this advice, the dealer, along with the manufacturer's delivery team, will be compelled to make the decisions without you. You will be responsible for accepting their decisions as well as all additional costs that attend them. If you are paying for your modular home COD, you must pay the balance due in full as soon as the modules are delivered to your site or the staging area.

After the modules are delivered, resist the urge to see what they look like inside. The modular dealer may be able to open a door or two to help you sneak a peak. But you should not attempt to cut open any of the protective coverings, since they are installed by the manufacturer to shield the modules against adverse weather conditions. If you or the GC attempts to do so and weather damage results, you will be liable for the cost to correct the damage.

Phase 15

- ☐ Remain at the site for the entire set
- ☐ Be safety conscious
- ☐ Ensure that the lender pays for your home
- ☐ Complete a walk-through with your dealer

For the same reasons that you should be present for the delivery of your home, you should be present for the entire set. Whatever problems arise, you have the right to be involved in the resolution, which is much easier to do if you are on-site.

Set day is very exciting for adults and even more so for children. It can also be very dangerous, especially for children but also pets. By all means invite your family and a few friends. Have a beer, if you like, but do not offer any alcohol to the crane or set crew. Keep any pets secure.

Do not go near the foundation, modules, crane, or set crew until the dealer or crew chief tells you it is safe. Be very cautious while walking around the site. When you are allowed to enter your home, you will find that none of the stairs will be in place, so be very careful and understand that if you enter the home, you are doing so at your own risk. Once you are in the house, you will find that it contains piles of uninstalled materials that can make exploring difficult and dangerous. Be careful!

If payment for your home is to be made with the assignment-of-funds method after the modules are set on the foundation, ensure that a representative from the lending institution is present at the agreed-upon time. The representative almost always inspects the home before delivering payment.

You and the dealer must walk through the house to complete his set-day inspection for all warranty work and material shortages. If you are hiring your own GC, have him complete the walk-through with you.

The GC's Construction Schedule

Your GC is responsible for meeting two construction schedules. The first, the pre-set schedule, includes those tasks he must complete to make your site and foundation ready for the delivery and set of your modular home. The second, the post-set schedule, involves those tasks he must do to finish the excavation on your property, button up your home, and construct your site-built structures.

Before beginning his work, the GC should have a preconstruction meeting with you. Even when a construction project goes well, it can be stressful to all concerned. It is not just that contractors have a habit of promising too much and delivering too little. It is also that customers without prior building experience have no idea what to expect. A wise GC helps his customers with their inexperience by preparing them for what is to come. With modular home construction, the greatest potential source of stress is with the GC tasks. The GC can help you have a more prepared, relaxed, and trusting experience by spending an hour or two with you in a preconstruction meeting. Ideally, you should meet a week or two before the GC begins his pre-set work.

Communications

Whenever possible, direct your important communications to the GC or his project supervisor. Since the GC is responsible for all aspects of the construction work, he will know all of your contract's details and the scheduled day-to-day plans and activities. The GC's employees and subcontractors will not know this information and may unintentionally mislead you.

You and the GC should agree how frequently you will speak with each other and whether it will be by phone or in person. There will be times when you must speak to him immediately, but you should respect the GC's need to work with few interruptions. (Keep in mind that one extra hour per day during an eight-week project is an extra 40 hours, which can cause your completion date to be delayed a week.) This is also true of subcontractors. Talk to them when you must, but do not jeopardize your completion date by chatting with them every chance you get.

Changes to your contract must be initiated with the GC. He will implement them with a change order signed by the two of you. The GC will instruct his subcontractors not to accept any changes directly from you, since he first must approve and document them. This is in the GC's interest, of course, since it ensures that he will earn a profit for the changes. This is also in his subcontractors' interest, since it ensures that they will be paid for completing the changes. But it is also in your interest, since it ensures that the full scope of work is taken into account in terms of both time and expense. For example, if you ask the plumber to relocate your vanity, you may also need the carpenter to do some repair work and the electrician to relocate the light above your vanity. If you work directly with the plumber, bypassing the GC, you will be faced with significant additional electrical and carpentry expenses.

Staying on Schedule

The GC should give you a written overview of his projected schedule. It may not be too detailed, but it should give you a sense of what he is trying to accomplish. If the GC experiences any delays, he can use the schedule to show you the impact on the remainder of the schedule.

The GC might build some extra time into his schedule to take into account the usual construction snafus. Or he might prefer to work on a tight schedule, with no room for error, and adjust the schedule each time something causes a delay. Most customers

say they prefer the second type of schedule, since it seems to promise a faster completion time. But most GCs say their customers typically handle a padded schedule a lot better, since there are fewer disappointments and the target date is usually hit. Ask the GC how he created his schedule.

Even if the GC builds in some make-up time, it will be helpful for you to understand the many things that can cause construction delays. Any of the following can change the start and duration of each construction task and, therefore, the targeted completion date:

- Inclement weather
- Delayed inspections or approvals by town officials
- Utility company service delays
- Unexpected problems during excavation
- Manpower delays involving the GC, his employees, or subcontractors
- Spill-over delays from other jobs
- Material back orders
- Custom or unusual features of your home
- Customer-initiated change orders that add significantly to the scope of work or delay other work
- Late payments to the GC
- The customer's late delivery of items he is responsible for, such as appliances
- Subcontractors that the customer hired do not complete their work in a timely fashion

The good news is no construction project has all of these delays. The bad news is that you are likely to experience at least a few of them. This does not mean your completion date will be delayed, since the GC may have budgeted some slack into the schedule or he may be able to make up the time. The GC is more likely to fall behind if one of the delays becomes extended because of circumstances beyond his control.

In construction, one delay can produce another delay. For example, the electrician might be ready to begin working on your home on his scheduled start date. If, however, the GC falls a week behind schedule because of a late inspection or a tardy subcontractor, the electrician may have to postpone the start of his work. If he is obligated to work on another home the following week, he may not be able to start on your home for two weeks. What could have been a one-week delay, then, has become a two-week delay. This type of situation happens with painful frequency.

In fact, this two-week delay can itself grow into a longer delay if another subcontractor, such as an HVAC contractor, cannot finish his work until the electrician is done. If the HVAC contractor is under contract with other projects, as virtually all subcontractors are, he could be in the middle of another job when your home is ready

Schedules and Delays

There are many types of scheduling situations other than straightforward delays that can cause misunderstandings and ill feelings between a customer and his general contractor. Here are five of them.

SITUATION 1

The GC does not initiate any work on your home the first seven to ten days after the set. You might conclude that the GC has dropped the ball and the entire project will be delayed. The GC, however, might be completing a few other homes that were started before yours. Completing the other homes first keeps him from having to jump back and forth among several homes at the same time. Once the GC starts work on your home, he will be able to make a concerted effort to complete it on schedule.

SITUATION 2

The GC gets ahead of schedule on one or two construction tasks. You might conclude that the entire project will be completed ahead of schedule. But there will likely be delays before your home is completed. Although it would be great if the GC completed your home early, it is more realistic for you to expect him to complete it on schedule.

SITUATION 3

The GC falls behind on one or more construction tasks. You might conclude that the completion date will not be met, but the GC's targeted completion date allows for some delays.

SITUATION 4

With three weeks left to the projected completion date, the GC appears to have only two weeks of work remaining. You might think that the entire project will be completed ahead of schedule. There are multiple small details, however, including inspections and punch lists, that must be completed during the last couple of weeks of the project. The crews completing these tasks may not be scheduled for another week. If the project is not completed early, you may feel the GC mismanaged the end of the project. The reality is he is right on schedule.

SITUATION 5

With two weeks to go, the GC appears to have three weeks' worth of work remaining to be done. You might feel that it is impossible to complete the work by the targeted completion date. For the past few weeks, however, the GC has focused on completing a few other homes that were started before yours. Once he turns his full focus to your home, he will be able to make an all-out effort to complete it on schedule.

for him. See the box on page 305 for examples of other types of scheduling problems.

Unless you are prepared for the almost certain ups and downs of construction scheduling, you are likely to be disappointed, frustrated, and angry: sometimes at the weather, town officials, or a utility company; sometimes at the GC or his subcontractors; and sometimes at your family or yourself. This could cause you to lose confidence and trust in the GC. The best thing you can do whenever you have a concern about your schedule is to discuss it with the GC.

If delays occur, try to be fair in judging whether the GC could have prevented them. Often it is best to withhold final judgment until the end of the job, when you can see how well the GC met his projected schedule. In light of the frequency of construction delays, try to keep your temporary living arrangements open-ended while your home is being built. Should your completion date get extended, you will save yourself a lot of trouble if you are able to stay where you are for the additional time. Avoid locking in a mover until you are reasonably sure your home will be ready. Consult with the GC before making final arrangements.

Hiring Some of Your Own Subcontractors

The GC's construction schedule assumes that you and your subcontractors complete your respective responsibilities on schedule. That is why you and your subcontractors need to know the GC's target dates. To facilitate this, the GC should meet with your subcontractors and you to discuss the schedule. The GC needs both you and your subcontractors to understand that he will not have time to oversee their work or teach them how to complete a modular home.

Your Responsibilities

If you have personally assumed responsibility for completing any of the GC's tasks, you need to do them on schedule. You must contact the telephone and cable companies to obtain service. You must open an account with either a gas or fuel-oil company, unless you are building an all-electric home. The GC should tell you when he needs these responsibilities completed.

You must coordinate with the GC for your appliance delivery. He will not want your appliances too early, since they could get damaged or stolen, or too late, because his subcontractors will not be there to hook them up. This same situation applies to any other items you are supplying that the GC will install, such as custom light fixtures.

Change Orders

Change orders that add new construction tasks almost always extend the completion date. If the additional work is requested late in your project, the GC may not be able to help you because of other commitments.

Timely Payments

You are responsible for ensuring timely payment for each GC invoice. Just as you cannot afford any delays on the completion of your home, the GC cannot afford any delays in receiving his payments.

Moving In before Your Home Is Complete

If the GC obtains the certificate of occupancy sooner than he projected, you may think he is sufficiently done for you to move in ahead of schedule. Do not do so unless you first obtain permission from the GC. He might ask you to wait because he still has to complete several tasks that cannot be done properly until the home is almost finished.

Do not move any of your belongings into your home without permission from the GC. If the GC agrees, he will ask you to use those rooms he has finished. If you intend to store your things in the basement, he must have already completed all of his work there. Since you are responsible for theft or damage, ask your insurance agent about your coverage.

Walk-Through Inspection

To ensure that all of the GC's work is satisfactory, you and the GC should complete a walk-through inspection one week before your anticipated move-in date. The two of you should use the walk-through to document your areas of concern. The GC can use the week to complete all of the listed tasks. He also should use the walk-through to demonstrate how your home works, discuss your maintenance responsibilities, and review the warranty procedures.

Change Locksets

Hire someone to rekey all of the locks just before you move in.

Final Payment

After the GC finishes his work, you need to pay him in full. If he does not complete a few items listed during the walk-through, you should hold back money until he does. Withholding 150 percent of the value of each item should be fair to both of you.

Scheduling and Timing of Construction Tasks

What follows illustrates how the construction schedule works on a modular home. The timeline is for a typical ranch-style home with a site-built, attached two-car garage and a rear deck. It includes some tasks that are done only under certain circumstances. The timeline assumes three things: The GC is installing few, if any, optional features on-site; the site work is relatively uncomplicated; and the GC and his subcontractors have no scheduling backlog. Although this type of house can be completed in three weeks after it is set on the foundation, a four-week schedule is shown to give the GC the extra week he might need because of typical construction delays. In addition, three weeks are scheduled before the house set to clear the land, excavate the site, and install the foundation.

Pre-Set Schedule

Three weeks before set day:
- Clear land of trees and brush

Two weeks before set day:
- Install rough driveway
- Drill well and submit water test, unless done earlier to obtain a building permit
- Excavate for foundation of house, garage, and deck
- Pour foundation footings of house, garage, and deck
- Inspect foundation footings of house, garage, and deck
- Install foundation walls of house and garage and pier supports for deck
- Install plumbing under the foundation floor
- Inspect plumbing under the foundation floor

One week before set day:
- Damp-proof the foundation
- Install footing drains around foundation
- Backfill
- Complete rough grade
- Pour foundation floor
- Build knee walls and install sill plate

Post-Set Schedule

First week after set day:
- Make sure home is weather tight
- Reorganize loose-shipped materials
- Remove temporary framing, where appropriate
- Pour foundation floor, if not done before set day
- Build basement stairs
- Build bulkhead stairs
- Air-seal interior marriage-wall seams
- Complete bolting and strapping of marriage walls, adjusting as necessary
- Get structural inspection
- Mount electrical panel box and meter socket and prepare for overhead or underground service

- Get electrical inspection and call utility company for electrical service (usually done by inspector)
- Complete rough electrical on house
- Get rough-electrical inspection
- Complete rough plumbing
- Get rough-plumbing inspection
- Complete rough-heating installation
- Get rough-heating inspection
- Install septic system
- Get septic system or municipal water and sewer inspection
- Begin rough carpentry of garage and deck
- Air-seal exterior marriage-wall seams of house
- Prepare exterior for siding

Second week after set day:
- Install electrical service (utility company)
- Complete electrical
- Get framing inspection of garage and deck
- Complete rough electrical on garage
- Get rough-electrical inspection on garage
- Complete garage and deck
- Install stairs at front door
- Connect waste line to septic system or muncipal sewer
- Connect water line to well or municpal water
- Complete plumbing, including appliance hookups
- Install built-in cooktop and oven
- Have fuel delivered for heating system
- Get gas inspection
- Complete heating system
- Get heating inspection
- Complete siding, soffits, fascia, and exterior decorative moldings on house, knee walls, and garage
- Complete modular adjustments and repairs

- Adjust windows and interior and exterior doors
- Repair drywall cracks
- Level floors
- Hang, tape, and paint drywall at marriage wall

Third week after set day:
- Complete both carpet and vinyl-floor installation
- Hang doors and install moldings at marriage wall
- Get final plumbing inspection
- Get final electrical inspection
- Get smoke detector inspection
- Insulate basement ceiling
- Complete miscellaneous tasks as needed (install dryer vent, vent fans, bathroom accessories; install toe kicks at exterior doors; tile around whirlpool tub; seal foundation penetrations)
- Complete finish grade
- Install gutters
- Do walk-through inspection
- Generate punch list

Fourth week after set day:
- Complete punch list
- Complete final cleaning of house
- Install window grills and screens
- Complete final touch-up
- Get final inspection
- Obtain certificate of occupancy
- Customer moves in

If you are building a house style other than a ranch, the GC needs to complete additional tasks.

For Raised-Ranch and Split-Level Designs

First week after set day:
- Frame split-level entry for landing, stairs, and close-off to basement

Second week after set day
- Finish split-level entry for landing, stairs, and close-off to basement

For Cape Cod Designs

First week after set day:
- Set and adjust interior stairs

Second week after set day:
- Install stair railings and trim
- Paint or stain stair system
- Build a close-off at top of the stairs if second floor is unfinished

For Two-Story Designs

First week after set day:
- Set and adjust interior stairs

Second week after set day:
- Install stair railings and trim
- Paint and/or stain stair system
- Install mechanical access panels
- Complete installation of finish flooring in rooms that contain floor access panels
- Complete installation of drywall in rooms with ceiling access panels

Some of the tasks listed above must be done in a particular sequence, while many others can be completed in a different order. In construction, those tasks that must be done in a specific order define what is called the "critical path." An example of a task that must be done in a particular sequence is the electrician's wiring of the heating system before the HVAC contractor can fire up the system. An example of a task that can be done at any time is the installation of the bathroom vent fans. The GC uses his professional experience to determine a sequence that works best for him. He is likely to change the sequence from project to project, depending upon the availability of his employees, subcontractors, and materials. If any delays occur, he adjusts the schedule accordingly.

Although the GC can complete the following four tasks when he chooses, it is best if he completes these tasks as soon as possible after the set:
- Bring electrical power to the house
- Pour the basement floor
- Facilitate access to the basement
- Hook up the heating system

Installing the electrical power and making the connections as soon as possible allows the GC's contractors to use their power tools without having to bring a generator. Installing basement and bulkhead stairs creates a safe and easy access to the basement for the electrician, plumber, and HVAC contractor. Pouring a basement floor, if it is not already done and one is being poured, makes it easier for the electrician and plumber to safely reach the basement ceiling, where much of their work occurs. It also enables the HVAC contractor to install the boiler or furnace in the basement. Hooking up the heating system as soon as possible during colder months enables drywall work, painting, and floor installation to begin. It also increases the effectiveness of all contractors, since they are more productive when they are comfortable. If you have selected hot-water baseboard heat, you need the plumbing hooked up as soon as possible to bring the water to the boiler.

During the first week after the set, the GC should complete the air sealing, bolting, and strapping of the interior marriage wall. This is particularly important if the building department requires a structural inspection before other work can be done to the marriage wall. Starting work on any site-built structures during the first week of a four-week schedule is important if the GC intends to have them built by the time your home is buttoned up.

Many of the other tasks listed in the proposed timeline can be executed in any sequence. Although the GC may have a preferred order, it is possible to do them early or late in the project and before or after other tasks. For example, the GC can repair the drywall cracks the first or the last day of the button-up. Since it is more efficient to fix them when he is finishing other drywall work at the marriage wall, he should wait until then to repair the cracks. He may also delay the repair to be sure a few more cracks do not appear as the house settles.

There are a few other tasks that can be completed at any time but are best done in a particular sequence. Another drywall example is the patching of the ceiling- and wall-access panels. A GC cannot finish his drywall work until all of the ceiling-access panels are closed up. For the sake of greater efficiency, some GCs do not even start the drywall work until this happens. For the same reasons, some GCs do not begin their flooring installation until the floor-access panels are closed. A risk with this approach is that the longer it takes the mechanical contractors to complete their work, the longer it will take the GC to close up the access panels, producing delays with drywall and flooring.

Another example of a task that can be done at any time but works better in a particular sequence is siding installation. The most efficient way is to side the house at the same time as you side all of the site-built structures, such as a garage. This means first building the garage and any site-built component that attaches to the home, such as a porch or chimney. If the site-built structures and attached components are delayed, the GC has to decide whether he also wants to delay siding the home. It should not be the GC's choice alone, however, as many siding contractors balk at having to set up their equipment, take it down, and set it up again.

Factors That Lengthen the Pre-Set Schedule

There are a few factors that can lengthen the GC's pre-set schedule. The GC's start date must be advanced if you need a well

drilled before you can receive a building permit. The GC must also start earlier if you have a complicated property requiring a lot of site work, such as a complex septic system, a long driveway, blasting of stone or ledge, large-scale wetland remediation, truckloads of extra fill, or extensive grading. Starting earlier enables the GC to complete the additional site work before your home is set. If the foundation is ready for the scheduled delivery date, his additional site work should not delay your move-in date. But it might not be practical to complete all of the additional site work in advance of the set. If the GC must complete it after the set, your post-set timeline could be extended.

If you are building in cold weather, expect the excavation and foundation work to take longer than it normally would. The GC must protect the ground and foundation from low temperatures, and all outdoor tasks take more time in the cold.

Factors That Lengthen the Post-Set Schedule

There are many factors that add time to the GC's post-set schedule. Some of these, such as the contractor's backlog of work, can add weeks, while others, such as the building department's inspections, might add but a few days.

Home Size, Style, and Features

The most obvious factor that can lengthen your post-set schedule is the size of your home. If you are building a larger home, the GC and his subcontractors need more time to complete it. Some tasks, such as the marriage-wall work, increase proportionally, with a 2,400-square-foot home requiring almost twice as much work as a 1,200-square-foot home. Other tasks, such as the installation of the heating system, might take only 40 percent longer because the larger home still needs only one heating unit and little additional time for installation. The complete system takes longer due to extra ductwork or plumbing.

The style of your home also greatly affects the duration of the post-set construction schedule. More complicated designs, especially those that require significant on-site construction, can add one to four weeks to your timetable. For example, if you are building a Whately 1 (see page 56), the GC needs more time to complete the drywall and stair railings in the vaulted foyer and the drywall in the master bedroom. Completing a chalet's additional framing, insulation, drywall, electric, heat, and stair railings takes considerably more time than closing off a traditional Cape Cod with an unfinished second story. State-of-the-art hybrid designs that are 75 percent modular and 25 percent site-built also require a lot more time.

When you select optional features for your home that must be installed on-site, the GC has to schedule additional time. This applies, for example, to hardwood and tile floors, a masonry fireplace, radiant heat, custom kitchen cabinetry, and tile showers. Some of these tasks can be completed while doing the standard button-up work. Others must be sequenced in a way that is certain to lengthen the post-set schedule. For example, if you are installing a radiant heating system on top of the sheathing on the first floor and tile and custom cabinetry on top of the radiant heat, you will need to complete them in that order.

You can certainly expect the GC to take longer when extra painting and staining are needed. This is especially true if the manufacturer delivered the home with all of the interior moldings, doors, and windows unfinished. If you have a lot of stair railings, as found in most vaulted foyers and chalets, you should expect the GC to budget extra time to finish the railings. The same happens if you opt to repaint all of the walls in custom colors, which will require at least two coats of paint.

On-Site Construction

The greater the number of site-built structures the GC is constructing, the more time he needs to complete your home. A garage, porch, deck, and mudroom, for example, each demand attention from the GC's crews. The post-set time is extended even further if excavation and foundation work for these structures cannot be done until after your home is set. Such work can interfere with set-day activities, especially for the crane. In this case, the GC will have to wait to start the framing of these structures until the foundation is ready, which could be in the second or third week.

A similar delay will occur if your property is too small to allow multiple crews and their materials to work productively at the same time. This can happen, for example, when the GC is digging trenches for the water, sewer, and underground electric services. Until that work is finished, it may not be practical for the framing crews to begin their jobs.

The GC will need additional time to complete a sizable unfinished space. For example, if you are having him finish the basement of a raised ranch or split level, the second story of a Cape Cod, or the front half of the second story of a three-box Cape Cod, you should expect his carpenters, drywall contractors, painters, electrician, HVAC contractor, and floor installers to take more time. If you are installing any plumbing in these areas, the plumber's schedule will be extended as well.

Crew Size

You might think that the GC could save time by using more and larger crews. You might also wonder why he cannot bring assembly-line thinking to his job site to knock off a hybrid design with lots of site-installed custom features and site-built structures. After all, the modular manufacturer builds each home, regardless of size, in about a week once it starts construction.

True, there are many construction tasks that can be spread out among several crews. The more the GC uses separate crews, the faster he completes the work. But there are limits to this strategy. The manufacturer is always building multiple homes at the same time in the same location with the same large crews. This is not the case with the typical GC unless he is building multiple homes in a subdivision, and all at the same time. More likely, he is building a few homes at a time, each in a different town, with subcontractors that have small crews.

In addition, it is not practical to have multiple crews working in a house at the same time. The modular manufacturer can sequence its crews like a finely tuned engine: One crew follows another with virtually no lag time. This is practical because assembly-line crews are located in one place. No GC building a few homes at different locations can pull this off efficiently or economically. His subcontractors are likely to be committed to projects other than the GC's, and they cannot make it work financially bouncing around from job to job. This does not mean the GC cannot do some construction tasks simultaneously with different crews. In fact, the projected timeline assumes he is doing just that. For example, it is very likely that he will be using different crews for the site work, mechanicals, and carpentry work, and these crews can complete at least some of their work alongside each other.

Some GCs use one button-up crew to do all of the carpentry tasks even though this lengthens the post-set schedule. These GCs feel they are better able to manage exactly what and when things are done if the same crew completes all of the framing and interior finish as well as the insulation, drywall, and painting. They select the crew knowing they can count on its quality and timeliness. Consequently, they sometimes produce a better job than another GC who uses several different crews. But this is not always true. Some GCs using multiple crews, each a specialist in its trade, deliver better quality and do so more quickly. Multiple crews speed up construction only if the GC uses them at the same time. If he uses them sequentially for one task at a time, he might need even more time than a GC who manages one crew well; the GC using one crew is better able to manage the schedule.

Cold Weather

Cold weather can lengthen a post-set-construction schedule. Installing a foundation after frost has set in the ground can make it impossible to pour the floor before set day. This in itself can add a few days to the schedule. It will add several more if the GC needs to melt the frost out of the ground after your home is set. Since pouring a post-set floor also delays the heating system installation, the GC must melt away the frost with a temporary heat source.

Cold temperatures have other adverse affects on the post-set schedule. Contractors are less productive when they are cold, and many kinds of construction tasks, such as drywall taping and painting, cannot be done when a home's temperature is below 50 degrees. If the GC is unable to get the heating system up and running quickly, he needs to extend the use of the temporary heat, which is less effective at warming the house as well as more costly.

House Touch-Up

If the modules experienced a rough delivery, your house may need some additional cosmetic work, especially to the drywall, as well as more door and window adjustments. Even if your home needs the average amount of cosmetic work, this part of the job could take longer than budgeted if you have asked for a higher level of finish than the manufacturer provided. For example, if you want every gap between the moldings and drywall caulked and this was not done by the manufacturer, it will lengthen the GC's schedule.

Inspections

Most building departments complete inspections in a timely fashion, but sometimes an inspector is delayed. This can throw off a construction schedule by at least the number of additional days it takes to get the approval. Construction also takes longer if your building department requires more inspections than those listed in the ranch timeline example. Such inspections usually occur because of some unique requirements of your project or your building department. For example, if you must add rebar to the poured-concrete foundation walls, the inspector might insist on seeing the concrete forms and rebar before the concrete is poured. The inspection itself might not take a lot of time, but it will delay the GC if the inspection cannot be performed within a reasonable time.

Contractor Backlog

A final factor that can lengthen a construction schedule is when the GC or his subcontractors have a substantial backlog of other work. A contractor who is overbooked cannot afford to build any slack time into his schedule, so if a construction problem at your home or with another project trips him up, he is more likely to fall behind schedule. An experienced GC who is preparing a construction schedule during a building boom takes this into account and builds in more make-up time than he normally would when times are slow. This allows him to recover from a delay without compounding it. It also means, however, that your post-set schedule will be longer than it would otherwise need to be.

Taking into account all of the potential factors that can lengthen a projected post-set-construction schedule, the four-week timeline can be extended by days, weeks, or even a few months. But however long it takes, your home will still be done faster than if you were having it stick-built.

By now it should be clear that your dealer, GC, and you have many responsibilities. You probably also realize that you must commit a lot of time and effort to these tasks if you are to remain on schedule. Most people who have built homes before can tell you that no matter how hard you work and how much you pray for good luck, you should be prepared for delays. After all, you cannot control all of the people you need to help you. But if you have selected the right modular dealer and general contractor, you will be able to count on them. This can make all the difference in the world.

Resources

Modular Manufacturers

You can obtain a list of modular manufacturers from either:

The Modular Building Systems
Association
717-238-9130
www.modularhousing.com
or
The National Association of Home
Builders
800-368-5242
www.buildingsystems.org/3.html

Both only list manufacturers that are members of their associations.

Training for General Contractors and Set Crews

Penn College Modular Housing Training
Institute offers a two-day course in
modular construction.
To register, call 570-327-4768.

Modular Industry Associations

The Modular Building Systems Association
offers membership to manufacturers, dealers, and other professionals associated
with the industry. You can reach them at
717-238-9130 or www.modularhousing.com

The Modular Building Institute also offers
membership to manufacturers, dealers,
and other professionals associated with the
industry. You can reach them at 434-296-
3288 or www. mbinet.org

Warranty Standards

A commonly used list of written warranty
standards is:

*Guidelines for Professional Builders and
Remodelers*, published by the National
Association of Home Builders. To order,
call 800-223-2665.

Books

Don O. Carlson, editor of *Automated
Builder Magazine*, published two very
good books on the building system
industry:

Don O. Carlson. *Automated Builder:
The Dictionary/Encyclopedia of
Industrialized Housing.* Carpinteria, CA:
Automated Builder Magazine, 1991.

Don O. Carlson. *Automated Builder:
How and Why to Buy a Factory-Built
Home.* Carpinteria, CA: Automated
Builder Magazine, 2001.

To order either book, call 800-344-2537.

Magazines

There are two periodicals that focus on
the building system industry:

Building Systems Magazine, published bi-
monthly by Builder Dealer Management
Services, Chantilly, VA. To order, call
310-356-4120.

Automated Builder Magazine, published
monthly by Automated Builder, Ventura,
CA. To order, call 805-642-9735.

Index

Note: Numbers in *italics* indicate illustrations and photographs.

Other Storey Titles You Will Enjoy

Be Your Own Home Renovation Contractor, by Carl Heldmann. Evaluate the structure of your building, estimate costs, negotiate loans, hire subcontractors, and save money with Heldmann's expert advice. Paperback. ISBN 1-58017-024-2.

Be Your Own House Contractor, by Carl Heldmann. Heldmann shows you how to get exactly what you want done on your new house, the way you want it done. Paperback. ISBN 1-58017-374-8.

Building Small Barns, Sheds & Shelters, by Monte Burch. Extend your working, living, and storage areas with low-cost barns, sheds, and animal shelters. Here is complete information on tools and materials to create everything from multi-purpose barns to various fencing designs. 248 pages. Paperback. ISBN 0-88266-245-7.

Deckscaping, by Barbara Ellis. Trellises, lighting, plantings, and furniture are all covered here with photographs, drawings and well-thought-out text. Paperback: ISBN 1-58017-408-6. Hardcover: ISBN 1-58017-459-0.

The Fence Bible, by Jeff Beneke. Easy-to-follow text, full-color photographs, and step-by-step illustrations make it easy to learn how to plan, design, build, and maintain every kind of fencing for every kind of environment. 272 pages. Paperback. ISBN 1-58017-530-9.

Fences for Pasture & Garden, by Gail Damerow. A good fence is essential for protecting livestock or gardens and this guide helps you select the best fence for your needs plus provides planning advice to ensure maximum efficiency of labor and materials. 160 pages. Paperback. ISBN 0-88266-753-X.

How to Build Paths, Steps & Footbridges, by Peter Jeswald. Transform any home landscape with paths, steps, or footbridges. Learn how to choose the best designs for your needs, define an outdoor living space, and realize your property's full potential. 248 pages. Paperback. ISBN 1-58017-487-6.

Outdoor Woodwork, by Alan & Gill Bridgewater. Enhance your landscape with these 16 attractive and useful projects for both the novice and the experienced do-it-yourselfer. 128 pages. Paperback. ISBN 1-58017-437-X.

Stonework, by Charles McRaven. Discover the lasting satisfaction of working with stone and learn the tricks of the trade with this collection of projects for garden paths and walls, porches, pools, seats, waterfalls, a bridge, and more. 192 pages. Paperback. ISBN 0-88266-976-1.

These books and other Storey books are available wherever books are sold and directly from Storey Publishing, 210 MASS MoCA Way, North Adams, MA 01247, or by calling 1-800-441-5700. www.storey.com